KT-551-530

Glass
Reinforced Plastics

Contributors

P. H. H. BISHOP, B.Sc.
Royal Aircraft Establishment, Ministry of Technology
R. W. CAHN, M.A., Ph.D., Sc.D., F.INST.P., F.I.M.,
Professor of Materials Science, University of Sussex
J. R. CROWDER, B.Sc., Ph.D.,
Building Research Station
M. A. DENNEY, F.P.I.,
Chief Chemist, British Aircraft Corporation Ltd., Weybridge Division
B. HARRIS, B.Sc., Ph.D.,
Reader in in Materials Science, University of Sussex
B. J. HAWTHORNE, B.Sc., A.R.I.C.,
Head of Plastics Laboratory, British Railways Board
A. C. JOLLY, B.Sc., Technical Service Manager,
Beck Koller & Co (England) Ltd.
G. KINGSLAKE, B.Sc., A.C.G.I., D.I.C.,
Chief Structural Engineer, Vosper Ltd.
J. MATTHAN, B.Sc., GRAD.P.I.,
Rubber & Plastics Research Association of Great Britain
R. G. B. MITCHELL, F.P.I.,
Sales Development Officer, Bakelite Xylonite Ltd.
R. H. NORMAN, M.Sc., A.INST.P., F.I.R.I.,
Manager General Physics Group, Rubber & Plastics
Research Association of Great Britain
R. M. OGORKIEWICZ, M.Sc. (Eng), A.C.G.I., D.I.C., C.ENG., M.I.MECH.E.,
Senior Lecturer in Mechanical Engineering, Imperial College of Science and Technology
P. C. OLIVER
Technical Manager, Osma Reinforced Plastics Ltd.
M. J. OWEN, B.Sc., M.S., Ph.D., C.ENG, M.I.MECH.E.,
Senior Lecturer, Department of Mechanical Engineering,
University of Nottingham
K. PARVIN, B.Sc., A.R.C.S., F.R.I.C.,
Technical Secretary, Scott Bader Co. Ltd.
R. PLEYDELL-BOUVERIE,
Designer, John West Design Group
I. W. REID, B.Sc., C.ENG., M.I.MECH.E.,
Technical Manager, Prodorite Ltd.
K. A. SCOTT, B.Sc., A.R.I.C.,
Manager Plastics Technology Group, Rubber & Plastics Research Association of Great Britain
M. H. STONE, B.Sc.,
Rubber & Plastics Research Association of Great Britain
W. C. WAKE, Ph.D., D.Sc. (Lond), F.R.I.C., F.I.R.I.,
Consultant, formerly Assistant Director (Research),
Rubber & Plastics Research Association of Great Britain
L. A. R. WARING, B.Sc., A.T.I.,
Technical Manager, Turner Brothers Asbestos Co. Ltd.
A. A. K. WHITEHOUSE, M. A., F.R.I.C., F.P.I.,
Director, Scott Bader Co. Ltd.
T. WILKS,
Managing Director, Viking Marine Co. Ltd.

Glass Reinforced Plastics

Edited by

BRIAN PARKYN

LONDON
ILIFFE BOOKS

ENGLAND

Butterworth & Co (Publishers) Ltd
London: 88 Kingsway, WC2B 6AB

AUSTRALIA

Butterworth & Co (Australia) Ltd
Sydney: 20 Loftus Street
Melbourne: 343 Little Collins Street
Brisbane: 240 Queen Street

CANADA

Butterworth & Co (Canada) Ltd
Toronto: 14 Curity Avenue, 374

NEW ZEALAND

Butterworth & Co (New Zealand) Ltd
Wellington: 49/51 Ballance Street
Auckland: 35 High Street

SOUTH AFRICA

Butterworth & Co (South Africa) (Pty.) Ltd
Durban: 33/35 Beach Grove

First edition 1970

Suggested U.D.C. number 678·5:666·189·2.

Suggested additional number 678·5:677·52

ISBN 0 592 05456 X

Printed in England by Page Bros (Norwich) Ltd
Norwich

Contents

PART 4—ENGINEERING DESIGN

Preface

It is now sixteen years since the publication of the first edition of *Glass Reinforced Plastics* edited by Philip Morgan. That book represented an important milestone in the history of plastics because it was one of the earliest comprehensive books to appear in any country covering the entire field of glass fibre reinforced plastics.

Since the appearance in 1961 of the third edition however, the reinforced plastics industry has grown enormously both in size and importance and it has now become a major industry in almost every part of the world. Whilst it was expected that the industry would develop rapidly in the United States, the United Kingdom and Western Europe, it is interesting to note the growth in Australia and the Republic of South Africa; in Brazil, Argentina and many other parts of South America; in India, S.E. Asia and Japan; in the U.S.S.R. and more recently in China.

The almost explosive growth of the reinforced plastics industry has been in part due to the development of better resin systems, curing systems and glass fibre reinforcement. New moulding techniques are now known, and as a result, particularly of the efforts of the British Plastics Federation, more reliable performance data on the finished product are now widely available. The reinforced plastics industry has 'come of age' and the increasing amount of work devoted to developing suitable specifications for reinforced plastics is evidence of this fact.

In addition to this, growth has been in part due to the very considerable literature devoted to resin formulations and to fabrication techniques now available. Much of the technical material in the earlier publications has become common knowledge while specialist books have also appeared which are able to provide more detailed information on resin chemistry, glass fibre and methods of application than is possible in a single volume.

There is however, a growing need for information on the design criteria and end uses of glass fibre reinforced plastics. It seemed appropriate therefore to produce a completely new book to replace both *Glass Reinforced Plastics* (Iliffe Books) and *Glass Fibre Reinforced Plastics* (George Newnes) which, while still covering the whole field in a comprehensive manner, places the emphasis on end products and design criteria about which there has been all too little published.

It is hoped that the specialist and non-specialist alike will find a wealth of new know-how and technology in these pages on what has become the most fascinating development of materials science in recent years.

Since the United Kingdom and many other countries have now officially adopted SI units, these have been used throughout the book with the equivalent English Units added in brackets where these are considered helpful. There are however, two deliberate exceptions. In the case of measurements relating to length, stress, weight, etc., of boat hulls, as for example in Chapter 3, English units have been used since they are widely understood internationally for this purpose, and furthermore, the Ministry of Transport Regulations are at present framed around English units. Secondly, where approximate weight quantities are referred to these are given in tons rather than tonnes or $kg \times 10^3$.

Finally, a word for the authors. Each has been selected for his particular knowledge of the subject. Some overlapping is inevitable in the preparation of a technical book in this way but where it occurs it will be found that the differing standpoints of each contribution is *per se* of great value and interest. The fact that the authors have kept to the time schedule necessary for a work of this complexity is gratefully acknowledged.

Brian Parkyn

Part 1

Introduction

1

Historical Background

BRIAN PARKYN

1. IMPORTANCE OF HISTORY

The importance of history is that it enables us to understand the times in which we live by placing them in their true perspective. The fact that history is often interesting is coincidental; indeed, were the interesting portions of history the only ones to survive, our knowledge of the past would be confined to a disconnected series of incidents illustrating the more salacious behaviour of our forbears. We should lack detailed knowledge of the evolution of most of the factors which go to make up the complex world of today.

The beginnings of many of these factors were quite unpretentious and their ultimate importance could not always be seen. In much the same way, the study of the development of glass reinforced plastics may serve to amplify knowledge and understanding of our present reinforced plastics industry.

2. GLASS FIBRE REINFORCED PLASTICS

Glass fibre reinforced plastics (GRP) are composites made by reinforcing a resin with glass fibre. Many other composites are known, ranging from wood, which is lignin reinforced with cellulose fibre, and reinforced concrete, to laminates made by reinforcing a resin with sheets of paper or fabric.

Glass reinforced plastics, however, have two particularly interesting characteristics. They can be given a mechanical strength greater than any earlier plastics materials, comparable on a weight basis with some metals. They can be moulded without heat or pressure, both of which are essential for shaping traditional plastics materials. It is a combination of these two characteristics which has made it possible for GRP to be used for moulding a type and size of end product impossible with any other plastics material.

The earliest commercial use of GRP was in the United Kingdom and the United States in 1942 for the manufacture of aircraft radomes. GRP boat hulls were made in the United States as early as 1946[1] and a seaplane float and similar mouldings had been made in Britain by 1947.[2] Up to that time the cost of the resin and the glass fibre seemed to preclude the development of any applications apart from specialised uses on aircraft, or for making military landing craft where the cost was of only secondary importance. The reason for this was that the principal form of glass fibre reinforcement consisted of fine woven glass fibre cloth made from an exceptionally fine glass filament.

It was then found that a coarser glass filament could be used without serious deterioration in the ultimate mechanical properties of the laminate. In addition to this, from 1947 onwards, much work was devoted to making a cheaper form of reinforcement than woven cloth. Diamond mat and swirled mat were developed, and in time this led to the now familiar non-woven chopped glass strand mat.

With the cold setting resin systems which became commercially available in 1947 the industry was ready to expand into many of the general applications of GRP which are now familiar throughout the world. The first translucent GRP sheeting appeared in Great Britain in 1949, followed by GRP boat hulls, GRP car bodies, and by the early 1950s by GRP lorry cabs. In 1952 the industry was already sufficiently established for the British Plastics Federation to set up a Reinforced Plastics Technical Committee which held biannual technical conferences. In 1959 the British Plastics Federation decided to establish a separate Reinforced Plastics Group to look after the interests of firms engaged in the industry. Since then they have been responsible for developing British Standards for the glass fibre reinforcement and the resins, and for filling in many of the gaps in the 'state of the art'. GRP is now firmly established and permeates almost every industry, and the extent of this is discussed in greater detail in succeeding chapters.

3. GLASS FIBRE

Glass fibres in one form or another have been known since about 1500 B.C.[3] In the Egyptian XVIIIth Dynasty small glass vases for holding cosmetics were made by taking a clay form, or some sand tied in a cloth bag and fixing this to a stick and rotating it in molten glass until the desired shape had been produced. Coarse coloured glass fibres were then heated and pressed into position on the vessels and combed to give a decorative effect. The core was then removed, and the outer surface of the glass polished.

The earliest use of woven glass fibre appears to be 1713 when Réne Réaumur submitted some glass cloth to the Paris Academy of Science which had been woven by the Venetian, Carlo Riva. In the 1860s a woven glass garment was presented to a Spanish princess, but it was found that the fibre was too coarse to allow the garment to be folded. Glass fibres of sufficient fineness and consistency for reinforced plastics were not available commercially until the 1930s.

4. POLYESTER AND EPOXIDE RESINS

Polyester resins can claim to be among the first of the many synthetic resins which are now the basis of the plastics industry. In 1847, 15 years before

Alexander Parkes introduced Parkesine, the great Swedish chemist, Berzelius, reacted tartaric acid with glycerol and formed a resinous mass poly (glyceryl tartrate). Since this was the reaction product of a polyhydric alcohol and a polybasic acid, it was a polyester resin, although it remained a laboratory curiosity for many years.

The next landmark in the history of polyester resins was the publication by Vorlander[4] in 1894 of his development of the first unsaturated polyester resins, the glycol maleates. This was followed by the work of W. J. Smith in 1901. He reacted phthalic anhydride with glycerol to produce poly (glyceryl phthalate), and this led to the development of the alkyd resins by G.E.C. and B.T.H. from 1913 onwards. By modifying this resin with various vegetable oils an excellent base for paints was produced having outstanding durability, and the commercial exploitation of alkyd resins for surface coating took place over the following 20 years.

The modern history of unsaturated polyester resins began with the filing of a patent application in 1922 by Carleton Ellis[5] and the subsequent publication of this patent in 1933. This covers the reaction products of dihydric alcohols and dibasic acids and acid anhydrides for use as lacquers. Hundreds of publications followed in the succeeding 10 years, but the important developments can be traced through several stages. Firstly, the formation of interpolymers of esters of dibasic acids with vinyl compounds by Dykstra[6] in 1934. Then followed the work by Bradley, Kropa and Johnston who prepared polymerisable compositions based on maleic anhydride[7–9]. This was followed by a further publication of Ellis[10] showing the copolymerisation of maleic polyester resins with monomeric styrene in the presence of peroxide catalysts. This patent is primarily concerned with the preparation of lacquers, and it is interesting to note that the benzoyl peroxide catalyst is referred to as a drier. Likewise the use of styrene is recommended to accelerate drying. This is an important patent since it is the first time that these compositions are suggested for use as moulding materials. Muskat[11] then showed that phthalic anhydride can be reacted with the maleic anhydride and glycol to reduce the tendency towards crystallisation and so improve the compatibility of the final resin with styrene. From this stage to the present day polyester resins have not changed much in principle. The changes which have been made mainly concern the proportion of reactants used in the polyesterification, and the actual nature of the reactants.

No account of the history of polyester resins would be complete, however, without mentioning the now classic work of Carothers on the theory of condensation polymers and polyesters[12, 13] in 1929, and Kienle's general theory of polymer formations based on polyester studies [14, 15] in 1930 and 1936.

The commercial development of unsaturated polyester resins began in the United States in 1941, when an allyl casting resin was introduced for use as a glass substitute. In this case unsaturation was obtained by using an unsaturated alcohol, allyl alcohol, instead of following the more usual practice of using an unsaturated acid such as maleic or fumaric. This was followed in 1942 by an allyl low pressure laminating resin, allyl diglycol carbonate, which was used for the manufacture of some of the first glass cloth reinforced resin radomes for aircraft.

In 1946 polyester resins were commercially available in the United States consisting of diethylene glycol maleate and styrene, and similar resins were soon manufactured in the U.K. At about the same time another type of polyester resin was also made in commercial quantities in England. This was the reaction product of methacrylic acid and phthalic anhydride with ethylene glycol[16]. It was copolymerised with n-butyl methacrylate and was used for some of the earliest glass fibre reinforced plastic mouldings to be made in England. With the commercial production of maleic anhydride and styrene in England, this type of resin has been almost entirely replaced by polyesters of maleic anhydride and ethylene or propylene glycol, with a saturated dibasic acid such as phthalic. These resins are mainly supplied as solutions in monomeric styrene and they represent to-day the bulk of unsaturated polyester resins used throughout the world.

Although the polyester resins by 1946 were contact resins, that is they could be reinforced with glass fibre and cured into mouldings and laminates with little or no pressure, they still had two drawbacks. They had to be cured at a temperature of at least 100°C, and they were air inhibited. These disadvantages restricted the size and complexity of what could be moulded. In 1946, however, it was found that polyesters could be 'cold cured' at normal room temperatures by the use of a tertiary amine such as dimethyl aniline as a promoter in the presence of benzoyl peroxide[17]. Many other cold setting catalyst systems have been developed subsequently, of which the most widely used is a cobalt soap with cyclohexanone peroxide or methyl ethyl ketone peroxide, first used commercially in 1947[18, 19].

The problem of air inhibition was solved about 2 years later by the use of a minute but controlled amount of wax in the resin[20]. Although this system is still in use, there are now several alternative ways of achieving the same result.

Epoxide resins can also be used with glass fibre for making reinforced plastic mouldings, but their use is not so extensive as polyester resins because of their higher cost and greater curing problems. W. H. Moss reacted glycerin dichlorhydrin with diphenylol propane and caustic soda and made the first epoxide resin in 1937[21]. This was followed by the work of P. Castan in Switzerland and others[22].

5. THE FUTURE

Although glass fibre has had no serious rival in reinforced plastics apart from asbestos, the growing demand by the aerospace industries for materials having greater specific strength and specific modulus led to the development in Britain of near perfect crystals of silicon carbide, silicon nitride and aluminium oxide (sapphire). These 'whiskers' have a much greater tensile strength and modulus than even the best glass fibres and by the early 1960s it seemed as if they might be seriously used for making reinforced composite materials. Concurrent with this work, boron fibres were developed in the United States for the same purpose.

In 1964 workers[23] at the Royal Aircraft Establishment, Farnborough discovered a process for making carbon fibre by carbonising under controlled conditions polyacrylonitrile fibre. This fibre which consists of a highly crystalline form of graphite is being used for making carbon fibre reinforced plastics and Rolls Royce use CFRP for the compressor blades on their RB211 jet engine. Carbon fibre has a strength and modulus vastly superior to glass fibre, and although it is at present much more costly, it will undoubtedly lead to a further development of reinforced plastics into even more exciting applications. In view of its heat stability, carbon fibre can be used for reinforcing ceramics, metals and glass, as well as plastics, giving to the engineer a completely new range of materials. Furthermore, carbon fibre gave rise to the first Parliamentary inquiry into reinforced plastics when in 1969 the Select Committee on Science and Technology studied the subject[24]. In just over 25 years, glass reinforced plastics have become a part of composite technology.

REFERENCES

1. Hicks, J. S. *Low Pressure Laminating of Plastics,* Reinhold, 1947.
2. Parkyn, B. and Hulbert, G. C. 'Twelve Years of Reinforced Plastics.' *Appl. Plast.,* London, April 1959.
3. Parkyn, B. *Jl. R. Soc. Arts* 1963 *CXl,* 205–23.
4. Vorlander, D., *Annalen,* 1894, **280**, 167.
5. Ellis, C., *U.S. Pat.* 1 897 977 (1933).
6. Dykstra, H., *U.S. Pat.* 1 945 307 (1934).
7. Bradley, T. F., *Ind. Engng. Chem.,* 1937, **29**, 440.
8. Bradley, T. F., *Ind. Engng. Chem.* 1937, **29**, 579.
9. Bradley, T. F., Kropa, E. L., and Johnston, W. B., *Ind. Engng. Chem.* 1937, **29**, 1270.
10. Ellis, C., *U.S. Pat.* 2 195 362 (1940).
11. Muskat, I. E., *U.S. Pat.* 2 423 042 (1947).
12. Carothers, W., *J. Am. chem. Soc.* 1929, **51**, 2548.
13. Carothers, W., and Arvin, J., *J. Am. chem. Soc.* 1929, **51**, 2560.
14. Kienle, R., *Ind. Engng. Chem.,* 1930, **22**, 590.
15. Kienle, R., *J. Soc. chem. Ind., Lond.* 1936, **55**, 229.
16. Hammond, R., *Brit. Pat.* 630 370 (1949).
17. Hurdis, E. C., *U.S. Pat.* 2 480 928 (1949).
18. Fraser, G. L., *U.S. Pat.* 2 516 309 (1950).
19. Simons, W. G., *U.S. Pat.* 2 537 375 (1951).
20. Parkyn, B. and Bader, E., *Brit. Pat.* 713 332 (1954).
21. Moss, W. H., *Brit. Pat.* 506 999 (1937).
22. de Trey, Gebr., *Swiss Pat.* 211 116 (1938).
23. Johnson, W., Phillips, L. N., and Watt, W., *Brit. Pat.* 1 110 791 (1968).
24. Report from the Select Committee on Science and Technology, *Carbon Fibres,* H.O.C. 157, 1969.

Part 2

End Uses

2

Building and Construction

R. G. B. MITCHELL

1. INTRODUCTION

Up to 1956 very few dwellings used GRP in their construction. In that year, however, the now legendary Monsanto 'House of the Future'[1] was built and paved the way for an upsurge of interest in the possibility of using GRP for structural applications. This chapter looks at some of the successful current applications and likely future uses of GRP in building.[2]

At the present time most British dwellings are still built of traditional materials, but there is a growing number of dwellings with GRP skinned external wall surfaces. An increasing number of houses, flats and bungalows contain cold and hot water storage cylinders made of GRP. Many industrial buildings have rooflights of corrugated GRP. These applications therefore, whilst modest enough, do show that a 'bridgehead' has been established.

Before examining the pattern of end uses that has emerged it will be important to define what we mean by the terms 'building' and 'construction'. This is because the widely-published statistics for house building draw attention to only a part of the whole building and construction industry activity, and they exclude some activities where reinforced plastics are being used.

The construction industry embraces the entire activities of building dwellings, offices, factories, warehouses, power stations, railway stations, municipal and all other buildings, as well as the construction of roads, bridges, docks, harbours, sea defences, dams, service installations and drainage systems. Buildings are, therefore, only a part of the construction industry activity and the term 'building' will be used in its literal sense in this chapter.

In 1968 the U.K. consumption of polyester resins for building applications was approximately 7 000 tons, yielding about 9 500 tons of GRP. By comparison, the construction industry consumes 18 million tons of cement a year and uses 7 500 million bricks. The industry employs between 6 and 7 per cent of the working population in the U.K. with an output of over £4000 million a year. It is a massive enterprise by any standards and the potential for approved materials is obviously attractive to the plastics industry.

2. PROPERTIES OF GRP

The physical and mechanical properties of GRP are discussed in later chapters. Properties of interest for building applications compared with properties of some other building materials are listed in Table 2.1[3–7].

Competitive installed cost and low maintenance costs, high strength, lightness in weight, translucency, good resistance to weathering and fire, and versatility of fabrication methods are all important properties.

Table 2.1

PROPERTIES OF GRP COMPARED WITH OTHER MATERIALS OF CONSTRUCTION

Property	*Units*	*GRP* laminate*	*Plywood (exterior grade)*	*Mild steel*	*Concrete*
Specific gravity	—	1.6	0.80	7.8	2.3
Density	kg/m^3	1 600	800	7 760	2 310
Actual strength					
Tension	$kN/m^2 \times 10^4$	7.6	5.8	55	
Compression	—do—	11.0	2.85	24‡	0.69§
Shear	—do—	6.9	1.2	38	0.069
Specific strength†					
Tension	$kN/m^2 \times 10^4$	4.7	7.2	7.1	—
Compression	—do—	6.9	3.6	3.1‡	0.30
Shear	—do—	4.3	1.5	4.9	0.03
Actual modulus					
Tension	$kN/m^2 \times 10^4$	0.55	0.80	20.7	—
Shear	—do—	0.28	0.034	8.6	—
Specific modulus†					
Tension	$kN/m^2 \times 10^7$	0.34	1.0	2.65	—
Shear	—do—	0.17	0.43	1.1	
Cost per m^3	$NP \times 10^4$	11.0	0.9	5.4	0.17
Specific cost†	—do—	6.9	1.1	0.7	0.07
Light transmission	per cent	85	Opaque	Opaque	Opaque
Maintenance	—	Translucent materials. Maximum useful life of 30 years without maintenance. Opaque materials. Require painting after 10–15 years and thereafter at similar intervals of time	Preservatives, varnishes, or paints necessary. Renewed application necessary according to type of finish	Corrosion protection essential and must generally comprise galvanising or priming followed by decorative finish. Repainting necessary	None except where decorative finishes are required

* The figures quoted are minimum design values for laminates made by the hand lay-up or contact moulding process with chopped strand mat reinforcement and resin content about 70 per cent.

† Specific properties are determined by dividing the figures for actual properties by the specific gravity.

‡ Yield stress.

§ Permissible stresses.

GRP can be fabricated by a variety of techniques so that different types of component can be made economically. For example, hot press moulding is particularly suitable for making large numbers of relatively small components like water storage cisterns. Hand lay-up is particularly suitable for making small numbers of large components like church steeples.

Not all properties will be needed for every application, but some of them will determine the suitability of GRP for specific applications where other materials do not possess the same combination of properties. Reinforced plastics are not necessarily substitutes for other materials. They may be used where traditional materials are used in a traditional way such as, for example, water tanks or cisterns. This is because they happen to perform as well or better than the traditional materials at the same or lower cost. In the long run, the greatest potential is in applications where traditional materials are not really satisfactory and where reinforced plastics will be used in a completely new way as, for example, in the manufacture of prefabricated bathroom units.

2.1. FIRE PERFORMANCE AND THE BUILDING REGULATIONS

The relevance of building regulations[8] on the use of plastics has caused considerable confusion and some understanding of the situation is necessary before any large scale use of GRP or, indeed, any plastics material for structural applications can be considered.

When a fire occurs, the main danger arises from spread either within the building or from one building to another. Regulations are imposed to limit this spread. In England and Wales, Parliamentary Acts for public health and safety are the enabling Acts for the Building Regulations 1965.[9] In addition, the London Building Acts[10] enable the Greater London Council to impose the London Building (Constructional) By-laws. These London By-laws are effective in the administrative area of the former L.C.C. Thus, buildings in the outer London Boroughs are subject to control by the Building Regulations 1965.

In Scotland the Building Standards (Scotland) 1963[11] are wider in scope than public health and safety requirements. For example, they impose controls on means of escape and electrical installations.

Buildings that are exempt from these Regulations include those for statutory undertakings, for example British Rail, Electricity, Gas, Water and National Coal Boards, British Road Services, the Crown and schools, where their special functions cannot necessarily be dealt with adequately by the standard Regulations. All the regulations and controls do, of course, take account of risks other than those caused by fire. For plastics the greatest risk is undoubtedly that caused by fire.

Controls to limit the risk of fire spread do so by assessing the use to which the building is put, its size, or the size of its parts, the degree of isolation between buildings or parts of buildings, and by laying down standards for the resistance of the structural elements and surface finishes to the action and spread of fire.

Tests which will influence the use of reinforced plastics and the behaviour of GRP when subjected to these tests are summarised in Table 2.2.

In practice, the most commonly applied yardstick for assessing the performance of a material like GRP is the surface spread of flame rating. There are many instances where GRP with a known rating has been permitted in applications where regulations have been waived in the absence of more realistic tests.

Table 2.2

B.S. 476 FIRE TESTS ON BUILDING MATERIALS AND STRUCTURES

Test Reference	*GRP Performance*
B.S. 476: Part 3: 1958 External fire exposure roof tests	The designation Ext.F. AA or Ext.S. AA is attainable with sheet made from special self-extinguishing polyester resins. These designations indicate the most exacting conditions covered by the test
B.S. 476: Part 4 Non-combustibility test for materials (replaces combustibility test in B.S. 476: Part 1: 1953)	GRP, in common with all plastics materials, is classed as 'combustible'
B.S. 476: Part 5: 1968 Ignitability test for materials	The designation 'X' identifies easily ignitable materials. The designation 'P' identifies not easily ignitable materials. GRP should achieve the 'P' designation, but claims have not been made because the Building Regulations do not yet incorporate this test designation
B.S. 476: Part 6: 1968 Fire propagation test for materials	The index of performance 'I' measures the amount and rate of heat evolved by a material subjected to fire in an enclosed space. The test was devised to distinguish between materials with a Class 1 surface spread of flame rating. The Index I is quoted together with the designation 'P' or 'X' of the ignitability test.
B.S. 476: Part 7 Surface spread of flame test for materials (replaces the test in B.S. 476: Part 1: 1953)	A Class 1 rating—surfaces of very low flame spread—can be achieved by GRP made from self-extinguishing polyester resins.
B.S. 476: Part 8. Fire resistance tests for elements of building construction (replaces fire resistance tests of structures in B.S. 476: Part 1: 1953)	Grading periods of 0.5–6 h are specified according to needs. GRP for elements of construction, such as walls, is normally used in combination with other materials like asbestos or concrete. These elements can be designed to achieve the specified grading

Basically, Building Regulations do not discriminate against any one material. In practice, however, Regulations are based on the behaviour of traditional materials and in the 'deemed to satisfy' schedules traditional forms of construction will obviously predominate. Architects, designers, engineers and the local authorities, who are empowered to interpret and to enforce Regulations, are therefore more likely to turn to traditional materials.

Local authorities may grant waivers for specific cases, but obviously there must be a strong incentive for this to occur. They cannot grant waivers where structural fire precautions are involved and such waivers for non-compliance must be referred to the Ministry. Fortunately, however, this inhibition on the use of new materials has been recognised. The Government have established an independent authority, the Agrément Board,[12] which examines new materials and new methods of construction to determine whether they can be recommended for general use by the construction industry. The Board works closely with the Building Research Station who may need to carry out detailed tests on materials and forms of construction for which British Standard specifications or codes of practice have not been published. For an acceptable product the Board issues a certificate which is valid for 3 years. This certificate carries the authority by which a specifying body can accept a new material or an innovation. The Board notify all local authorities about the certificates.

In addition to the encouragement for innovation being made through the Agrément Board, the British Plastics Federation Building Group have commissioned the Rubber and Plastics Research Association to study the behaviour of plastics in fire, with the objective of establishing really meaningful tests. This work is being done with the close co-operation of the Fire Research Station, but final results which can be incorporated into Building Regulations will take many years to achieve.

One of the problems is that little is known about the behaviour of traditional building materials which are combustible, for example, timber, and tests for combustible materials must be meaningful for all of them. If the work is satisfactorily concluded it will provide another way of getting plastics accepted without causing unnecessary risk.

2.2. WEATHERING[15]

The performance of plastics on external surfaces of a building is also difficult to assess[13–16]. Performance of GRP in any given situation is affected by the method of manufacture as well as by the type of polyester resin and the glass fibre reinforcement which is used. Artificial weathering tests can give misleading results and there is no universally accepted test method which can be correlated with natural weathering.

The practical effects of weathering on GRP surfaces include discoloration (or reduced light transmission of a translucent material), surface crazing, exposure of the glass fibre reinforcement and loss of mechanical strength. It is possible, however, for a translucent rooflight material to give 30 years satisfactory service. If opaque material is resurfaced after 10–15 years, there is no reason why the material should not survive the life of the building.

The processes of degradation of GRP are very much slower than with many traditional materials. For example, GRP weathers much better than most timber, and rooflight sheet does not embrittle with age like cement asbestos sheet. Good quality GRP mouldings can be said to require very little maintenance, but they cannot be compared with materials like stone, brickwork or slate.

3. METHODS OF MANUFACTURE

GRP components in building applications may be made by any of the conventional techniques, including hand lay-up, cold press moulding or hot press moulding, and filament winding[17,18]

3.1. HAND LAY-UP

The laminate is made by consolidating by hand layers of brush or spray applied polyester resins and glass fibre reinforcement in an open mould. The mould itself is usually made of GRP. After treating the mould surface with parting agents, a gel coat of pigmented resin is commonly applied. The gel coat improves the appearance of the finished laminate by obscuring the glass fibre pattern, and it also improves weather performance. Laminates made by this technique will only have one surface that is smooth and accurate in contour. The process is particularly suitable for making large components and small numbers of mouldings.

3.2. COLD PRESS MOULDING

Laminates are made by consolidating resin and reinforcement in matching tools, usually made of concrete with a polyester or epoxide resin facing. Simple clamping rigs, combined with the weight of the moulds themselves, provide the required pressure. The process is particularly suitable for making large mouldings required in moderate quantities, i.e. 2 000 or more.

3.3. HOT PRESS MOULDING

Mouldings are made from resin and glass fibre 'preforms' in heated steel moulds mounted between the platens of a hydraulic press. The preforms are made by depositing the glass fibre on a male mould and coating them with a binder to retain their shape. The process is suitable for long runs of small components, such as water storage tanks and cisterns[19]. Mouldings made by this process have a higher glass fibre to resin content than mouldings made by hand lay-up or by cold press moulding. This results in improved strength and better fire performance.

3.4 FILAMENT WINDING

This process consists of winding continuous filaments or rovings on a mandrel after they have been passed through a resin bath. Whilst generally used for making cylindrical pipes and tanks on a collapsible mould, the process is also used for cladding rigid PVC and spun concrete pipes.

3.5. CONTINUOUS PROCESSES

Corrugated rooflight sheet can be made by hand lay-up, but it is usually made by various continuous processes[20]. Basically, they comprise the deposition of strands of chopped rovings onto a film of regenerated cellulose which is supported by a moving conveyer belt, coating with a spray of resin and then consolidating. Corrugations are formed before curing and removing from the conveyer. GRP profiles, such as angles and channels, can also be made by continuous processes[21,22].

3.6. WHICH METHOD TO SELECT

The actual method of manufacture will be determined by the number of components which are needed, their dimensions, the properties required and the costs of manufacture.

Many of the more exciting new applications for GRP components involve the use of large mouldings which are generally made more economically by hand lay-up. Press moulding gives better consistency of physical and mechanical properties which, in turn, may influence fire and weather performance. The designer must, therefore, determine the method of manufacture of a component at an early stage in design.

4. APPLICATIONS

In 1968 U.K. consumption of polyester resin was about 30 000 tons. Approximately 7 000 tons or 23 per cent of the resin was used for building applications. Some of the resin is used in paste form for adhesives, coatings and fillers. The balance is reinforced with glass fibre. Rooflight sheet is the major end use for glass fibre reinforced polyesters and took about 5 000 tons of polyester resin in 1968. Applications are summarised in Table 2.3.

4.1. ROOFLIGHT SHEET

Reinforced plastics provide a versatile medium for lighting many types of building. The material is available in a variety of forms, ranging from the mass-produced corrugated sheet to the special effect designed and probably made in a studio. Other building materials do not possess the same useful combination of properties. The light transmission (usually about 85 per cent) lightness, toughness, weather resistance and, where appropriate, self-extinguishing properties and the cost of GRP, make it particularly suitable for roof and vertical lighting of factories, warehouses, loading bays, railway stations, sports arenas and swimming pools[23].

Corrugated sheet is the most commonly used form of GRP for rooflights. It is made in a range of profiles which match corrugated steel, aluminium

Table 2.3

PRINCIPAL BUILDING APPLICATIONS FOR GRP[14]

Application	*Form in which supplied*	*Advantages over traditional materials*	*Disadvantages*
Rooflights	Corrugated or flat sheet	Stronger than glass, lighter in weight and easier to install	Light transmission less than that of glass. Deterioration of light transmission on ageing
Domelights	One piece components moulded or fabricated	Light in weight, easy to install	——do——
Domes and other roof structures	Modular components, single or double skins	Light in weight, easy to erect	Stiffening with metals or timber may be necessary
Internal partitions	Corrugated or flat sheet	Convenient in use. Special decorative effects can be easily incorporated	Limited use because of cost
Cladding	Flat or profiled unsupported sheet or as surface 'skin' to concrete or asbestos	Lightness, range of decorative effects, versatility for individual designs	Fire performance limitations and the effects of prolonged weathering
Sectional buildings	Modular components, often double-skinned with 'sandwich' construction	Lightness and ease of erection	——do——
Bathroom units	Assembled modules	——do——	No specific disadvantages
Tanks and cisterns	One or two piece press mouldings	Lightness, no corrosion, low thermal conductivity	No specific disadvantages
Pipes and ducts	Continuous profiles or as cladding on concrete or PVC pipe	Improves strength of concrete pipes and protects them from chemical attack. Increases temperature range for PVC pipe	No specific disadvantages
Window frames	Assembled press moulded components or as sections for cladding timber	Reduces maintenance associated with most other materials	Less suitable than timber for non-standard dimensions
Concrete moulds	Mouldings generally made by hand lay-up, but sometimes by press moulding	Lightness. Gives concrete of high quality and excellent finish. Provides a new medium for architectural designs on concrete	Low stiffness means that additional support is often necessary

and asbestos sheeting, thereby allowing an interchange of materials to give the desired degree of lighting.

Translucent GRP is used in two other forms. One is the domelight which is fabricated from either flat or corrugated or specially moulded sheet. Domelights are used on industrial, commercial and public utility buildings. GRP is particularly suited for this application because the structure can be self-supporting and can be designed so that it can be secured to the roof or the upstand without the need for a supporting framework. The other form of rooflighting is when the GRP spans the entire area to be covered by forming both walls and roof. Examples are swimming pool canopies, storage buildings, and radar domes. The structures may be assembled from mass-produced small modules or by joining large mouldings made by the hand lay-up techniques. Joints are usually made by bolting through flanges and sometimes a metal framework is incorporated to give additional stiffness to the structure.

For improved thermal insulation, double-skinned components can be made in flat or corrugated form where the two skins are joined through flanges or webs of GRP. Metal grids have also been used as a core. By combining colour tinted translucent panels, special decorative effects for church windows are possible.

4.2. INTERNAL PARTITIONS

Ordinary corrugated sheet is used for partition glazing in industrial and commercial buildings. More decorative forms incorporating special effects are made for domestic use and for hotels and places of entertainment. An example of such a panel is a double-skinned construction which envelops a core of honeycomb or similar structure to give a decorative effect.

4.3. CLADDING

GRP is used in two ways. Firstly, for cladding other structural materials, and secondly as an integral part of either a structural or a non-loading bearing wall panel.

For cladding structures of concrete or brick, sheet is used in the profiles made for rooflights and vertical lighting, and also in special profiles such as 'ship lap'. Particular advantages of GRP are the ease of fabricating large panels to minimise joints, and the infinite range of coloured and textured surfaces that can be made at acceptable cost. Fixing devices and joints must, however, accommodate the movement caused by the different thermal expansion between GRP and other materials, such as concrete and metals.

GRP is used in a variety of ways in combination with other materials. Applications range from a simple spandrel panel design to a complex storey height component. In these panels the GRP is invariably the exterior skin. Panel construction may be a simple sandwich panel. A cellular plastics core is often used to give thermal insulation. The inner skin may be GRP or another type of sheet material.

Figure 2.1. Three-storey dwellings constructed from GRP composite loadbearing panels (Courtesy of William Old (Resiform) Ltd.)

Figure 2.2 22-storey G.L.C. dwellings clad with Indulex GRP skinned panels (Courtesy of BP Chemicals (U.K.) Ltd.)

GRP can impart an acceptable finish to asbestos sheet. There are several types of GRP skinned asbestos panel made and a good example is the structure shown in *Figure* 2.1[24]. Here the panels comprise an industrialised building system of novel design. Each panel embodies a loadbearing framework suitable for three-storey buildings. For high rise structures the panels can be used as a cladding.

Moulded GRP shells backed by lightweight concrete and an inner skin of plaster have been successfully used for cladding high rise dwellings as shown in *Figure* 2.2[25]. The panels were developed for the Greater London Council to clad high rise steel frame buildings. Design requirements included the need for lightness in weight, a one-piece surface, dimensional accuracy, low maintenance, a one-hour fire resistance and a satisfactory rating for surface spread of flame.

In the selection of GRP two points of particular interest are that (1) the press moulding was chosen to achieve the required physical properties, including the required surface spread of flame rating, and (2) that the GRP and the lightweight concrete core are complementary materials used to achieve the overall design needs. The panels in the illustration measure 2.6 m high by 2 m wide, with a maximum depth of 0.15 m. Completed panels weigh approximately 250 kg. The average density is 400 kg/m^3 compared with dense concrete at 2400 kg/m^3.

A novel way of using GRP for cladding buildings both externally and sometimes internally is the mural which, like the church window, is usually designed and made in the artist's studio. These panels are often very large and of striking design and they have drawn attention to the use of plastics to achieve effects that could not be created with any other material.

4.4. SECTIONAL BUILDINGS

Lightness in weight of GRP mouldings, combined with desirable physical and mechanical properties, ease of fabricating large components and simplicity of jointing, has meant that the design of sectional buildings is a logical application. Such structures are particularly suitable on sites where access is limited or where the ground cannot support traditional structures without excessive cost for foundations. Examples include railway trackside relay rooms, living accommodation and laboratories at Antarctic bases, lighthouse towers, desert accommodation, petrol stations and even the weekend chalet.

Sections will often incorporate a cellular plastics core to give thermal insulation. The sandwich construction gives improved stiffness. Other design features employed to give adequate stiffness include folded plate designs (*Figure* 2.4), curved sections, ribs and flanged joints. (*Figure* 2.5)[26]. Ribs and joints may also be reinforced by incorporation of timber or steel sections. Such additional stiffening, however, requires very careful design at the joint to ensure even stress distribution, as concentration of stresses may cause premature failure. The incorrect use of steel which is so much stiffer than GRP can, in fact, cause stress concentration and premature failure in adjacent areas of the laminate.

Figure 2.3. Weekend chalet made from 36 GRP sections of double-skinned sandwich construction. The building has a diameter of 8 m, *a floor area of* 25 m^2 *and weighs less than* 2 tons (Courtesy of Oy Polykem AB, Helsinki)

Figure 2.4. Stores building constructed from 'foldedplate' GRP sidewalls. Each module measures 1.2 m *wide by* 6 m *high. The complete structure measures* 15 m *by* 14 m *and is roofed with translucent GRP units* (Courtesy of Scott Bader Co. Ltd.)

Some of the largest reinforced plastics structures made have been for roofing buildings made of other materials. GRP church steeples and spires result in considerable economies because of their lightness in weight and ease of installation. They are used as replacements on traditional stone structures which may not support the weight of a traditional steeple or spire without expensive structural alteration, and they are used on new buildings which can be designed for the lightweight GRP structure. *Figure* 2.6 is an outstanding example of a GRP steeple.

4.5. BATHROOM UNITS

There are two and a quarter million dwellings in the U.K. without bathrooms. Of these homes one and a half million are unfit and will be replaced in the near future. This means that there is a potential market of threequarter million homes where a bathroom can be installed. Dwellings without bathrooms are usually too small for internal re-design, but a shell or pod containing the bathroom, wash basin and possibly W.C. can often be sited externally with access from a doorway made in an outside wall[27]. GRP is a very convenient structural medium for this type of component and *Figure* 2.7 is a good example of such a unit.

4.6. TANKS AND CISTERNS

Smaller mouldings for both hot and cold water storage are good examples of applications for press moulding. Tool costs can be amortised over the long production runs expected for these standard components. The hot water cylinder is a particularly interesting development (*Figure* 2.8). Thorough testing has shown that, with the use of a special type of polyester resin, the cylinder withstands the effect of temperature and pressure, yet it is competitive in price with copper cylinders. Furthermore, GRP is not subject to the world market fluctuations of price and availability of copper.

4.7. PIPES AND DUCTS

GRP is used for cladding spun concrete pipes and for skinning rigid PVC pipes. Concrete pipes clad with helically-wound glass fibre rovings impregnated with polyester resin have improved strength and they can be laid without having to be embedded in concrete. They are particularly suitable for effluent disposal in estuaries and areas where concrete may be attacked chemically, or where ground movement can be expected. If the pipes do not have to be buried in concrete, some degree of flexibility can be designed for at the joints.

GRP skinned PVC pipes are generally made for installation in chemical plant. The GRP not only provides greater mechanical strength at normal temperatures, but protects the PVC from distortion caused by hot fluids.

Ducts are used for cable laying or cable support. GRP ducts have been used for railway electrification. In domestic dwellings GRP cable and meter covers have been adopted where meters are installed for external reading.

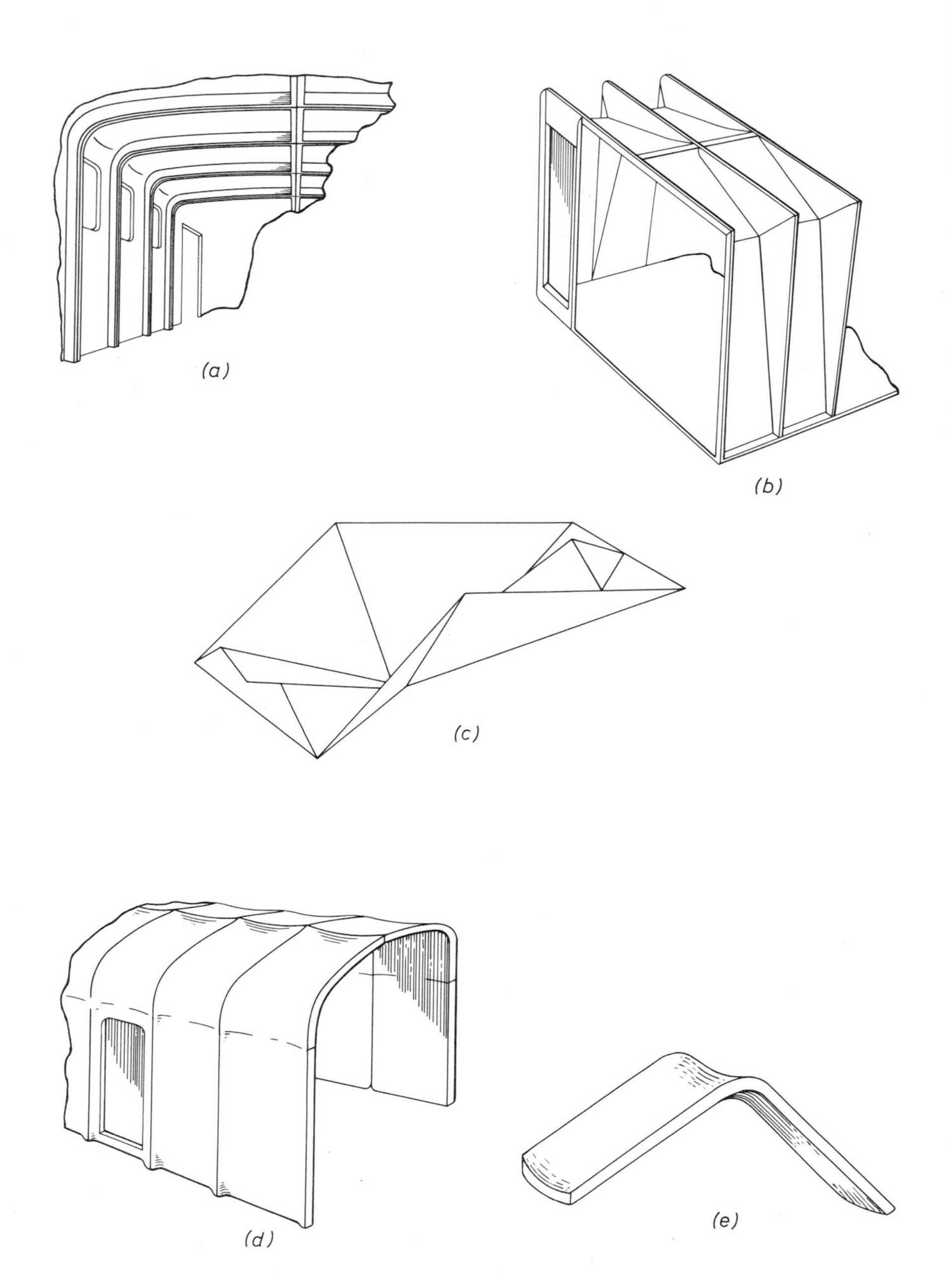

Figure 2.5. Illustrates (a) the use of stiffening ribs (b) the folded plate or rectilinear form (c) a folded plate roof structure (d) the curvilinear form and (e) one of the sections in more detail

Figure 2.6. 24 m *high GRP steeple comprising base, clock section, bell tower and spire with weather vane. The structure is an exact replica of the original Wren design for St. James's Church Piccadilly. The total weight of the GRP structure is less than* 10 *tons* (Courtesy of Bakelite Xylonite Ltd.)

4.8. WINDOW FRAMES

A number of designs have appeared in both GRP and rigid PVC in the last 10 years. Both materials offer the freedom from maintenance which is costly with timber and with steel. GRP window frames could be expected to last 15 years without maintenance. A GRP frame will not suffer from warping or distortion and should provide a good weather seal. Unfortunately, the high cost restricted market penetration until the emergence of a recent design which is only marginally more expensive than timber[28]. This design is based on a full assessment of the cost stiffness relationship combined with the high quality and good appearance made possible by a modified press moulding technique. Standard sections can be assembled for modular sizes and for purpose made frames.

Figure 2.7. GRP bathroom unit weighing only 105 kg *for Cleethorpes council house. The unit is assembled from two mouldings joined horizontally. Plan area is* 1.7 m *by* 1.9 m (Courtesy of Rollosrank Ltd.)

Figure 2.8. GRP hot water cylinder designed as equipment to the B.S. 699 grade 3 copper cylinder which is capable of withstanding a test pressure of 110 kN/m^2 *and operating under a maximum* 9 m *working head. Dimensional area* 0.9 m × 0.45 m *and nominal* 130 litres *capacity* (Courtesy of Osma Plastics Ltd.)

4.9. CONCRETE MOULDS

Formwork accounts for 30–40 per cent of the cost of a concrete structure. It is essential that this money is well spent. The appearance and the quality of the concrete surface is determined by the formwork. The surface finish cannot be better than the finish of the forms or moulds which comprise the formwork. The term 'form' is used for site-poured concrete and the term 'mould' is used for pre-cast concrete.

Today, engineers demand high quality concrete surfaces free from surface blemishes and cracks to obtain optimum strength, and the architect and designer want surfaces that can achieve the appeal of Portland stone.

GRP moulds and forms yield concrete of very high quality[29]. The best results are obtained with the use of the correct grade of mould oil or chemical release agent which also improves the life of the mould. Prime costs are generally higher than costs of timber moulds and less than the cost of steel moulds, but the cost per unit depends upon the number of units required and the complexity of the mould.

Where artists' original designs are required on a concrete panel, GRP is likely to be the cheapest mould[30] material for a single use and, indeed, it can be said that GRP has given rise to a freedom of design for large concrete structures not practicable in any other medium. (*Figure* 2.9.)

Figure 2.9. Concrete panel cast in GRP mould (Courtesy of William Mitchell, Design Consultants)

In purely functional applications, GRP moulds and forms are ideal for casting coffered floor slabs, bridge columns, ducts and pre-cast beams, posts, stairways and paving slabs.

A particularly interesting development is the use of GRP as permanent formwork. Examples include duct linings for power station cooling water and sewage treatment. Externally located permanent formwork is used for column cladding and for building panels.

4.10. MISCELLANEOUS APPLICATIONS

Other uses for GRP in building and construction include road signs, lamp columns, garage doors, porch roofs and canopies, eave flashing, roof vents, ventilation cowls, etc. In general, these applications are viable because GRP gives low maintenance in use, is light and easy to install and can be fabricated by a variety of techniques for large or small production runs.

5. CONCLUSIONS

The unique combination of properties which assured the acceptance of GRP for rooflights, and which is still the largest single building application, has been followed up by designs for domelights, roof domes and complete building structures. In the future, GRP will continue to be successfully exploited for moulded components in increasingly sophisticated household needs for heating, ventilating, cooking, plumbing, sanitation and power supply. GRP has established itself for concrete moulds and forms, and this application is likely to grow as the demand for better quality and better finished concrete grows.

GRP for external applications will also increase as the building industry is forced to divert more of the labour force from maintenance to new construction. At present one-third of the total labour force is devoted to maintenance of existing buildings. An obvious way to reduce this wastage is by the use of new materials which need less maintenance than the materials they supersede.

GRP is establishing itself in structural applications as performance data is recorded for the benefit of future designers. At the beginning of this chapter the 'Monsanto House of the Future' was mentioned. It was dismantled in 1967 because it was considered to have served its purpose, and not because of deterioration.

The restraining factor on the wider use of GRP is undoubtedly the lack of information on behaviour in fire, the problems associated with improving the performance in fire, the difficulty of specifying realistic tests and the difficulty of interpreting existing tests. The solution of so many inter-related factors is inevitably long term. There will be no sudden 'breakthrough'. A 'code of practice' cannot be written until 'practice' is established, but advancement there will be and as each reinforced plastics structure is erected it will breed ideas for others.

REFERENCES

1. Monsanto Plastics 'House of the Future'. *The Architect Bldg News,* 9th October, 1957
2. Plastics in Building Structures. *Proc. conf. Plastics Institute,* June 1965. Pergamon Press
3. *F.R.P. Design Data.* Fibreglass Ltd.
4. 'The Evaluation of Published and Unpublished work on the Long-term Properties of Reinforced Plastics.' *Br. Plast. Fed.* No. 44, 1960
5. 'Fatigue and Creep in Reinforced Plastics.' *Br. Plast. Fed.* No. 64/1 1967
6. Owen, M. J. *Trans. J. Plast. Inst.* **35**, 115, 353, February 1967
7. Owen, M. J. and Smith, T. R. *Plastics and Polymers,* **36**, 121, 33, February 1968
8. Fitt, R. A. *Guide to the use of plastics under the Building Regulations* 1965. Polyplan Ltd., Leicester. Nov. 1967
9. 'The Building Regulations 1965.' *Statutory instrument No.* 1373, H.M.S.O.
10. *London Building Acts* 1930–1939, *Constructional By-laws.* Greater London Council, March 1965
11. 'The Building Standards (Scotland) Regulations 1963.' *Statutory instrument No.* 1897 (S.102). H.M.S.O.
12. *The Agrément Board and its Work.* The Agrément Board, Hemel Hempstead
13. 'Applications and Durability of Plastics.' *Digest* 69, Second series — Building Research Station
14. 'Plastics in Building.' *Overseas Building Note* 108 — Building Research Station
15. Crowder, J. R. 'The Weathering behaviour of Glassfibre reinforced polyester sheeting.' *Miscellaneous Papers* 2 — Building Research Station
16. Crowder, J. R. and Matthan, J. *Plastics* **33**, 372, 1135 October 1968
17. *Cellobond Polyester Resins.* BP Chemicals (U.K.) Ltd.
18. *Polyester Handbook.* Scott Bader Co. Ltd.
19. 'Plastics Hot Water Cylinder.' *R.P.A. Journal,* June 1967
20. 'Corrugated Plastics Sheeting.' *Br. Plast. Fed.*
21. Morris, D. R. *New Scientist,* 29th November, 1962
22. Kannelby, G. 6*th Int Reinforced Plastics Conf.,* British Plastics Federation, November 1968
23. 'In Four Days—A roof fit for a swim.' *Contract Journal,* 2nd May, 1968
24. '225 flats by Resiform Construction.' *Industrialised Buildings, Systems and Components,* September 1968
25. Blake, H. V. 5*th Int. Reinforced Plastics Conf.,* British Plastics Federation, November 1966
26. Makowski, Z. S. *Systems, Buildings and Design,* October 1968
27. 'More help for Councils on Twilight Housing.' *Coun. Equipment and Building News,* June 1968
28. Hughes, G. 'Low cost window frames using DMC.' 24*th S.P.I. Reinforced Plastics/Composites Division Conference Proceedings,* February 1969
29. 'Moulds and Form Faces for Concrete.' *Plastics in Building,* Vol. 1, No. 4 Bakelite Xylonite Ltd.
30. 'Glass Reinforced Plastics for Moulds.' *Concrete Building and Concrete Products,* January 1967

3

Boat Hulls

T. M. WILKS

1. INTRODUCTION

Boats, that is vessels up to about 15 m, have for all time and in nearly all places been traditionally built in wood. This practice has persisted right up to the middle 1960s when GRP can be said to have become the primary boat building material. Experimental hulls made in the late 1940s were amongst the largest GRP mouldings to be made.

This revolution has been caused by the low cost of equipment, tools and machinery necessary for production together with the very marked comparative lack of skill required to produce good products. These facts also ensure that there is no lack of small firms willing to start up and that the industry will remain highly competitive.

Furthermore, most boat hulls are not so adversely affected by the low rigidity of GRP as many other structures. This is due mainly to the need for an outside skin capable of withstanding the high point loadings of groundings on uneven, and perhaps rocky, surfaces or of lying alongside a quay or another vessel while afloat without sustaining damage. Such a skin will be far in excess in thickness of that required to take any calculated hull bending loads, and indeed in most cases is thick enough to only require intermittent framing for adequate panel stiffness. For this reason also there does not seem to be a case for using epoxy resins in preference to polyesters.

The corrosion and weathering resistance of GRP are naturally advantageous in boat hull construction, this being particularly so in the case of ships' lifeboats. These are subjected to heavy marine atmosphere weathering and intermittent industrially polluted environments with only minimum maintenance. Needless to say, for these reasons GRP is almost exclusively used today for constructing ships' boats.

In some cases it is necessary to produce many identical hulls, such as those of one-design class racing sailboats. Before the advent of GRP no material existed which could be shaped using tools of a cost low enough to give economy for the numbers involved. These one-design hulls had to be craftsman built largely depending on accurate hand shaping of a great number of small parts. Now, however, every year new racing classes of GRP boats appear, offering very substantial price savings over the comparable wooden boats they aim to replace.

In the design of ships, aircraft or hovercraft, or indeed of most dynamically loaded structures, it is useful to consider the loads applied and then design mathematically a structure to withstand them with an adequate safety factor.

It is possible to use such a theoretical design procedure with a GRP hull and in some special cases it may be useful. However, a GRP boat hull will most probably be produced in large numbers where final design of lay-up, choice of thickness, and type of shell stiffening, will be found by experience gained on a prototype. Prototype design will almost certainly be made using comparative strength to a previous design.

Let us now look in more detail at GRP hull structure, and, for this purpose, divide it into four boat types.

2. SHIPS LIFEBOATS

Ships boats ranging from 16 ft to 36 ft in overall length, i.e. approx. 5 m to 10 m, must conform to rules laid down by the International Marine Consultative Organisation (I.M.C.O.) and the regulations, inspection, and certification, of the country under whose flag the ship will be sailed. In the United Kingdom marine safety is the concern of the Board of Trade who have accepted GRP ships lifeboats since 1955.

Manufacturers seeking Board of Trade approval must satisfy the surveyors that the workshops are suitable and proper quality control available. Temperatures must be kept between 15°C and 22°C and records kept of actual temperatures and humidities. Resins must be recommended by their manufacturers as suitable for boat hulls and for working ambient temperatures between 65° and —32°C. Glass reinforcements must contain not more than 1 per cent alkali, calculated as Na_2O. Laminates must pass carefully controlled self-extinguishing tests.

Apart from fulfilling the design rules governing lifeboats built of any material, the Board require a theoretical maximum working stress of 0·5 ton/in^2 (approx. 0·8 kN/m^2 × 10^4) on the GRP structure design. A typical stress calculation for a 28 ft (8·4 m) long 70 person capacity lifeboat is given in Table 3.1.

The tests for prototype lifeboats are laid down in the Board's instructions to surveyors 1959, which state 'Tests for GRP lifeboats shall be as follows:

(*a*) A 100 per cent overload test with the lifeboat suspended by the lifting hooks. Where alternative positions are provided for the lifting hooks, the test is to be made with the lifting hooks at the maximum span, unless otherwise instructed by the Chief Ship Surveyor. The lifeboat is to be loaded incrementally and measurements at full load, 25, 50, 75 and 100 per cent overloads should be recorded as follows:

(i) Deflection of keel amidships.

(ii) Change in length as measured between top of stem and stern posts.

Table 3.1. CALCULATIONS FOR MOMENT OF INERTIA AND STRESSES IN GUNWALE AND KEEL

Formula for calculating stresses due to bending moments $p = MY/I$

where p = intensity of stress exerted on gunwale or keel
M = Bending moment
Y = Distance of section from neutral axis ($N.A.$)
I = Moment of inertia of the section about its neutral axis

Formula for calculating bending moment [from M.O.T. survey of lifesaving appliances supplement No. 1 (1963) *para* 6]

$BM = WL/6$

where W = Total davit load—weight of lifeboat
L = Distance between lift hooks

	Description of item	*Section*	*Area* in^2	Y ft	AY in^2 ft	AY^2 in^2 ft^2	k ft	AK^2 in^2 ft^2
Items	Gunwale	18 × 0.562	10.12	2.875	29.10	83.65		
situated	Shell at side	12 × 0.250	3.00	1.60	4.80	7.68	1.00	3.00
above	Side seat	14 × 0.310	4.34	2.125	9.22	19.60		
assumed	Cleading	15 × 0.250	3.75	1.50	5.63	8.44	1.25	5.86
neutral	Shell at bilge	13 × 0.250	3.25	0.50	1.63	0.81	1.10	3.93
axis	Lower side seat	12 × 0.125	1.50	0.80	1.20	0.96	0.80	0.96
	Cleading	10 × 0.187	1.87	0.45	0.84	0.38		
			27.82		52.42	121.52		13.75
Items	Shell at bilge	22 × 0.250	5.50	0.375	2.06	0.77		
situated	Bottom shell	27 × 0.370	9.99	0.80	7.99	6.39		
below	Cleading below NA	9 × 0.187	1.68	0.33	0.55	0.18	0.75	0.95
assumed	Half keel	10 × 0.750	7.50	1.40	10.50	14.70	0.50	1.88
neutral axis			24.67		21.10	22.04		2.83
			52.50		31.32	143.56		16.58

NOTE: K = Vertical height of section
NOTE: *Gunwale section area allows for a* $\frac{3}{4}$ *in. dia. hole for rowlock*
Area = 52.50 in^2: $AY_A - AY_B$ = 31.32 ft in^2: AY^2 = 143.56 ft^2 in^2: AK^2 = 16.58 ft^2 in^2

Position of actual NA about assumed $NA(h) = AY/\text{area}$ = 31.32/52.50 ft 0.597 ft
$\therefore$ Y to keel 1.38 + 0.597 = 1.977 ft
$\therefore$ Y to gunwale 2.76 — 0.597 = 2.163 ft
Moment of inertia about assumed $NA = AY^2 + (AK^2)/12$ = 144.94 in² ft²
Correction to MI due to transfer of NA_A to $NA = A \times h^2$ = 18.71 in² ft²
$\therefore$ Moment of inertia $= MI \,.\, NA_A$ — correction $= a$ = 126.23 in² ft²
Total moment of inertia $= a \times 2$ = 252.46

Bending moment:
W = 164.10 cwt, L — 27 ft
$BM = W \times L/6 = (164 \times 27)/6 = 730$ cwt ft

Stress in gunwale:
$p = MY/I \quad \therefore p = 6$ cwt/in²
$\therefore$ Stress in gunwale = 0.3 ton/in²

Stress in keel:.
$p = MY/I \quad \therefore p = 5$ cwt/in²
$\therefore$ Stress in keel 0.2 ton/in²

Important: Stress in gunwale or keel not to exceed 0.5 ton/in²

Boat weights:

Boat with fixed equipment	=	45.00
Skates	=	0.70
Fuel	=	1.65
Provisions	=	6.71
Loose equipment	=	2.90
70 persons at 165 lbs	=	103.12
		160.08 cwt

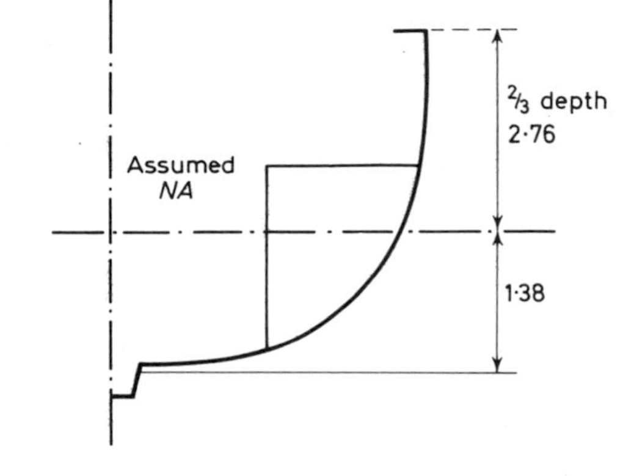

Table 3.2. RESULTS OF OVERLOAD TESTS OF THE 28 FT 70 PERSONS LIFEBOAT

Total davit load:			
	Boat with fixed equipment	=	45.00
	Fuel	=	1.65
	Provisions	=	6.71
	Persons	=	103.12
	Skates	=	0.70
	Loose equipment	=	2.90
	Total davit load	=	160.08 cwt
	Total davit weight	=	160.08
	Less boat with fixed equipment	=	45.00
			115.08 cwt (weights to be added)

% *Overload*	25%	50%	75%	100%	150%
Total davit load	160.08	160.08	160.08	160.08	160.08
Overload	40.02	80.04	120.06	160.08	240.12
	200.10	240.12	280.14	320.16	400.20
Less boat with fixed equipment	45.00	45.00	45.00	45.00	45.00
Weights to be added	155.10	195.12	235.14	275.16	355.20

Strength test results

Load	*Wt. cwt*	*Keel deflection*			*Gunwale spread*			*Length decrease*	*Gunwale drop*		
		For'd $\frac{1}{4}$	*Amidships*	*Aft* $\frac{1}{4}$	*For'd* $\frac{1}{4}$	*Amidships*	*Aft* $\frac{1}{4}$		*For'd* $\frac{1}{4}$	*Amidships*	*Aft* $\frac{1}{4}$
Zero–normal		$\frac{1}{8}$	$\frac{7}{16}$	$\frac{3}{8}$	$\frac{11}{32}$	$\frac{3}{32}$	$\frac{7}{16}$	$\frac{7}{16}$	$\frac{1}{32}$	$\frac{1}{32}$	$\frac{1}{32}$
25%		$\frac{11}{32}$	$\frac{5}{8}$	$\frac{9}{16}$	$\frac{9}{16}$	$\frac{1}{8}$	$\frac{11}{16}$	$\frac{3}{4}$	$\frac{1}{32}$	$\frac{1}{32}$	$\frac{1}{32}$
50%		$\frac{15}{32}$	$\frac{7}{8}$	$\frac{5}{8}$	$\frac{3}{4}$	$\frac{1}{8}$	$\frac{7}{8}$	1	$\frac{1}{16}$	$\frac{1}{16}$	$\frac{3}{32}$
75%		$\frac{15}{32}$	$1\frac{1}{8}$	$\frac{13}{16}$	$\frac{7}{8}$	$\frac{7}{32}$	$1\frac{1}{32}$	$1\frac{1}{8}$	$\frac{3}{32}$	$\frac{3}{32}$	$\frac{3}{32}$
100%		$\frac{7}{8}$	$1\frac{3}{16}$	$\frac{7}{8}$	$\frac{31}{32}$	$\frac{7}{32}$	$1\frac{1}{8}$	$1\frac{5}{16}$	$\frac{3}{32}$	$\frac{3}{32}$	$\frac{1}{8}$
150%		$\frac{7}{8}$	$1\frac{1}{4}$	$\frac{7}{8}$	$1\frac{1}{16}$	$\frac{7}{32}$	$1\frac{5}{16}$	$1\frac{3}{8}$	$\frac{1}{8}$	$\frac{1}{8}$	$\frac{5}{32}$
Permanent set		—	$\frac{1}{4}$	$\frac{1}{8}$	$\frac{3}{8}$	$\frac{1}{8}$	$\frac{3}{16}$	$\frac{1}{4}$	—	—	—

(iii) Change in breadth over gunwale.

(iv) Change in depth measured from gunwale to keel. The keel deflection and change in breadth at (i) and (iii) above should not exceed 1/400 of the lifeboat's length when the boat is subjected to 25 per cent overload and the results at 100 per cent overload should be in proportion to those obtained at 25 per cent overload.

(*b*) A drop test into water from a height of 7 ft 6 in (keel to surface) with the lifeboat loaded to represent its service condition.'

The results of the overload tests on the 28 ft 70 person lifeboat are given in Table 3.2.

The drop test described in (*b*) may at first sight appear to be nothing very startling. However, it cannot be dismissed lightly. Lifeboats are comparatively flat bottomed over about half their length amidships. This causes high acceleration, of the order of 6*G* or even 8*G* over the bottom area on impact with the water surface. Accelerations at the gunwales during the same drop do not exceed 2*G* although here they will have a more gradual build-up and decline. This shows quantitatively what might be expected, namely that the bottom acts as a buffer for the rest of the structure, and in so doing will deflect in the order of 3 to 4 in in a 26 ft boat. The GRP shell can easily take such deflection if so designed. Care, however, must be taken that no rigid fitting is positioned close enough to be damaged by the deflecting GRP.

The reinforced plastic lifeboat can truly be described as a major breakthrough in ship's lifesaving equipment. This is because with the use of closed cell structure polyurethane foam as a buoyancy material instead of metal air tanks, the need for both inspection of the hull structure for rot and corrosion and the testing of air tanks for tightness has been eliminated. This alone results in big savings incurred by the weather resistance of GRP.

3. DINGHYS

Because of their small size, boats under this heading were in many cases the subject of very early experiments in GRP production. In some cases the boat was of secondary importance to the proving of a technique or even a resin or reinforcement. There are now, however, hundreds of large and small firms manufacturing good GRP dinghys.

Open rowing or outboard boats and canoes are usually about $\frac{1}{8}$ in (3·2 mm) thick. Anything less than this becomes too lacking in stiffness to be acceptable without corrugation, and even with these or other forms of added stiffening a shell thinner than $\frac{1}{8}$ in will be very liable to puncture on grounding. Skegs and bilge rails can be moulded in one with the hull but must be protected with metal or wood cappings to prevent rapid wear. In most cases these items are separate, reasonably easily replaceable parts. Some manufacturers put a craftsman made wooden finish to a bare GRP hull shell, making gunwales, thwarts, bottom boards, and all other components in wood. Others make practically the whole boat in two mouldings and show great ingenuity in producing attractive looking designs.

In some earlier designs positive buoyancy was produced by sealed com-

Figure 3.1. Because the heeling forces are transmitted to the hull by the shrouds which are in the forward part, and the C.G. *of the crew is aft, there will be a high torsional stress in the hull* (By courtesy Beken & Son Ltd.)

partments formed in areas between two skins. This is not a good method and has now largely been dropped. Water will always find its way into a built-in tank sooner or later and water slopping about in a buoyancy tank is a difficult thing to laugh off in front of a disillusioned owner. In most cases buoyancy is provided by blocks of rigid polyurethane foam or separate PVC air bags strapped beneath the thwarts.

Sailing dinghys, particularly racing class boats, can be quite highly stressed but the stresses are calculable. Such boats are subjected to heeling forces which can only be resisted by the weight of the crew being laterally displaced from the centre of buoyancy of the hull. Because the heeling forces are transmitted to the hull by the shrouds (laterally disposed rigging wires) which are in the forward part, and the centre of gravity of the crew is aft, there will be a torsional stress in the hull of the order of 500–2000 ftf/lb, depending on the size of boat and number of crew. *Figure 3.1.* The low rigidity of GRP would be a serious problem in such a boat if it was what is called 'open', meaning that no decking is fitted. This makes the structure an open channel very unsuitable for resisting torsion. Fortunately most such vessels require fore and side decking of generous enough width to reduce the stresses to manageable limits.

4. FAST MOTOR BOATS

Under this heading it is proposed to discuss what is generally known as a 'planing' hull. That is, a true surface vehicle practically all the weight being taken by the dynamic lift generated by the reaction between the hull bottom and the water. This takes place at speeds of about

$$\text{Knots} = 2(\text{waterline length in ft})^{\frac{1}{2}}$$

and above, and can be felt as a distinct transit from one state to another like an aircraft becoming airborne. Such a hull will be subjected to severe vertical accelerations while moving over waves. *Figure 3.2.* The quantitative value of these accelerations vary of course for different speeds, wave shapes, and hull shapes but can easily be more than a human being can tolerate even when specially padded and positioned. Very few boat owners will intentionally continuously subject the hull to such treatment, but the occasional inadvertent 'slam' is inevitable and the structure must be designed to withstand it.

The machinery will be relatively powerful and heavy and the fuel tanks and batteries are likely to be of large capacity. The main problems in the structural design of this type of hull therefore, are the bottom strength and stiffness and the attachment of machinery and equipment.

The most usual way of overcoming both these problems is to provide a deep 'egg box' structure, the fore and aft members of which are well fastened to the bulkheads and transom, preferably with conventional through fastenings. This obviates the need to provide a stressed joint between the bottom of the boat and the members of the 'egg box' and consequently any need for through fastenings in the hull. Usually the 'egg box' is made up of marine grade mahogany plywood. This is fitted to the boat bottom and simply 'glassed in' using *in situ* laminating with polyester or epoxy resin and glass fibre strips. By completely laminating on to the surface of one or more of the compartments thus formed, built in fuel and water tanks can be very easily made. Machinery and other fittings can then be attached in a conventional way to the wooden structure.

Needless to say such a structure can be made in any suitable material, steel, aluminium or GRP and fitted to the GRP hull in the same way. Alternatively, 'sandwich' structure hulls for fast boats can be made with relatively thin skins of the order of $\frac{1}{8}$ in using plastic foam or balsa wood cores. The success of such structures depends on the quality of the joint between skin and core and the correct choice of core density. This of course applies to sandwich

Figure 3.2. A hull subjected to severe vertical accelerations while moving over waves (By courtesy Beken & Son Ltd.)

construction in any field and the use of it in boat hulls will provide outstandingly light construction. However, for any but highly specialised craft, great care must be taken due to the very high local pressures occurring in normal everyday handling of boats due to grounding, coming alongside etc.

5. SLOW MOTOR BOATS AND LARGER SAIL BOAT HULLS

Hulls of this type, which while moving as well as at rest, rely on the displacement of water of a weight equal to their own to stay on the surface, 'float' in other words, constitute the major part of the worlds boat weight and value. The number of types and uses under this heading are of course almost unlimited, but structurally they will pose much the same problems.

The hull skin will probably be a mainly chopped strand mat laminate for the most part unstiffened by any regular framing. Such stiffening as is necessary will be provided by bulkheads, engine bearers, tanks, lockers or other furniture. Regular framing is sometimes necessary in flat areas of the hull in which case formers or cores of any material are usually used layered on to the inside of the partly laid up hull, the final layers being laminated over them to form a corrugated inside surface.

This method of providing stiffness is not so often used now as in early days of GRP boat hulls. The reason is interesting in that then it seems to have been thought necessary to copy the appearance of conventional wooden boats. These had what are called timbers; approximately square-sectioned wooden members running round the hull whose function was to hold the planks to which they were through fastened together, edge to edge, and so make a watertight structure. Secondly it was found that less rigidity than was thought necessary proved adequate in most cases.

The highest predictable stresses on this type of hull are probably those imposed by the mast and rigging of sailing boats. All modern rigs use a wire stay to support about half the sail area and a mast to support the other half and to provide the compression member of the whole rigging system. *Figure 3.3.*

Figure 3.3. All modern rigs use a wire stay to support about half the sail area and a mast to support the other half and to provide the compression member of the whole rigging system (By courtesy Beken & Son Ltd.)

This results in quite high continuous concentrated loads. For instance, the forestay and backstay of a 6 ton displacement 30 ft long sloop will have about 1·5 tons tension in the forestay when sailing in normal conditions.

This means that there will also be 1·5 tons tension in the backstay. In addition, there will be some 0·5 ton in the windward shrouds, heeling the boat over. These forces together result in 3·5 ton point loading at the foot of the mast and there are others which will bring the total to something like 4 tons under normal conditions.

The flexibility of GRP is a nuisance in racing sailboats of this type because really tight and straight forestay wires are essential for optimum performance. GRP hulls sag under these loads and it is necessary to prestress the hull by tightening the rigging while the boat is at rest to a far greater extent than with a wooden or steel hull.

6. CONCLUSION

To sum up the subject of GRP and boat hulls it can be said that this is the right material at the right time. The huge expansion in yachting is being helped by the availability of GRP and the availability of GRP has undoubtedly contributed to the expansion. This state of affairs has in turn resulted in the revolution now taking place in all boatbuilding from conventional materials to GRP.

4

Chemical Plant

I. W. REID

1. INTRODUCTION

During the last decade the range of application of reinforced plastics in the chemical process industries has widened very considerably. The continuing build-up of knowledge on long-term chemical resistances, short- and long-term mechanical properties and the problems associated with high temperature operation have led to a position where chemical plant purchasers *know* that if they purchase from the more experienced designers and fabricators of this type of equipment, their problems will have been thoroughly investigated and equipment they purchase will be built to the highest specification and be tailored to the precise requirements of the specified duty.

Storage tanks up to 75 000 gal (340 m^3) capacity, 25 ft (8 m) diameter by 25 ft (8 m) high, are in service in this country as are process vessels of 6 ft (2 m) to 10 ft (3 m) diameter working under full vacuum conditions or internal pressures as high as 45 lb/in^2 ($0{\cdot}03\ kN/m^2 \times 10^4$) and chimney stacks over 300 ft (90 m) high. Pipelines of lengths measured in miles are in service under extremely arduous chemical, pressure, temperature and external loadings (soil or tidal) conditions.

With this present range of application, and bearing in mind future probable developments, it is absolutely essential that the designs of such equipment are to the highest possible standards from both chemical and mechanical viewpoints.

2. MATERIALS

Whilst this chapter refers specifically to glass fibre reinforced thermoset resins, it should be pointed out that thermoplastic (especially PVC and polypropylene) liners are frequently used in conjunction with glass/resin reinforcement for strength requirements, in applications where the particular chemical properties of these thermoplastics show an advantage.

2.1. GLASS REINFORCED RESIN CONSTRUCTION

By far the largest proportion of chemical plant equipment manufactured in glass/resin construction uses polyester resins of various types (isophthalics, bisphenols etc. and modified versions of these types) and reinforcement in the form of rovings, chopped strand mat, woven rovings and fabrics together with corrodent contact surfaces of surface 'tissue' or 'veil'.

The volume of fabrication of chemical plant involving glass fibre reinforcement with other chemically resistant resins such as hot cured epoxides, phenolics and furanes is at the present time on a very small scale. The majority of fabricators use only 'E' type glass, but 'A' type glass is often specified for some special applications where its particular chemical properties are required. All glass surface tissues used for chemical industry application are of 'C' type glass. Synthetic fabrics such as 'Courtelle' or 'Dynel' are used in applications which require their good abrasion resistance and specialised chemical resistant properties.

2.2. THERMOPLASTIC LINED CONSTRUCTION WITH GLASS/RESIN REINFORCEMENT

The thermoplastic liners are generally hot formed from sheet and welded together by hot gas/filler rod techniques. Special methods are used to achieve the high bond strengths between liner and glass/resin reinforcement which is necessary especially for equipment operating at elevated temperatures. Both PVC and polypropylene have extremely low elastic moduli and for 'dual' laminates of this type, the thin (approx. 3 mm maximum thickness) liners are not normally considered as contributing towards strength requirements in design calculations.

3. METHODS OF MANUFACTURE

3.1. MOULDS

Ideally, GRP chemical plant should be manufactured as one piece mouldings, thus having optimum mechanical and chemical properties. Since this type of manufacture would necessitate the use of complex dismantable formers or 'melt out' formers, chemical tanks, vessels etc., are normally fabricated from separately moulded components such as ends, cylinders and branches, as is generally the procedure with metal.

Pipe sections and barrels of closed cylindrical vessels are normally made on either vertical or horizontal mandrels which may be of a simple collapsible type or have a built-in taper to facilitate withdrawal. Materials for these formers can be mild steel GRP or wood. End closures, whether dished or conical are normally made on one piece moulds. 'Blow off' points are generally included in such moulds. Branches, manway bodies and other small fittings as are required on chemical process equipment are most frequently made on mild steel moulds since the required numbers off are generally fairly high.

3.2. LAMINATING METHODS

The reinforcement and resin may be applied by any one of a fairly large number of methods, including filament winding, tape winding and automatic spray-up, hand lay-up and the manual application of woven rovings. In the case of hand lay-up applications, the resin is either deposited by gun or by bucket/brush technique. As in the case of moulds the most suitable moulding method largely depends on the anticipated numbers off, standardisation of size of items, and shop handling facilities.

3.3. COMPONENT ASSEMBLY

Assembly of components is generally satisfactorily achieved by offering two items together (either butt or lap joints), and applying carefully controlled internal and external integrating overlays on to abraded surfaces of the two items. This technique will give a chemically satisfactory joint with a joint efficiency of 1 or more as determined by the design engineer, and has been used in high-pressure hot duty vessels to the satisfaction of the major insurance companies.

4. LAMINATE PROPERTIES—MECHANICAL

4.1. SHORT TERM

Table 4.1 gives a summary of the mechanical properties of special interest to the designer of process plant. There are several further aspects of interest to the designer of reinforced plastics in this field, especially in relation to those items which are either large or working at high stress levels. The use of special thermoset resins which have high chemical resistance and thermal stability and are therefore fairly brittle presents considerable difficulties. The low strain to failure of some of these resins means that there is a serious risk of damage to the internal gel coat during manipulation in the fabrication shop, during installation, or during service, unless the vessel is designed to accommodate this. Table 4.1 shows in column 6 the wide variation in craze characteristics of different types of polyester resin.

It should be pointed out that the long-term effect of many corrodents, liquid or gaseous, is to further embrittle the resin. High temperature operation and ultraviolet exposure also adversely affect this characteristic and can lead to flexibility losses of up to 100 per cent.

Table 4.1

TYPICAL PROPERTIES OF LAMINATES

Type of glass reinforcement	‡ *Resin type*	*Typical glass content by weight, %*	*Specific gravity (approx.)*	*Laminating thickness/layer of reinforcement required to justify* lb/in² *figures*	*Craze tensile strength* lb/in² (kN/m² × 10^4)	*Ultimate tensile strength* lb/in² (kN/m² × 10^4)	*Tensile modulus* lb/in² (kN/m² × 10^7)	*Shear strength* lb/in² (kN/m² × 10^4)	*Poissons ratio*
Chopped strand mat (1 oz)	A	30–35	1.5–1.6	0.025	12 000 (8.3)	17 000 (11.8)	0.8 × 10^6 (0.53)	25000 (17.3)	0.30
	B	30–35	1.5–1.6	0.025	8 000 (5.5)	17 000 (11.8)	0.8 × 10^6 (0.53)	25 000 (17.3)	0.30
	C	30–35	1.5–1.6	0.025	15 000 (10.4)	17 000 (11.8)	0.8 × 10^6 (0.53)	25 000 (17.3)	0.30
Balanced 16 oz woven roving,	A	50–55	1.7	0.020	30 000 (20.8)	50 000 (34.5)	1.4 × 10^6 (0.97)	58 000 (40.0)	0.15
parallel laminated	C	50–55	1.7	0.020	37 000 (25.6)	50 000 (34.5)	* *	* *	0.15
Progressively orientated	A	*	1.7	0.020	16 000 (11.0)	27 000 (18.6)	1.0 × 10^6 (0.69)	30 000 (20.8)	0.25
balanced 16 oz woven roving	C	*	1.7	0.020	20 000 (13.9)	27 000 (18.6)	* *	* *	0.25
2/1 biased 16 oz woven roving—major axis	A	50–55	1.7	0·020	36 000 (25.0)	60 000 (41.5)	1.8 × 10^6 (1.24)	* *	†
	C	50–55	1.7	0.020	45 000 (31.0)	60 000 (41.5)	1.8 × 10^6 (1.24)	* *	†
minor axis	A	50–55	1.7	0.020	18 000 (12.5)	30 000 (20.8)	1.0 × 10^6 (0.69)	* *	†
	C	50–55	1.7	0.020	22 000 (15.2)	30 000 (20.8)	1.0 × 10^6 (0.69)	* *	†
Uni-directional 21 oz woven	A	55–60	1.8	0.030	42 000 (29.0)	70 000 (48.5)	2.0 × 10^6 (1.38)	* *	†
roving	C	55–60	1.8	0.030	52 000 (36.0)	70 000 (48.5)	2.0 × 10^6 (1.38)	* *	†
Filament wound rovings	A	60–65	1.8–1.9	0.015	*	80 000 (55.0)	3.0 × 10^6 (2.08)	* *	†

* Not available.

† Resin A is a general purpose polyester resin to B.S. 3532:1962 type A; resin B is a heat-resistant polyester resin to B.S. 3532: 1962 type C; and resin C is a medium high flexibility 'Bisphenol A' based polyester resin.

‡ Values vary widely with directions of test.

4.2. LONG TERM

Figures 4.1, 4.2 and 4.3 show typical fatigue, creep and strength-retention characteristics of different laminates. Considerable evaluation of these properties has been done in the U.S.A. and is now in hand in Great Britain on laminates of the types being discussed. It is generally on the basis of these published results, of up to 10^8 cycles and 5 years continuous loading, that vessels operating at stress levels of approximately 10 per cent of the short-term ultimate are considered adequately safe.

Examination of the characteristic curves indicates that:

1. Both tensile and flexural fatigue strengths with both random reinforcement and bi-directional reinforcement tend to level off at between 20–25 per cent of their short-term ultimate strengths at 10^7 cycles (*Figure* 4.1).

2. At a tensile loading of 50 per cent of the short-term ultimate strength, the total creep after 10^4h is in the order of 0:005 in/in. Even though glass reinforced polyester resins exhibit extremely low strain to resin fail characteristics it is seen that the laminate creep over extended periods will be negligible when the working stress levels are in the order of 10 per cent of the short-term ultimate (*Figure* 4.2).

3. The tensile strength retention of laminate continuously loaded to a level of 50 per cent of the short-term ultimate strength will sustain this load for 10^4 h. As with the creep characteristics, at a working stress level of approximately 10 per cent of the short-term ultimate it is expected that strength retention would be entirely adequate. (*Figure* 4.3).

A general but most interesting observation from the above long-term data is that random chopped strand glass mat reinforced laminates are shown to have fatigue strength retention, creep resistance and continuous-loading strength retention similar to laminates reinforced with woven continuous fibres.

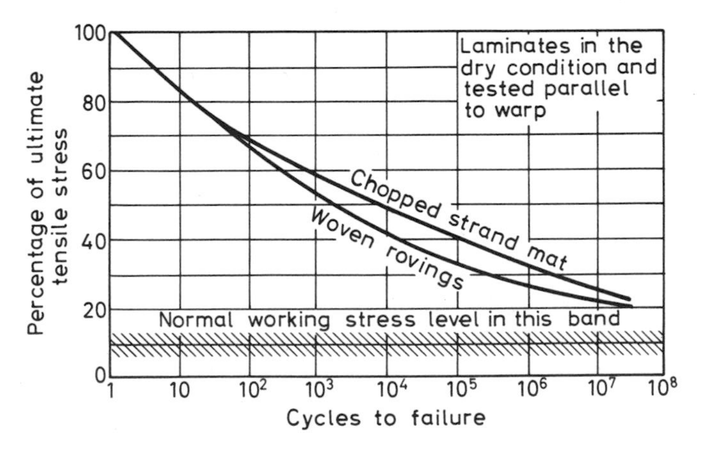

Figure 4.1. Tensile fatigue strength of polyester glass fibre laminates
(By courtesy of Gibbs and Cox Inc. *Marine Design Manual 1960*. McGraw-Hill Book Company, New York and London)

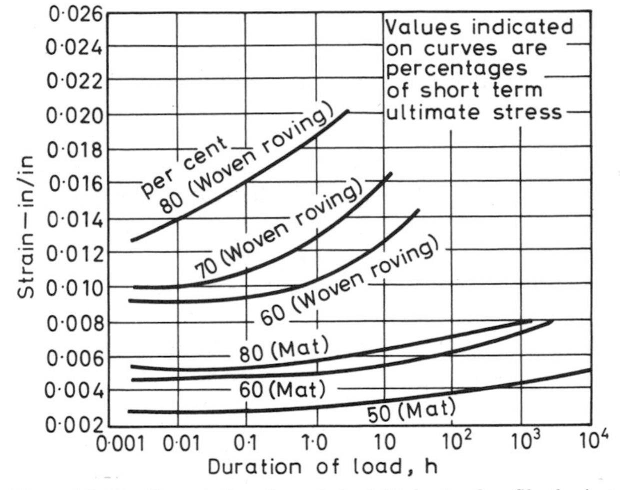

Figure 4.2. Tensile creep of continuously loaded polyester glass fibre laminates
(By courtesy of Gibbs and Cox Inc. *Marine Design Manual 1960*. McGraw-Hill Book Company, New York and London)

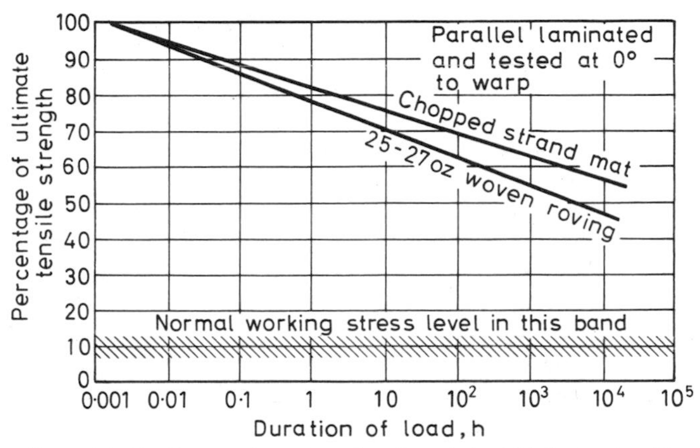

Figure 4.3. Tensile strength retention of continuously loaded polyester glass fibre laminates

(By courtesy of Gibbs and Cox Inc. *Marine Design Manual 1960*. McGraw-Hill Book Company, New York and London)

5. LAMINATE PROPERTIES—CHEMICAL

This aspect has been treated in greater detail in Chapter 17. It is important however, to ensure that an item of equipment be equally chemically resistant *all over* its corrodent contact surface.

It is good practice to have a glass surface tissue (or synthetic fabric such as 'Dynel') liner backed up by a *minimum* of 0·8 kg/m^2 (3 oz/ft^2) of resin rich chopped fibre reinforced laminate before any woven roving, continuous filament or similar reinforcement is introduced into the laminate. For really severe duties a minimum of 1·3 kg/m^2 ($4\frac{1}{2}$ oz/ft^2) is recommended. Large items are often fabricated with a cheaper 'general purpose' resin backing up this resin rich liner.

Joints between components (barrels to end closures, branches etc.) should be made to meet the same chemical resistance requirements. That is, joints should be internally overlaid, where chemical duty requires such a procedure, with 0·8 kg/m^2 to 1·3 kg/m^2 chopped fibre together with a surface tissue or similar finish. Such an approach should lead to a joint construction which is satisfactory chemically. This type of joint should be externally reinforced to meet *mechanical* requirements.

Since most of this type of equipment will be in service in a corrosive environment, care should be taken to ensure that a continuous chemically resistance finish is applied also to *all* external surfaces.

It is not generally recommended that fillers other than those required to give thixotropic, fire retardant, anti-static and ultraviolet absorbent properties be added to resins for use in building chemical plant and when additives are used it is important to incorporate the absolute minimum.

6. TEMPERATURE CONSIDERATIONS

With the steady increase in the chemical resistance of thermoset polyesters at high temperature, problems associated with laminate softening due to temperature are becoming of critical importance.

The normal heat distortion test on resin castings is of considerable value to the designer. This test is a very good discriminator between different resins and through service experience can be used to determine the maximum recommended working temperature for a *laminate* of a given resin.

Obviously, if two resins are equally attractive in chemical resistance, usability, availability and cost for a given application, it will be preferable to use the resin with the higher heat distortion point. The vessel might require steaming out, or even be required for another higher temperature duty at a later date. Companies fabricating in GRP for the chemical process industries use five types of resins:

1. Low heat distortion point (approx. 60°C), medium chemical resistance resin.
2. Medium heat distortion point (approx. 100°C), high chemical resistance resin.
3. Medium/high heat distortion point (approx. 120°C), very high chemical resistance resin.
4. High heat distortion point (approx. 140°C), high chemical resistance resin.
5. Special resin for specific duty (e.g. foodstuffs or petroleum spirit.) Heat distortion point to suit requirements.

7. WORKING TEMPERATURE *v.* HEAT DISTORTION POINT

It is accepted practice for fully stressed laminate applications that the resin or resins used in the construction should have a heat distortion point at least 10°C greater than the maximum working temperature of the equipment in service. Very occasionally higher temperature conditions could probably be accommodated provided the temperature does not rise above approximately 10°C *over* the resin heat distortion point, but these occasional conditions should be fully investigated from all aspects.

In the case of lower stressed applications, ducts, stacks etc., there is a wide background of success with operating conditions considerably above (60°C or more!) the resin heat distortion point. Such conditions can only be permitted if a very full analysis of mechanical operating conditions (vacuum, pressure, vibrations, wind loads etc.) is available, if the chemical corrosion problem is not severe, and the designer has a successful background knowledge of such items of equipment.

Care must be taken in applying these 'rules of thumb' in cases where the corrodent may permeate the matrix of the laminate over extended periods and hence cause degradation and loss of strength. Higher temperature differentials between operating temperature and resin heat distortion point are required under these conditions.

8. DESIGN

Vessel designers working with the more conventional materials such as mild steel, stainless steel and aluminium, have one major advantage over their counterparts designing in GRP. The metallic construction designer is able to purchase his materials in sheet or section form with guaranteed mechanical properties in the form of a mill certificate. Through the mill certificates, analysis of the materials in the plate or section, the chemical performance can be fairly accurately predicted.

The reinforced plastics designer in his design has to allow for all known factors affecting the materials during service life (fatigue, creep, chemical corrosion, ultraviolet degradation etc.), and also for the unknown problems associated with the fact that the materials of construction are manufactured on the shop floor, often by semi-skilled labour. It must be pointed out also that full inspection techniques for this type of material have not yet been satisfactorily evolved. Non-destructive tests are as yet in their infancy, and even destructive test results can only be really associated with the cut-out from which the testpiece was taken, and not from an area on the same vessel, distant from this cut-out. Obviously 'test-plates' made by laminators *may* have no relationship whatsoever to an on-job laminating.

8.1. FACTORS OF SAFETY

From this preamble it will be readily understood that factors of safety on ultimate stresses at present have to be set at a higher level than with more proven materials. It is accepted practice to vary the factor of safety with duty, operational cycles if any, vibration conditions, inspection facility in the particular works involved, and even the very ability and background of the design engineers in individual fabricating companies. Factors of safety as low as 7:1, and as high as 20:1 are in general use. A figure of 10:1 is widely used but is not always soundly based.

It will be understood how full inspection and non-destructive testing facilities are urgently required when one considers that it is quite possible with 'conventional' GRP laminated by hand techniques, to achieve tensile figures of 60 000 lb/in^2 (42 $kN/m^2 \times 10^4$) to 65 000 lb/in^2 (45 $kN/m^2 \times 10^4$). The application of a factor of safety of only 8 brings the design stress down to approx. 7500 lb/in^2 (5 $kN/m^2 \times 10^4$).

The practical determination of correct factors of safety by individual firms for specific process equipment is a matter of real importance and has to be handled with the highest possible ethical standards.

Figure 4.4. 13 ft 6 in (4.1 m) *diameter* $\times \frac{1}{3}$ *mile* ($\frac{1}{2}$ km) *long polypropylene lined, glass/resin reinforced fume extraction system handling acidic/organic fumes from a battery of rayon spinning plants. Volume of throughput* 13 $m^3 \times 10^3$/min (Courtesy Prodorite Limited)

8.2. SOME OTHER DESIGN ASPECTS

Subsequent to the nomination of particular design stresses, tensile, shear, bearing etc., and the relevant elastic moduli, the design itself is a result of a design based on limiting the strain in the gel coat liner as much as stress in the laminate, together with a sound background of experience and an understanding of the laminate construction required for the chemical operation of the process plant.

The experienced designer will, however, very often find himself nominating thicknesses of vessel shells, covers etc. for stability requirements rather than for limitations on gel coat strain or laminate stress. This is obviously necessary

Figure 4.5. Sulphur dioxide stripping columns. Unit on left of illustration is 6 ft 0 in (1.8 m) *i.d.* 23 ft 6 in (7.1 m) *overall height. Unit on right of illustration is* 6 ft 9 in (2 m) *i.d.* × 21 ft 3 in (6.5 m) *overall height. Both vessels, manufactured from bisphenol polyester resin, reinforced with composite of woven glass rovings and chopped strand mat were designed and tested to* 100 kN/m^2 *positive pressure and to* 35 kN/m^2 *vacuum*

with horizontal cylindrical vessels, rectangular vessels, large covers for vessels and ductwork, and especially where internal vacuum conditions exist.

A fundamental problem which faces the reinforced plastics designer is that the conventional strength measurements, lb/in^2, kN/m^2 etc., are meaningless unless read in conjunction with a definition of the amount of glass reinforcement necessary in the laminate, and in the required orientation and correctly worked 'wet-out' to give reliable and reproducible lb/in^2 or kN/m^2 figures under test conditions. There are two approaches to this problem: Firstly to design such units with mechanical properties as lb/in or kg/cm width/unit layers of reinforcement, and secondly to design lb/in^2 or kN/m^2 units and then translate back into units of glass layers of reinforcement, the translation being based upon the correlation of lb/in or kg/cm per layer units and lb/in^2 or kN/m^2 units achieved on test. Table 4.1 refers.

9. INSPECTION

At this stage of the development of process equipment in reinforced plastics,

Figure 4.6. Site built pvc lined process vessel 20 ft (6.1 m) *diameter* × 28 ft (8.5 m) *deep; fitted with agitator equipment* (Courtesy Plastics Design and Engineering Limited)

it is almost entirely the responsibility of the manufacturer to ensure that the designer's requirements with regards to reinforcement layout and number of layers is met. Even on pressure vessel applications, it is at this stage of the development of the industry, often the case that the insurance company's surveyor has to be advised by the fabricator's shop inspector what particular aspects to investigate. This position throws a great responsibility on the fabricator's own inspection system.

Very close attention must be paid to factors affecting the chemical integrity of the equipment (pinholes, air inclusions, glass/resin ratio, under-cure, dry glass on corrodent surface), and to those factors which affect the mechanical integrity of the equipment (amount and disposition of reinforcement, correct design radii, minimum thickness requirements etc.).

Even closer inspection will be required in cases where site work is involved. Wet or cold conditions may seriously affect the cure of the resin used under such atmospheric changes.

The most satisfactory results are achieved by having a *high* standard of *on-job inspection*. Inspection of completed items is of great importance but it is not the ideal solution.

Sections of laminate at branch cut-outs etc. can be analysed to show reinforcement content and reinforcement/resin ratio, items can be visually inspected (pigmentation of main body of laminate is *not* recommended), and thicknesses can be measured; but as mentioned earlier, these only give confirmation at the position of the cut-outs, and a few inches from this position the quality could be totally unacceptable. Thus, again, it is stressed that full inspection during manufacture is the only real solution for chemical process plant in reinforced plastics.

10. INSURANCE CERTIFICATION REQUIREMENTS

The major insurance companies in Great Britain at present call for a factor of safety of approx. 10:1 on design stresses which must also be fully proved to the certifying body. In some instances, on critical duty vessels, a factor of safety of 10:1 of *burst* pressure/working pressure has been required, but in more recent years the competency of the major fabricators in this country has been well established and such extreme factors of safety are not now required.

Smaller fabricators may well have difficulty in proving their designs for high duty vessels to the necessary certifying bodies without either carrying out an extensive series of prototype tests or overdesigning the vessels considerably to give adequate performance on test. This approach though acceptable is generally not economical.

11. BRITISH STANDARDS

A British Standard for reinforced plastics tanks and vessels for use at low pressures is now (at the time of writing) at a fairly advanced stage. The scope of this chapter is not sufficient to detail the design recommendations of this work, but when published, the standard will be of considerable value to the designer of GRP process equipment.

The detail design of pressure vessels will not be covered by this British Standard but details of design techniques, stress raisers etc. for this type of vessel are available from other sources. This data is also applicable to many lower duty applications, such as storage tanks and ductwork.

12. SOME FURTHER APPLICATIONS

12.1 PRESENT

Apart from the more conventional tanks and process vessels discussed above and the more generally accepted application for GRP in chemical plant

Figure 4.7. Soap splitting installation operating at temperatures up to 105°C. *The fume extraction system and general pipework are also constructed in glass/bisphenol resin laminate* (Courtesy A.P.V. Kestner Limited)

these materials are being used in increasing amounts for other related applications. Examples of these include underground storage tanks; linings for mild steel and concrete tanks; pressure and gravity discharge chemical transportation road tankers; mixer impellers, fans, pump bodies and impellers, valve bodies, filter plates, etc.

12.2 FUTURE

It is expected that during the next few years the industry will consolidate its position in chemical process applications. With the new generation of materials becoming commercially available, it is probable that a considerable increase in range of sizes, temperature and chemical environment, and of operating pressures will be developed during this period.

REFERENCES

1. Reid, I. W., *Proc. Instn mech. Eng.*, 1964, *Symposium on Plastics and the Mechanical Engineer,* Paper 7.

5

Rail Transport

B. J. HAWTHORNE

1. INTRODUCTION

The plastics industry is past the stage where the attitude prevailed that plastics in general, and glass reinforced plastics in particular, were suitable for everything, and it is evident that a more rational approach now exists which recognises the limitations of these newer materials. Railway requirements are very often different from other users and should be borne in mind when novel applications are being considered as well as where past uses are being examined.

During the past decade every major railway organisation in the Western Hemisphere has gone into the 'red' and most have contracted continuously; for example, the number of main line coaches used by B.R. has dropped in the last 7 years from 55000 to 28000 and during this period only 3800 new coaches have been built. However, it is evident that the vigorous attempts now being made to stabilise the railway industries will offer scope for the use of plastics. The developments of high speed, lightweight rolling stock and the need for fast commuter transport in and out of cities are likely to bring about increased utilisation of plastics. There are several reasons why plastics generally are likely to be used to a greater extent than at present but the most important one is that, with high speeds and rapid acceleration, the weight of railway vehicles is becoming critical.

Before the railway applications of plastics are examined, it is expedient to reiterate briefly the advantages and disadvantages of GRP with respect to railway uses. The first and perhaps most advantageous property is the ease with which three dimensional curves can be fabricated compared with metals, particularly where only short or medium production runs are required. Secondly, the high strength to weight ratio and toughness of GRP can help to reduce tare weights and at the same time can give a tough and durable product, for example a carriage door. Thirdly, their non-corrosive properties can help to reduce maintenance, which is a very considerable advantage when one takes into account the many millions of pounds of railway expenditure on this item.

The disadvantages of GRP which have limited their adoption include inadequate design data, the lack of stiffness, the lack of information on their fatigue performance and finally the inconsistency of the standard methods of production which gives rise to inconsistent products. These disadvantages can be serious when one considers the major criteria which have to be met when applying plastics in railway rolling stock. Stresses in vehicles can be very high and complex, for instance, deceleration forces can be up to 7 *g* compared with $1\frac{1}{2}$ *g* in the case of road vehicles; railways generally own and maintain their rolling stock and a life of at least 15 years is required (until recently the figure was 30 years). The problem of the fire resistance of organic materials to be used in passenger vehicles travelling at high speed has to be faced.

It will be evident from the following pages that the above criteria have in the main resulted in applications of GRP to non- or semi-structural applications. There have been attempts to use them for complete structures, notably passenger coaches and containers, but at present the main structural interest, in Europe at least, is in the construction of non-metallic tank wagons and small trackside buildings. Although the railway industries have not parallelled the marine industry in finding in GRP a combination of properties which exactly meets their requirements, substantial quantities of material are used in a number of different environments.

2. GENERAL ROLLING STOCK USES

Typical non- or semi-structural uses of polyester glass fibre are end wall sections of diesel railcars and locomotives, doors, various ducts, rectifier trays, floors in toilet compartments, corridor ceiling panels. In the U.K., British Railways use GRP rodstock coated with butyl rubber as underbridge, tunnel and section insulators on the 25 kV overhead power system. On the civil engineering side, the use of polyester rooflight sheeting for station canopies and passenger shelters is widespread. The reason for this is that the maintenance costs of a good quality sheet can be substantially lower than those of patent glazing over a period of years.

The utilisation and awareness of GRP on the part of British Railways began in 1954 with the production of canopy ends for diesel railcars. This was a straight substitution of plastics for metal and it resulted in a reduction in cost of 50 per cent. From this application, a number of other uses developed. Having made a single skin for the outer layer it was logical to do the same thing on the inside of the vehicle and for many years now interior ceiling panels and carriage end walls have been produced in GRP. The carriage end walls have a simulated wood finish which is obtained by using a clear fire-retardant gel coat over a tissue printed with a wood grain pattern. On the Southern Region of B.R. the canopy ends of electric traction coaches are manufactured in sandwich form and the space between the skins is used to carry electric cable ducting and the connections for the warning horn.

The Netherlands Railways have also used GRP extensively for walling applications and they are now experimenting with thin GRP sheets for the sidewalls of motor coaches of multiple train sets. As a rule these train sets are painted every 8 years but the Netherlands Railways have started a test programme in which GRP sheets, 2.5 mm thick and made in the required colour, are glued to the walls of the vehicle at the time of overhaul.

Also, Netherlands Railways have used glass fibre reinforced polyester for the production of gearcases on motor coaches and by 1967 some 500 had been made. They are less expensive than steel and dimensionally more accurate than welded steel sheet. Furthermore, repair is easier and the sound transmission is less. These cases, which were developed by Netherlands Railways and are manufactured by them, have been so successful that the introduction of a similar gearcase in locomotives is planned.

In Sweden, the State Railway has been experimenting since 1966 with a sandwich structure for roof panels. These panels, which are 60 mm thick, are made of two GRP skins with a core of rigid PVC, and are being evaluated on eight double-deck coaches. The Swedish State Railway is also evaluating toilet floor units made from GRP having a surface containing carborundum. The Netherlands and German Federal Railways are already using a similar item and in both cases it is a one-piece moulding. This one-piece construction facilitates cleaning and prevents water spillage corroding the vehicle underframe.

Carriage water tanks are another example of an application which utilises the strength of GRP together with its chemical resistance. These tanks are used by the French and Netherland Railways and are being evaluated by the Swedish State Railways and British Railways. Their cost is lower than that of a copper tank and condensation troubles are eliminated. The French Railways have commented that even when the water freezes there is no damage to the tanks because the properties of the GRP tolerate the great forces exerted by the freezing water. B.R. have had a test tank in service for 5 years which has an average wall thickness of 6 mm and a capacity of $0{\cdot}36\ m^3$. Pipe connections were made by bonding in the fittings and the tank was constructed in two halves so that a GRP baffle system could be inserted before the top was bonded in place. These tanks have to withstand not only severe surging forces as the train brakes, but also the filling systems frequently operate at a high pressure. When the test tank was removed for its final inspection, no damage had occurred apart from a little superficial crazing around some of the pipe connections. There was no evidence of any mould growth and the tank was in a very good condition. It is likely that this type of tank will be adopted as standard by B.R.

Previously mention was made of the fact that the moderate modulus of elasticity and light weight of GRP structures can be utilised in carriage door construction and it is worth examining this application in a little more detail. The light weight of a GRP door minimises inertia forces while the moderate modulus allows the door to absorb energy elastically rather than by deforming as would tend to be the case with metals. For instance, B.R. have used GRP sandwich construction for carriage doors for some years and there are at least 40000 of these doors in service. Although the B.R. door costs about 10 per cent more than a steel and wood door of traditional design, it weighs only 14.5 kg compared with 42.5 kg for the traditional door. This difference in weight is reflected in the markedly superior performance of the GRP item. In a severe test, which simulated the impact which a door receives when it flies open and slams backwards against the carriage wall, the traditional door withstood 70 slams while the test on the GRP door was stopped at 10 500 slams.

Figure 5.1. Cold moulded seat shells fitted out in a coach

Figure 5.2. Bodyside quarters mounted on a vehicle underframe

A cast aluminium door, which weighed 60 kg withstood 1000 slams before it was so badly deformed that it would have been useless for service.

Another B.R. application which has proved to be successful is that of seat shells. In 1965, the process of cold moulding was developed and adapted[1] to the manufacture of seat shells. Three sizes of seat were required in quantities of 1000 of the single seat and 2000 of each of the double seats. Suitably shaped preforms 16 oz/ft^2 (4.8 kg/m^2) were manufactured simply by fusing the thermoplastic binder around the outer edges of the eight laminates of glass mat in a small rig conforming to the profile of the appropriate seat. These seats are attached to the coach floor by a simple metal support bolted through the seat pan and timber blocks are screwed to the sides of the shell to provide a base for the arm rests. The cushioning material used is fire-retardant, flexible polyurethane foam (*Figure 5.2*). Following the success of this project, one of the double seats was adopted as a standard and 6000 have been manufactured in pre-preg material. It is expected that the next generation of coaching stock, which is now being designed, will have similar seat units.

3. COACH STRUCTURES

Although a design scheme has been drawn up for a coach based on glass reinforced epoxide[2] resin and a standard coach underframe has been fitted out with a GRP body and roof[3], as yet no railway organisation has adopted a plastics coach as standard. Czechoslovakian State Railways are operating a number of electric locomotives having GRP panels bonded to a metal framework, and this is the nearest thing to a bodywork structure which is in use in a railway.

A design scheme[2] was submitted to B.R. in 1958 for a coach structure based on a sandwich structure of glass reinforced epoxide resin and polyurethane foam. The structure was designed to carry a vertical load of 12 tons and a compressive load between side buffers of 200 tons. It also had to withstand a maximum drawbar pull of 30 tons. At the time the cost of the structure was 1.8 times greater than that of a standard steel structure. This cost, which was based on a production run of 100 coaches, was one of the reasons why the scheme was not pursued at the time. The other reasons were the lack of realistic information on the long-term fatigue properties of the sandwich structure and on the fire risk. It was also thought that there would be considerable difficulties to be overcome in the manufacturing process before the design scheme would work.

In 1963, B.R. built[3] two vehicles consisting of a standard underframe with a GRP sandwich structure for the coach bodies and roofs. One of these coaches was completed and put into limited service while the other was retained for test purposes. The object of the exercise was to study the production feasibility of large one-piece roof mouldings and at the same time to evaluate, practically, coach body construction techniques. The coach which was put into service showed several interesting features, not least amongst which was a marked decrease in the sound level. Also the total structure was 3.5 tons lighter than a similar steel coach (this compares with an estimated weight saving of 12.5

tons for the all-plastics coach referred to previously). At this juncture it should be pointed out that one of the potential advantages of plastics is that of freedom from corrosion. On B.R. the cost of painting rolling stock is something like £0.8 m a year and a large part of this goes towards combating corrosion. The major advantage of light weight could bring useful savings in traction costs. For example, compared with the conventional mild steel coach the estimated cost saving in 1958 for the all-plastics coach was approx. £3000 over 30 years. By comparison, an all aluminium structure would save about £1000 over the same period.

The component parts of the coach bodies were constructed using a sandwich structure consisting of outer skins of polyester glass fibre 6 oz/ft^2 (1.8 kg/m^2) with a core of rigid polyurethane foam. At regular intervals, the foam was interrupted to allow the inclusion of GRP reinforcing webs which tied the two outer skins together, so that the risk of buckling failure was minimised.

The bodyside quarters were moulded as separate units (*Figure 5.3*) and the roof, which was a one-piece moulding 19.5 m long and 2.6 m wide, had a cross-sectional thickness of 102 mm at the centre reducing to 57 mm at the sides. The developed width of the roof was 3 m.

The construction of the sandwich structure fo both the roof and the bodyside quarters was achieved by using a closed mould technique. A box-like mould was manufactured having either a hinged or detached lid so that when the lamination process was complete the box could be closed to provide a smooth surface on all external surfaces. The layers of polyester glass fibre were put down on the inside faces of the box and slabs of moulded polyurethane cut to a suitable size. The GRP webs which tie the skins together were then laid upon the faces of the foam. The blocks of coated foam were placed over the wet laminate in the base of the mould and a laminate laid-up on the lid of the box which was finally closed to complete the process. It will be evident that it was essential that the resins used had a variable, but controllable, pot life so that good bonding could be achieved between the various parts of the sandwich. During the manufacture of the sandwich, items like steel tapping plates, nut plates, communication channels, etc. were inserted and accurately located without any difficulty.

The coach body was assembled by bolting and bonding the component parts both to each other and to the underframe. For reasons of safety, four steel crash pillars were incorporated at the ends of the coach. When finished and completely fitted with doors and interior furnishings the coach body weighed 6250 kg which was 3560 kg less than an equivalent steel structure.

While a second coach was being built for destructive test purposes, the first coach was put through a series of running trials. The feature which emerged from these was the high degree of comfort which the coach provided. Not only did the sandwich structure provide good thermal insulation, but also it reduced the noise transmission to a remarkable extent. The measured reduction in the noise was found to be 10 dBA.

The other coach when completed was not furnished in anyway but was tested[4] to obtain design data for future work. In order that the properties of the plastics structure alone could be determined the underframe and crash bars were removed for the principal tests. A number of subsidiary tests were

Figure 5.3. Inverted GRP coach in test rig

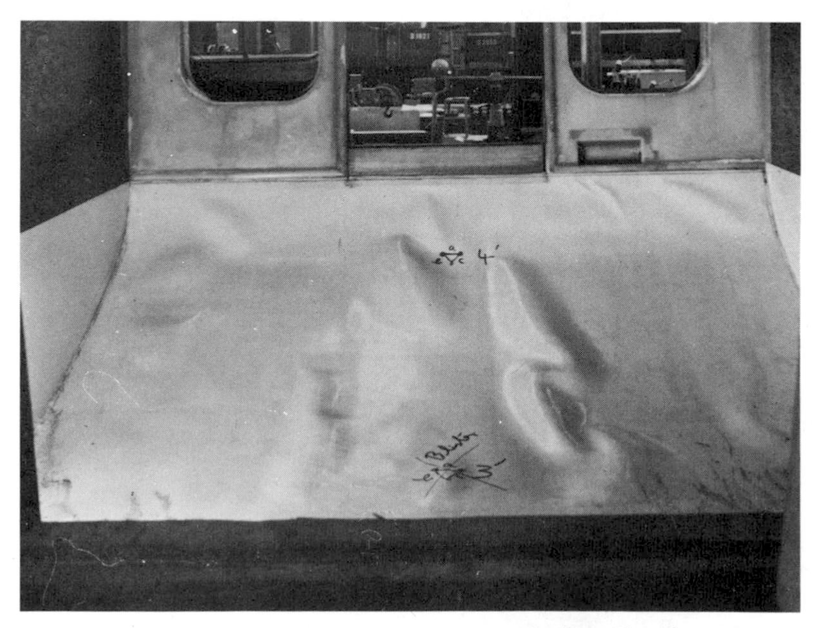

Figure 5.4. Compartment interior; 30 ton end loading on the roof

Figure 5.5. Compartment exterior; 30 ton end loading on the roof

carried out but the main test involved the measurement of the strain developed when an incremental series of compressive end loads was applied to the roof section (*see Figure 5.3*).

The roof structure consisted of a longitudinal spine, made up of rigid PVC foam box section wrapped with GRP, with transverse ribs of GRP running from the spine to the edges of the roof. The spaces between the ribs were filled with rigid polyurethane foam which was used as a former for the ribs and the whole was skinned with GRP on both sides to complete the sandwich structure.

Under compressive end loading the central spine behaved in a reasonably elastic manner, but uneven strain distribution occurred between the transverse webs because of local buckling of the GRP skin. At 25 tons applied load the maximum recorded strain was 2.7×10^{-3} and the mean recorded strain was 1.34×10^{-3}. The mean figure was compatible with the overall change in length of 29 mm. When the compressive loading reached 30 tons, the roof buckling became acute (*see Figures 5.4* and *5.5*) and no further loads were applied.

After these end loading tests, samples were cut from the coach roof and bodysides so that the mode of failure could be examined more closely. It was found that the premature buckling was due to local variations in the composition or thickness of the skin. These were manufacturing faults and it was estimated that the roof should have been able to withstand an end compression load of at least 50 tons. A higher performance still could almost certainly be obtained if a stronger core material was used and a positive effort made to utilise the core as part of the structure. In the original design concept of the coach, the foam core was intended to serve only as a spacer between the skins and as a former for the transverse webs. The webs themselves were intended to prevent buckling.

In conclusion, it should be pointed out that the only criterion determining the choice of material for coach construction is the cost of achieving the construction and performance required. While it seems likely that a coach could be manufactured using GRP in combination with steel to obtain the necessary stiffness, the adoption of GRP would call for a complete change in manufacturing technique and a heavy investment in new equipment and new skills. On the other hand, a switch from steel to aluminium involves a far smaller degree of change of manufacturing technique and, moreover, aluminium is a more consistent material so that a stressed skin design can be adopted with confidence. For these reasons it looks as though if a change is made from steel construction, aluminium is more likely to be adopted than GRP in the immediate future.

None the less, the tendency towards higher average speeds is making it necessary to reduce coach weight and this is going to favour plastics for interior fittings and semi-structural components generally. For example, one of the B.R. designs which is in an advanced state of development, is that of the High Density Multiple Unit Stock, which as its name suggests, is intended for use on short distance commuter routes and the costs of traction are such that one can spend up to fifty pence per pound of material to save a pound of vehicle weight. These commuter vehicles will have lightweight seat shells moulded in GRP and for the sliding doors a lightweight GRP-foam construction is being considered.

4. TANK WAGONS

However, although in the case of coach building the merits of GRP are more or less matched by those of aluminium, in the case of tank wagons the situation is different. Here there is a need both for light weight and specific corrosion resistance and a number of European railways are experimenting with various types of tank wagon or tank containers.

The Hungarian Railways have about 10 GRP-lined steel tanks which are used for the transport of hydrochloric acid, but more interesting developments have taken place elsewhere, particularly in France. The French National Railways have been experimenting since 1963–64 with a fleet of ten GRP tanks, lined with either PVC or polyethylene, for conveyance of chemicals. The tare weight of these wagons is some 9.2 tons compared with 11.2 for a conventional design, and they have survived impact tests at 15 km/h. The tank is bolted to the wagon underframe by means of continuous strips along each side of the tank. The authorities responsible for dangerous goods traffic at present have given approval to these wagons on an annual basis and so far approval has been granted each year.

The French are also using GRP containers for the transport of milk. These containers, which have a capacity of 43 m^3, have been designed for easy transfer on and off specially designed rail vehicles. The container consists of a filament wound core covered with a normal laminate to make an inner shell 4 mm thick; the outer shell is 8 mm thick GRP and there is a 50 mm thick polyurethane foam core between the inner and outer shells. The end covers are made up of an inner facing 5 mm thick, a polyurethane foam core 50 mm thick and a GRP skin 9 mm thick. These covers are glued to the cylindrical core. The core insulation is foamed *in situ*. This container has been designed for an over-pressure of 196 kN/m^2.

The Italian State Railway authorities are experimenting with a compartmented tank wagon for carrying fuel. The tank is 7.3 m long and is made up of three separate cylindrical units bolted and bonded together. To mount the tank on its underframe, steel T sections are bonded into the sides of the tank, which is bolted through the stem of the T onto the vehicle mounting. The tank wall thickness is 10 mm and the capacity is 34 m^3. This construction showed a weight saving of about 2 tons compared with a standard tank wagon.

To complete the review of tank wagons, mention should be made of two being evaluated by the German Federal Railways. One has been in use since 1960 for the conveyance of wine and fruit and has survived two accidents without serious damage, both of which could well have severely damaged a steel or aluminium tank. This tank is made from a sandwich structure of GRP and rigid PVC foam and has a capacity of 23.5 m^3 and weighs 980 kg. A comparable insulated steel wagon weighs 3680 kg and has a capacity of only 20.5 m^3.

The other tank wagon, which has been built under the aegis of one of the major German resin companies, is being used for the conveyance of liquid adhesives and other chemical liquids. The tank, which has a wall thickness of 10 mm and an overall loaded weight of 38.3 tons, has successfully withstood impact tests.

5. REFRIGERATED CONTAINERS

While the case for GRP tank wagons has yet to be proved there is no doubt that synthetic materials have established themselves for use in refrigerated containers.

British Railways freightliner insulated containers have been made from a GRP polyurethane foam sandwich held in a metal frame to conform with I.S.O. lifting requirements and the most recent production has been based on GRP-faced plywood for the outer layers of the sandwich. In addition to the British usage of GRP foam sandwich construction for these containers, similar products have been manufactured for French, Italian, Dutch, German and South African railways. The obvious advantages of the synthetic materials are the absence of metallic connections, which could serve as thermal paths, and freedom from corrosion combined with ease of cleaning. At present the main disadvantages of GRP as a skin material is that, at an economically sensible thicknesses, it can be punctured and this is the reason why B.R. are now producing insulated containers having GRP faced plywood as the skin material. In one particular design of polyester glass fibre container manufactured for B.R. in a small quantity, the container was progressively constructed from the outside inwards so that the inner surface was relatively rough. These containers were used for conveyance of meat and, although there is a stringent cleaning process using hypochlorite, some containers developed a mould growth in the surface. This was overcome by re-sealing the surface, but it is obvious that a smooth inner surface is essential for any type of container carrying foodstuffs, and GRP coated plywood has just such a surface.

An example of an insulated wagon used by South African Railways has been described in detail recently[5]. This wagon is 12.2 m long and is used for the high-speed transport of freshly caught fish from the Cape coast to the inland areas on the Reef. The wagon consists essentially of a laminate 8 oz/ft^2 (2.4 kg/m^2) as the inner skin, a laminate 4 oz/ft^2 (1.2 kg/m^2) for the outer and a foam core. An interesting point is that the internal dunnage bars, which are 102 mm and 10 mm thick, are formed in GRP by cold-press moulding. In essence the method of manufacture is to make the inner and outer skins separately, the inner skin having a considerable number of both horizontal and vertical half-round reinforcing ribs built in, as well as having a considerable number of steel tapping plates accurately located in it. The cavity between the skins is filled with polyurethane foam using a frothing technique. The

Table 5.1

	Standard refrigerator wagon	*Sandwich structure refrigerator wagon*
Length of wagon over buffers (m)	11.74	12.54
Effective floor area (m^2)	22.2	25.4
Effective loading capacity (m^3)	46.0	53.0
K-value (kcal m^{-2} h^{-1})	0.35	0.20
Tare weight of wagon empty, without ice (kg)	16 000	
Tare weight with ice (kg)	21 000	
Weight empty with cooling plant but without fuel (kg)		15 500
Weight with fuel (kg)		16 050
Load capacity (kg)	15 000	19 950

pressure generated is dealt with by the ingenious method of using the moulds to reinforce the skins. In other words the inner skin on its mould is lowered into the inverted outer half in its mould and foam is injected through the roof cavity.

Another design of refrigerated van, which has been described in detail by the Dutch manufacturers[6] is being evaluated by Interfrigo. Once again the structure is made up of a sandwich of polyurethane foam skinned with GRP. To make the sandwich strong enough it was necessary to tie the skins together with 'hat' sections which, in addition, made up suitably sized cavities for froth foaming the insulation *in situ*. After the first wagon of the test series had been in service for a short time it was found that in some places delamination occurred between the skin and the 'hat' sections. This was caused by the temperature gradient of up to 80°C in summer between the inner and outer skins and was overcome by putting horizontal stiffeners on the outer skin to prevent buckling.

The glass content of the skins varies between 5 and 9 oz/ft^2 (1.5–2.8 kg/m^2) depending on the location and foam of 40 kg/m^3 density was used. The practical advantages of this structure over a standard refrigerator wagon are shown in Table 5.1:

6. CONCLUSION

The applications discussed above have been related to rolling stock of one form or another, but it should not be forgotten that glass reinforced plastics are used widely on the civil engineering side. Not only are large quantities of standard items like rooflight sheeting used, but also complete buildings to house relay equipment have been made for a number of years. In addition, the use of GRP trackside boxes and cupboards to house switchgear and signalling equipment is commonplace. Apart from reducing condensation these components require less maintenance than their metal equivalents. Finally, it is fair to say that glass reinforced plastics have made a useful technical and economic contribution to the railways during the past decade, and this material will undoubtedly be used increasingly in the future.

REFERENCES

1. Scollay, I. J., *et al.* Paper 3, B.P.F. *Fifth Int. Rein. Plast. Conf.*, Nov. 1966.
2. Bristol Aeroplane Plastics Ltd., 'Railway Coach in Reinforced Plastics', 1958.
3. Cleathero, B. E. Monocoque Construction in Transport: *Trans. Plast. Inst., Lond.*, No. 64, p. 287.
4. British Railways Research Department, *Report No.* 588.
5. Turner, P. E. Insulated Wagons for S.A. Railways; *Reinf. Plast.*, Oct. 1968.
6. Wethmar, A. Paper 3, *Symposium: Some Transport Applications of Reinforced Plastics*, Feb. 1966.

BIBLIOGRAPHY

Office for Research and Experiments of the International Union of Railways: Question B.68: Enquiry Reports, 1963 and 1967.

6

Road Transport

A. C. JOLLY

1. INTRODUCTION

It has been said 'The Car Industry is quite objective in its approach to materials, and does not really mind if the end product is made from compressed grass cuttings so long as they do the job at least as well as previously used metals and at a lower cost.'[1]

The principal advantages of GRP over standard materials include light weight, lack of corrosion, high strength–weight ratio and good surface finish. In addition, complex contours and shapes can be obtained without the need for expensive tools and variation of shape can be made at any time with minimum cost. GRP is easy to repair and results in much lower stresses due to thermal expansion than with steel and other metals due to a lower modulus. Tensile strength can be high depending on the type of reinforcement and its method of application and the reinforcement can be laid in the line of maximum stress. Pilot production runs on prototypes can be carried out at low cost, and GRP can be made translucent.

It is also possible to obtain built-in flame resistance by using carefully formulated resins. The current specifications however, do not adequately cover the requirements and an S.M.M.T. (Society of Motor Manufacturers and Traders) working party is at present trying to establish an acceptable level of fire retardancy in GRP vehicles. This will involve a critical review of the specifications dealing with this matter such as B.S. 3532, B.S. 2782, A.S.T.M. D635 and B.S. 476, Part 1.

Two of the main disadvantages of GRP with regard to vehicle manufacture are that the labour content is high and the speed of production is slow compared with the more conventional methods involving metal. As a result, GRP is found to be most suitable where production rates of 50–350 vehicles per week are acceptable. The high labour content can be turned however to advantage. Many developing countries raise tariff barriers and call for a high local labour content. In some instances even the complete assembly of the bodyshell from components shipped from the U.K. is not sufficient to

Figure 6.1. British Leyland Motor Corporation Minis. GRP bodies made in Chile (Courtesy of Pressed Steel Fisher Ltd.)

Figure 6.2. Body moulds for Mini produced in Chile (Courtesy of Pressed Steel Fisher Ltd.)

satisfy the restrictions imposed. Manufacturing overseas 'on site' is the only solution if the market outlet is to be maintained and under these circumstances the economics for using GRP are often acceptable. Furthermore, GRP produced by hand lay-up, spray and cold-press techniques, has proved to be very convenient in view of the low capital outlay required to establish production and the need only for semi-skilled labour. Examples of this are the Reliant production of the Anadol in Turkey and the GRP versions of the British Leyland Minis and M.G. 1300 in Chile.

2. DEVELOPMENT OF CAR BODY DESIGN

As an example of the way a GRP car body is designed and manufactured it is interesting to follow through the development and production by Lotus Cars Ltd of the Lotus Elan.

Initially, a quarter scale model is made in modelling clay and all mould and manufacturing problems sorted out at this point. The next stage is to make a full-size pattern, in this case of aluminium. From the full-size pattern the GRP master body mould is made with stiffened metal internal framing. At this stage the final shaping and fitting is carried out and the whole surfaced to a high standard using a furane resin. The pattern is then released for production and body and component tools made from it. Each mould tool is approved before being released and from then on is subject to routine servicing as and when required but is expected to yield one moulding per day if a large section is involved such as the body, and two mouldings a day where smaller non-essential parts of an internal nature are required.

The production of the car is then commenced with wax and PVA release agent coatings being separately applied to the moulds. This is followed by two gel coats and then pre-tailored glass mat, issued in car sets to the laminators, is impregnated with resin by the hand lay-up method. After curing, the mould tool is broken down into pieces, each moulding being mechanically jacked away from the mould surfaces. The same procedure is applied both to components and to the main body shells. Mechanical attachment points are then bonded into the body and components to facilitate chassis fitting, door hanging and so on. On passing through the fettling shop, all edges and holes are trimmed and drilled and surplus material removed. Doors, boots, bonnets and lamp-housings are then fitted to the body and inspected. After this all surfaces are flatted to give a good key for painting. The body is then removed to the paint shop and coated with polyurethane primer, flatted, coated with a nitro-cellulose primer surfacer, flatted and then finished off with a number of coats of synthetic enamel top coat. Between each stage the finish is cured at 130°C. The assembly of the complete car is then carried out and the car finally inspected.

3. WORLD SURVEY

The production of GRP car bodies, commercial vehicle cabs and other applications of GRP in road transport is now well established throughout the world. Whilst attention should be drawn to the interesting growth of this market in Australia, South Africa, Japan, Brazil, Argentina and most European countries, the following specific examples are of particular interest.

3.1. ITALY

The construction of refrigerated and non-refrigerated truck bodies as well as tanks for the transportation by road of milk, beer, fruit juices, wine and many chemical products accounted for over 1000 tons of GRP in 1967. In some of the GRP trucks the sandwich construction is used with expanded polyurethane as the core. Filament winding techniques are used for tanker production up to a capacity of 60 m^3. Among other applications are GRP bodies for some sporting cars, buses and roofs for public transport vehicles and also containers[2].

3.2. U.S.A.

Transportation ranks as a major user of GRP and was about 12 per cent of the total 1967 consumption of 250000 tons. This is about £10 to £12 per vehicle based on an estimated production of 9.5 million cars, trucks and other commercial vehicles. The projected growth rate of GRP in the transportation industry is about 15 per cent a year. For many years the Molded Fiber Glass Company Inc. have been foremost in the United States in manufacturing GRP car bodies, and they have over 65 presses at Ashtabula, Ohio, devoted exclusively to this end. The body panels are press moulded with an isophthalic polyester resin using glass fibre preform on the continuous strand 8600 mat. Their Chevrolet Corvette was the first plastics bodied car to be made in quantity production anywhere in the world. There is a growing trend in the car industry towards producing small cars and different body types. At present there are more than 30 known new applications of GRP in cars manufactured during 1968. These include the bumper extender for Ford which is expected to be produced in quantities of 200000 this year, the highest volume of any GRP parts specified; 16 major parts in the Ford Shelby G.T. 500 soon to be produced with a planned volume of 9000 units; Chrysler hard top package tray; the completely re-designed Chevrolet Corvette with an entire GRP body, which it has had for the past 15 years (matched moulds are used for the body panels); and the timer belt housing for the Pontiac Tempest. In addition, it is expected that about 10000 tons of glass fibre reinforced thermoplastics will have been used in U.S.A. cars during 1968.

New low-shrink premix polyester resins is another development of interest to U.S.A. car manufacturers. Combined with new moulding techniques these resins produce GRP parts with a surface finish as smooth as steel when it comes from the mould. Sanding, putty rubbing and other normal GRP surface preparations are eliminated. Another innovation is the one-piece 1968 Falcon bumper extender made from these new resins and moulding techniques with a tooling cost saving of 25 per cent.

Not all the 1968 transportation advances of GRP are in cars. For example,

International Harvester is introducing a new front end truck design in GRP and White is designing the production of 5000 tilt cabs equipped with GRP panels. In addition, the John Deere Co. now fits a GRP hood on its garden tractors[3].

3.3. CHILE

For the reasons discussed earlier in this chapter the GRP versions of the British Leyland Mini and M.G. 1300 are being manufactured in Chile after development by Pressed Steel Fisher Ltd. The design is based on a moulded plastics shell incorporating a steel foundation frame to improve the torsional and beam strength properties. Additionally, there is provided a light steel frame ('birdcage') about the door apertures and roof areas to improve stability and as protection in roll-over contingency. It must be emphasised that this stiffening has not constituted a chassis. Thirty-one moulds are used for the Austin Mini and twenty-seven for the M.G. 1300.

3.4. TURKEY

In addition to manufacturing GRP cars in England, the Reliant Motor Co Ltd has established licensees in many countries. A large operation exists in Turkey where production is now up to 4000 vehicles per annum with regard to one particular car, the Anadol. This is a modern four to five-seater saloon

Figure 6.3. Anadols with complete GRP bodies coming off the assembly line (Courtesy of Otosan Otomobil Sanayii AS., Istanbul

with a GRP body. Its method of manufacture fulfils the criteria previously mentioned with regard to situations of this kind, namely it involves a high labour content, low foreign currency content and relatively low capital expenditure. The major moulding items include the main body shell, front bulkhead, rear bulkhead, bonnet lid (outer), bonnet lid (inner), boot lid (outer), boot lid (inner), front door (r.h. outer *and* inner), front door (l.h. outer *and* inner), rear door (r.h. outer *and* inner), rear door (l.h. outer *and* inner), plenum chamber, air ducting (to radiators), fascia (2 parts), glove box and lid, petrol tank cover, steering column nacelle, crash pad and five inspection covers.

3.5. ISRAEL

GRP has been used in the car field in Israel for some years, particularly in the Sabra and Sussita vehicles manufactured by Autocars Ltd of Haifa—another Reliant licensee. They have been coming off the production line at the rate of 3000–4000 per year. Use is also being made of GRP extensively by a Leyland assembly plant manufacturing an all-GRP lorry cab. Other uses are in buses for both front and rear doors and seats, and police jeep bodies and roofs for the Dechevaux model of Citroen. Sidecars for motorcycles and scooters have virtually replaced metal ones and are standard. In terms of tonnage therefore, road transport is by far the largest and most developed outlet for GRP in Israel[4].

3.6. CZECHOSLOVAKIA

Skoda are involved in using GRP for vehicles and have achieved a cost saving of up to fifteen times over conventional materials. GRP has been used for the bodywork of electric locomotives, buses, trams and track vehicles in general. The hand lay-up method is used for about half the total use but it is expected that spray-depositing systems will supersede it[5].

4. UNITED KINGDOM

4.1. CARS

The largest concern producing GRP cars in the United Kingdom is the Reliant Motor Co. Ltd. whose Regal three wheel car has been manufactured for many years. If consists basically of 21 mouldings ranging from major items like the outer and inner shells and air ducting for the main body, the boot floor, the bulk head stiffener and minor items such as inspection covers. The four wheel saloon car known as the Rebel consists of 34 GRP mouldings whilst the very sophisticated and attractively designed Scimitar consists of 65 mouldings. The main moulding methods used by Reliant are spray gel coat and hand lay-up techniques. Some smaller parts however, such as battery boxes, engine covers and roof stiffeners are press moulded. Fire retardant resin is used, the gel coat is self-coloured and the body is finished off finally by a painting process.

Another firm specialising in GRP cars is Bond Cars Ltd. whose amalgamation with the Reliant Motor Company Ltd. has recently been announced. The moulds for their range of cars are made from GRP. The main body of the 2-litre and 4S '1300' Equipe consists of only four mouldings—the top section,

Figure 6.4. Full range of Reliant GRP vehicles (Courtesy of Reliant Motor Co Ltd.)

two sides and one underscoop. The top and two sides are bolted together and this is layed up. The underscoop is layed up and then bolted to the sides and top. These are then bonded together. The bonnet of the two models consists of five mouldings; the bonnet top and sides, two wheel arch return flanges, one rear flange and one underscoop. The wheel arch flanges and rear flange are bolted to the main bonnet section and layed up. The underscoop is layed up, bolted to the main section and both are bonded together. The Equipe boot lid is made from one mould with removable flanges to facilitate removal of the moulding. Eight more single piece moulds are used for ancillary pieces.

Lotus Cars Ltd. produce three high performance cars; the Elan, the development of which has already been described; the Elan Plus Two and the Europa. Their components are made entirely of GRP and the production methods used are similar to those of the Elan.

The London taxi cab is a high specialised vehicle and has to be designed and built to a detailed specification which is laid down by the public carriage office to whom all design proposals, detailed drawings and material specifications must be submitted for acceptance. When used in the Metropolitan Police area, it can be granted a licence up to 10 years of age and since it can do approximately 30000 miles per year it is important that the body and chassis will withstand this severe service. For this reason the Wincanton Transport and Engineering Co. Ltd. used GRP for their 'Winchester' London taxicab.

It was decided to make the body sectional, the two halves being the roof (from waist upwards) and the lower body extension (wings, scuttle, interior, floor, etc.). The lower body section incorporates all the cab interior floor and seat boxes, the engine compartment and the boot interior. Because of this the complete lower section is self-coloured and only the outer skin of the body is painted. The boot lid, bonnet, all doors, all exterior panels and head lining are

Figure 6.5. The Bond Equipe 2-*litre G.T.* (Courtesy of Bond Cars Ltd.)

in GRP. Hand lay-up is used for all parts but with the cold pressing of some interior panels at the main stage. Fire-retarded resins are used throughout as specified by the carriage office.

4.2. INVALID TRICYCLES

Three-wheeled vehicles for invalids are made in the U.K. in collaboration with the Department of Health and Social Security. Considerable numbers have been made in GRP for many years by A.C. Cars, Invacar and Frank Tippen. The A.C. Car version has a complete body including door panels produced from 15 self-coloured GRP mouldings. They have an output of approximately 50 bodies per week and achieve this by employing six complete sets of moulds.

4.3. COMMERCIAL VEHICLES

GRP has shown itself to be extremely attractive to commercial vehicle builders and during the past 15 years it has been widely used in the United Kingdom for this purpose. Dennis Brothers Ltd. for instance, have an output of between 500 and 600 GRP cabs per annum. The hand lay-up technique is mainly used because of the unsuitability of the spray process for mouldings incorporating complicated returns of sections. Spray-up is used for simpler

Figure 6.6. Pax V 15 ton *g.v.w. commercial vehicles. The cabs are one-piece GRP mouldings* (Courtesy of Dennis Bros. Ltd., Guildford)

types of bodies such as tanks. The design has been rationalised to enable major mould components to be utilised for various types of vehicle. For instance, the front of the cab includes front doors which are identical on single cabs, double crew type cabs as well as fire appliances. The cab mouldings are supplied self-coloured thereby saving painting costs. Minor scratches can be polished out, again saving painting. Fodens, British Leyland and ERF are amongst others using GRP in their vehicles.

GRP moulded roofs have also been used for some time. Originally their great attraction was translucency, but other qualities have since been recognised, such as the fact that being moulded in one piece they are leakproof and they are quicker and easier to fit. It is interesting to note that with confectionery vans the roofs are usually painted out in livery colours so that they are no longer translucent.

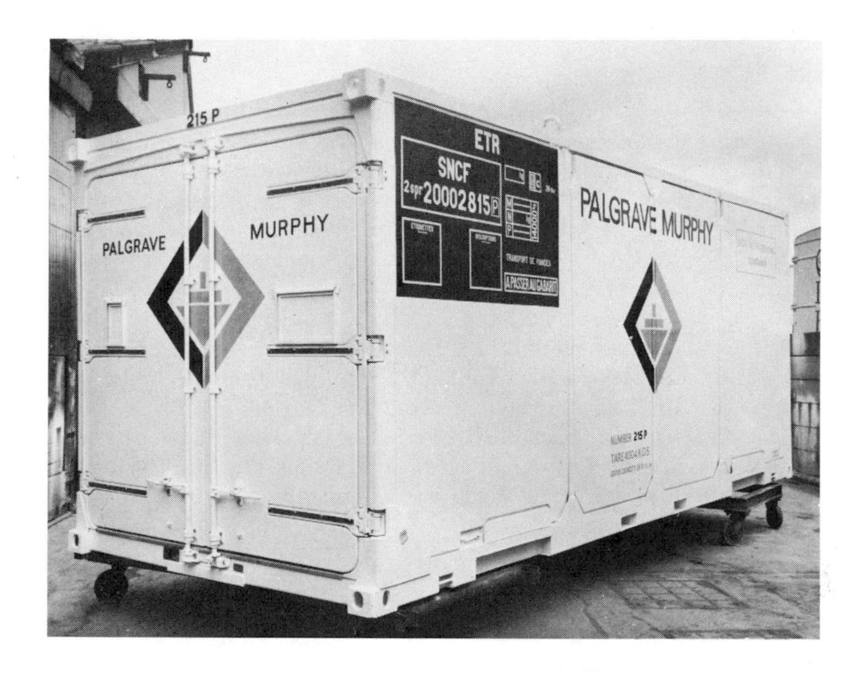

Figure 6.7. One-piece moulded GRP insulated container for the transport of hanging carcase meat. Refrigerated with liquid CO_2 (Courtesy of Mickleover Transport Ltd.)

Figure 6.8. One-piece moulded GRP insulated icecream van (Courtesy of Mickleover Transport Ltd.)

4.4. CONTAINERS

The foremost use of containers by the transport industry is to standardise and rationalise its function of moving goods[6]. A container or trailer has to be strong enough to carry the loads contained and to withstand severe handling damage, but at the same time it has to be as light as possible in order to increase the payload to the maximum. Basically two types can be distinguished, namely the 'dry' or non-insulated and the 'insulated' or refrigerated. Standardisation has been brought about concerning sizes and tests resulting in both an ISO standard and the British standard B.S. 3951, 1967. This requires that all containers will be 8 ft (2.4 mm) wide × 8 ft (2.4 mm) high overall and be capable of being handled by standard lifting methods. Most will be 20 or 40 ft (6 or 12 m) long with arrangements to include 8 ft (2.4 m), 10 ft (3 m) and 30 ft (9 m) also when required. Nowadays, both the dry and insulated types of containers utilise a basic frame design which takes all the load. The use of GRP in the dry or single skin containers is in the coating of both

Figure 6.9. A fitted Avonaire Elite as supplied to most of the police forces throughout the U.K. (Courtesy of Mitchenall Bros. Ltd.)

sides of plywood. The refrigerated containers make greater use of plastics since the outer and inner skins are of GRP with foamed plastics such as polystyrene, PVC and polyurethane as the insulating core. Since methods of lifting have changed, special corner castings in metal are mandatory and side-wall loads and dynamic loads have greatly increased. The whole design of containers has radically altered during the past three years[7]. Work has also been carried out to overcome the contamination of certain foodstuffs by monomer and certain solvent vapours.

4.5. OTHER APPLICATIONS

The introduction of GRP into the motor cycle and scooter industry nearly 20 years ago was mainly due to weight-saving and the need to obtain an inexpensive weather resistant streamlined fairing. These are usually self-coloured, the shape being a useful aid to rigidity. They can be designed to give either partial or almost complete weather protection. Many petrol and oil tanks for motor cycles are also made of GRP in the United Kingdom.

Other important uses of GRP in road transport include ambulances, caravans, mobile shops and indeed wherever a custom built or semi-custom built vehicle is required.

5. THE FUTURE

The introduction of carbon fibres with their exceptionally high modulus and strength[8] will provide a challenge to glass fibre as a reinforcement but will probably be used in conjunction with the latter for some time because of cost. Nevertheless, carbon fibre reinforced plastics have already been used in the body of the Ford GT 40 which won the 1968 Le Mans Grand Prix.

Much still needs to be done to improve the self-extinguishing properties of GRP used in road transport. Greater gel coat durability is also needed and a means of adequately backing it with a sprayed-on bubble resistant filler or finely chopped strand. Resins which could withstand conventional stoving temperatures, to enable paint finishes comparable with those used on metals, is another area for more research and advancement[9]. A good survey of the restraints and incentives relevant to polymers in transport generally has been published[10].

Meanwhile much work is being carried out on other materials which are in direct competition with metal[11]. The chief contenders for the high production car body structure are cast urethane monomer, thermoformed sheet, rotational moulded thermoplastics and injection moulded thermoplastics. Though the latter process might be considered too costly and inflexible it should be remembered that GRP has a relatively high material cost and labour content thus limiting its use to applications of low production rate. Press moulding while more efficient, involves subdivision of the body shell into a considerable number of pieces necessitating extensive assembly and finishing operations. Thus the more extensive use of GRP in the transport industry will probably be in tanks and containers where mechanical strength is at a premium, rather than in conventional car bodies.

REFERENCES

1. *Rubb. Plast. Age,* **49**, No. 8, (1968) 724.
2. Rial, E., B.P.F. *6th Int. Reinf. Plast. Conf.,* London, November 1968.
3. Menton, J. D., B.P.F. *6th Int. Reinf. Plast. Conf.,* London, November 1968.
4. Eden, A., B.P.F. *6th Int. Reinf. Plast. Conf.,* London, November 1968.
5. Bohumil, H., B.P.F., *6th Int. Reinf. Plast. Conf.,* London, November 1968.
6. Hawthorne, B. J., Plastics Institute Symposium '*Whither Reinforced Plastics,* London, May 1968, paper 6.
7. Hayes, C., Plastics Institute/I.R.I. Conference *Polymers in Transport,* London, February 1969, Paper No. 1.
8. Watt, Johnson and Phillips: 'High Strength High Modulus Carbon Fibres', *Engineer, Lond.* **221**, 815 (1966).
9. Adams, A. A., Plastics Institute Symposium *Marketing Reinforced Plastics,* London, February 1969, Paper No. 6.
10. Sharp, H. J., Plastics Institute/I.R.I. Conference *Polymers in Transport,* London, February 1969. Paper No. 7.
11. Hill, A. C., Plastics Institute/I.R.I. Conference *Polymers in Transport,* London, February 1969, Paper No. 6.

7

Cladding and Sheeting

J. R. CROWDER

1. INTRODUCTION

Since it was first offered commercially in Great Britain in 1952, corrugated glass-fibre reinforced polyester sheeting has accounted for a large proportion of the output of reinforced plastics. It has been estimated that in 1969 about 3×10^6 m^2, weighing over 5×10^3 tons, of this sheeting was produced in the United Kingdom. Although production has not kept pace with the growth curve of the plastics industry as a whole, it still constitutes 18 per cent of the total usage of GRP; this proportion is similar to the quantities used in the two other major fields of application, namely, marine and land transport.

Most of the corrugated sheeting is used by the building industry and the greater proportion has until recently been used in conjunction with other corrugated sheet materials, in particular asbestos cement. Translucent and transparent sheeting usually manufactured to conform to BS 4154[1] makes up the bulk of the market but a small proportion is in the form of opaque sheets for vertical or roof cladding. More usually, when the material is used for cladding the external walls of buildings, it is made up in panels which incorporate other materials as well as the plastics. Other clear plastics materials, notably acrylics, and PVC with or without incorporated wire mesh, are also available in corrugated forms. The relative merits and different usages of the materials are discussed later.

Corrugated translucent sheets were the earliest form of GRP available to the general public. The formability of the material and its translucent nature when cured suggested its use as a lightweight alternative to framed glazing or very heavy corrugated reinforced glass. It has additional advantages of toughness and ease of handling or cutting on site. It is possible to produce sheeting to meet the requirements of building regulations for any roof situation, but some restrictions, discussed later in this chapter, apply to its use in vertical surfaces.

Almost all the corrugated and flat sheets and cladding panels are made with polyester resins; styrene, or mixed styrene and methyl methacrylate monomers are added to effect cross-linking during curing. Other materials, particularly polyester cloths and film, and polyvinyl fluoride films have been incorporated in surfaces to provide special properties. Various lacquers are occasionally applied to surfaces during or after sheet manufacture to improve surface

Figure 7.1. Blue tinted corrugated translucent reinforced plastics sheeting was used to clad this 68 m *diameter sports dome in Perth, Scotland* (Courtesy of British Industrial Plastics Ltd.)

properties. Choice of resin and of the glass reinforcement used with it play a large part in determining the quality of each individual product, and the method and care in manufacture exert an overriding control.

2. METHODS OF MANUFACTURE

Although a large proportion of corrugated GRP sheeting is now produced by mechanised processes, hand lay-up is still successfully employed by a number of firms in the U.K. Mechanically made sheets are usually more uniform in quality, though it is possible to produce excellent quality sheets by hand. Manual production is particularly useful for moulding to unusual profiles which may be difficult or costly when made on a continuous laminating machine. With more than fifty profiles in regular use in the U.K., smaller hand moulds are often more economic than the long formers required for continuous production. Manual or only partly mechanised processes are also used for the production of most cladding panels.

Figure 7.2. Longitudinally corrugated glass reinforced polyester sheeting is produced in continuous lengths on this machine
(Courtesy of British Industrial Plastics Ltd.)

2.1. HAND LAY-UP

In this method, chopped strand mat is almost exclusively used, although fabrics such as woven rovings can be employed. Manufacturing details vary but basically the technique is the same[2]. A warm dry room is normally used for the manufacturing process.

A sheet of dry regenerated cellulose film, larger all round than the glass fibre mat to be impregnated, is laid out on a flat table top and may be secured against slipping. The polyester resin, thoroughly mixed with catalyst and accelerator, proportioned to give the required setting time, is then usually spread uniformly over this film and a piece of the chopped strand mat of appropriate size is laid on it. The mat may be placed on the cellulose film first and the selected resin applied to the mat, usually by sprinkler, taking care to ensure uniformity of distribution.

Most of the currently used mats are themselves bonded with powdered polyester resins and a soaking period of 5–10 min is allowed for this binder to dissolve and for the mat to become completely impregnated. The mat must be dry and, in order to improve and control the rate of wetting, the impregnation table may be warmed to a constant temperature. Once all the occluded air has escaped, a second cellulose film is placed on to the impregnated mat and, by using squeegees or rollers, any air bubbles are worked outwards from the centre and escape at the edges. Great care must be taken not to squeeze out too much resin at this stage or the optimum glass/resin ratio will not be obtained. Excess resin is held between the overlaps of cellulose film surrounding the impregnated mat and may be reclaimed for adding to the next mix of catalysed resin. The edges of the sandwich are bent over and taped to prevent the return of air bubbles during shaping.

The sandwich is then transferred to a former, usually made of metal, of the required profile, taking care to place it symmetrically. Starting from the centre of the former, and working to the sides, it is shaped to the corrugation contours with soft cloth pads, taking care not to develop any wrinkles.

Contact pressure only is used when curing, and sometimes light tubing of suitable size is laid along the corrugations. Loaded formers are then stacked vertically, separated by bars to allow air to circulate and avoid undue exothermic development of heat.

When the sheets are sufficiently cured to be handled without permanent distortion, they may either be stacked to mature at room temperature for a week or two, or they may be 'post-cured' at somewhat higher temperatures for considerably shorter periods. Completion of cure is usually determined very empirically, by the 'ring' of the sheet when it is tapped. The sheet is trimmed to size before despatch, although the cellulose film may be left for removal when the sheet is installed.

2.2. MECHANISED PRODUCTION

Mechanisation of manufacture is usually geared to continuous production. Machines in current use have the same basis as hand-forming processes but with many refinements. Fully continuous methods for production of corrugated roofing sheets have been developed in several countries and are the subject of numerous patents[3–6]. The various systems are often used under licence and full details are not available.

Processes are of two main types, one using chopped strand mat as the reinforcement, the other incorporating strands chopped continuously from rovings. In a process typical of the former type, catalyst and accelerator are mixed with resin, either in batches or continuously, using a ratio pump or other monitoring device. The mix is fed on to a moving carrier film which may be of regenerated cellulose or polyester ('Melinex'). Polyester film is often used advantageously in the continuous process and, being stronger, it is able to sustain the stresses imposed by the forming process and the weight of glass and resin. Because it is relatively expensive, this film must be re-used several times and is continuously rewound at the end of the production line.

The resin content of the sheet is controlled by a metering device and doctor-blade. Glass fibre reinforcement, in the form of chopped strand mat, is fed into the moving layer of resin, which is readily absorbed and, in wetting out the mat, displaces most of the air. At this stage, the moving film may be led over a smooth table which provides sufficient space to permit coating with additional resin or the introduction of surfacing tissue. The upper release film of cellulose or polyester is then applied by adjustable rollers to form a sandwich, at the same time removing any remaining air. When the correct resin content has been obtained, this sandwich is led on to continuous formers for the appropriate corrugation profile, starting at the centre and working towards the sides, so that there is no stretching of the impregnated mat. By the time it has all been brought into shape, it is passing into the curing oven and the resin is beginning to gel. For some distance in the oven, it passes along heated cauls, which ensure retention of the correct profile as well as providing the heat necessary for cure of the resin. The temperature is brought up to between 60° and 90°C and in the period of up to 30 min, depending on the process, that it takes to pass through the oven, the sheet is cured. No post-cure is usually necessary and the release film is removed from above and below the sheet before edges are trimmed continuously with abrasive disc saws to give the required width. A disc saw, set in a platform that moves with the sheet, cuts it to the required length when the moving front of the sheet trips a control switch. Such a machine requires three or four operators, but gives a uniform product of consistent quality and to closer tolerances than are possible with hand lay-up methods.

A continuous process developed by Montecatini[4] in Italy, has been described[7] and differs from the above in many details, but primarily in that the continuous sheet produced is cross-corrugated. Considerable lengths of this may be rolled to permit advantages in handling and transportation but it is used in the U.K. only on a very small scale. This is because of the requirement, that until recently, for most building installations, sheeting could be used only in chequerboard patterns in conjunction with sheets or other corrugated materials. The variety of corrugation profiles required to match these other materials would make the cost of a complete set of formers for the Montecatini process too high in the U.K. (more than 50 profiles). Also, the variety of lengths required, in small increments, would increase the difficulty.

Another continuous process[5] makes use of continuously chopped strands distributed through a forming hood. The strands effectively form a mat on a number of evenly spaced nylon threads, which carry the sheeting through the process until it is strong enough to be self-supporting. A film of flexible and rapidly-cured resin may be applied to the moving cellulose release film before the glass fibre and the bulk of the resin are introduced. The resin itself has good weathering qualities but deterioration may occur on weathering if the glass fibre strands are close enough to the surface to be readily exposed. In consequence, this very thin gel coat improves the weathering qualities; also, without affecting the overall resin content the resin/glass ratio of the interior of the sheet may be reduced by passing it between squeeze rollers; these also help to drive out excess air and ensure thorough impregnation. The sandwich is then shaped on cauls, in a similar manner to that described above and after passing through the curing ovens, the product is trimmed and cut to the required lengths. Since glass rovings are less costly than glass mat, and can be distributed to a desired width with a minimum of waste, sheet made by this process is potentially cheaper than sheet produced from mat.

It should be noted that the longitudinally corrugated sheet produced by two of the continuous processes described, is available in considerable lengths. In recent years, as well as being used in the familiar chequerboard pattern, lengths of this sheeting have been employed to provide the complete roof-span of many sports buildings, in particular swimming pool covers, and have permitted construction of some unusual arc-shaped roofs.

2.3. DECORATIVE SHEETING

A variety of decorative sheets has been produced, often to special order, and usually by hand or semi-continuous processes. They may be either flat or reeded, translucent or opaque and often incorporate printed fabrics, as well as glass fibre reinforcement. Care must be paid to the distribution of resin between the cloth and glass mat to give a balanced sheet, otherwise distortion may occur on demoulding or during use. Also, inks and dyes must be chosen so that they neither interfere with cure of the resin nor are affected by any of the resin components; they must also be stable to light.

Figure 7.3. The cladding panels of this curtain wall building consist of dark green pigmented GRP backed with asbestos cement sheeting (Crown copyright)

2.4. CLADDING PANELS

Panels for the facing and cladding of framed buildings have been regarded for some years as a potentially large market for reinforced plastics but have not yet achieved the expected major breakthrough. They have an advantage in being suitable for production in relatively small numbers so that an architect is able to obtain an individual finish for a building and still have some variation in the sizes of panels used. The result is that small production runs are normal, and manufacturing processes are carried out in small shops either by hand lay-up or simple 'spray-up'. Contact moulding techniques are used, and mould surfaces are normally textured to make any surface imperfections, arising either from manufacture or from weathering, less obvious. The mould surface is treated with release agent, and then a thin film of pigmented,

Figure 7.4. The cladding panels on this building in North-west London are faced with GRP which also serves as permanent formwork for the lightweight concrete insulation and serves as a medium for detailed edge design (Copyright Sam Lambert)

catalysed resin is applied as a gel coat. When this begins to gel, appropriately tailored glass mat is laid over the surface and impregnated with a further supply of activated resin. The wet mat is rolled to eliminate trapped air and to ensure even and complete impregnation. Any framework for the panel or backing such as asbestos cement pegboard may then be placed in position and built into the structure, using further layers of glass mat and resin, until the desired thickness is achieved. The panel is allowed to cure at room temperature until it can be handled, and then cure is completed either in an oven or over a longer period without further heating. Similar results can be achieved using either spray-applied resin on mat, or spray/depositor applied resin and chopped glass rovings. More skill is necessary to achieve the same evenness of finish and resin/glass distribution, but where large numbers of panels have to be produced, considerable saving in labour may be achieved in this way.

Considerable care is necessary in the design of such panels to ensure that the laminate is balanced to minimise warping, though panels exposed on the sunny side of a building, particularly if dark in colour, will be subject to a wide temperature range. Firm bonding to a rigid backing is necessary to avoid excessive deformation due to thermal movement produced by solar heating. Great care is also necessary in designing the edges of panels so that they can be fixed in suitable frameworks which can provide allowance for thermal movement as well as positive fixing and exclusion of rain-water.

Cold press moulding is a relatively new technique which may have been used for the manufacture of building panels but one set of panels, produced for a block of flats in north-west London, was manufactured by a hot press moulding technique using matched metal dies. The expense of mould making is justified by a building design which permits multiple use of near-identical panels and by the expectation of further large orders for similar panels. In this instance it is notable that the GRP material was used in conjunction with lightweight concrete, asbestos cement sheeting and a steel framework and it provides an external finish, permanent formwork for the concrete and a suitable medium for design of edge details and incorporation of window units. Use with other materials in such composite units has wider potential in cladding panels than reinforced plastics used by themselves.

3. MATERIALS AND PROPERTIES

Most of the corrugated sheeting used in the U.K. is made to meet the requirements of BS 4154: 1967. This standard specifies, among other properties, dimensions, tolerances, strength and light transmission and diffusion of sheets. For most corrugated sheeting, light transmission is the characteristic most important to the user. By proper choice of resins and control in their cure, up to 85 per cent transmission of daylight may be obtained, although usually with some diffusion. Use of methyl methacrylate incorporated in the styrene monomer for cross-linking provides a cured resin whose refractive index matches more closely that of the glass than a resin cross-linked with styrene alone. Resin manufacturers aim, by achieving this match, to give the maximum degree of clarity and light transmission to a cured sheet.

If colour, or a high degree of diffusion are required, they are best achieved with carefully chosen dyes and pigments added to the resin. For many applications, fire retardant treatments are necessary, either integral in the resin structure or in the form of additives, to achieve appropriate gradings in standard fire tests. Light stabilisers must be incorporated, particularly with fire-retardant sheeting, to minimise the degradation and discoloration produced by weathering.

Sheeting and panels are usually reinforced with E-glass. It has better resistance to the action of water and weather than the cheaper alkali glass but, more particularly, its refractive index (approx. 1.55) can be matched by polyester resins to give good light transmission. The glass fibres are usually treated with a keying agent, most commonly of a silane type, and mats are almost invariably bonded with powdered polyester resin, which gives good wetting-out properties as well as sufficient strength for handling during the moulding process. Depending on the corrugation profile, chopped strand mat of between 400 and 600 g/m^2 is used and, with proper impregnation, ensures the requisite stiffness for the 'load deflection' test in BS 4154[1]. For some government or public authority applications, stiffer corrugated sheets may be required and for these and for cladding panels, this is usually achieved by increasing the number of layers of impregnated reinforcement.

3.1. WEATHERING QUALITIES

Most corrugated sheeting is used, in the first place, for its light transmitting qualities and the most important change that normally occurs on weathering is loss of transmission. Three principal effects combine to reduce transmission: the action of solar radiation, particularly in the ultraviolet region, causes yellowing and eventual degradation of the resin so that transmission of visible light is much reduced; water diffusing into the sheet breaks the bond between glass fibres and resin and because of change in refractive index at the interface, more of the incident light is reflected; at a more advanced stage of breakdown, erosion of resin from the surface exposes glass fibres which collect dirt, which may eventually reduce the light transmission to as little as 13 per cent of the original[8]. The transmission characteristics of two sheets of different qualities, but more typical of current products, are shown in Table 7.1, which indicates the effect of composition and manufacture on weathering qualities: Various techniques have been suggested for improving the weather resistance. The resins most badly affected by light and showing the greatest tendency to yellow are those that incorporate fire-retardants, and manufacturers incorporate the minimum of such additives compatible with obtaining the appropriate grading in BS 476 Pt. 3[9], or other standard requirement. A surface rich in resin weathers better than one in which the fibres are close to the surface. In practice, such a surface may be obtained by the use of a gel coat, this being normal practice for cladding panels, but less practicable for corrugated sheeting. In the latter case, a surfacing tissue of glass filaments has a considerable beneficial effect[8]; polyvinyl fluoride surfacing film has also been used to advantage, though sheeting becomes relatively more expensive.

For the best quality sheets, with a minimum of fire-retardant additives, estimates of life expectancy in the U.K. have been offered of 30 years or more; but poor quality sheets, particularly of 'self-extinguishing' grades, may last for no more than 10 years before their light transmission becomes unacceptably low[10]. With the recent realisation that for many building applications in the U.K., sheets do not require the maximum amount of fire-retardant additives,

Table 7.1 *Change of light transmission on weathering*

	Percentage of original transmission after period of U.K. weathering							
Years Exposed	0	1–3	4	5	6	7	8	9
Sheet No:								
I—Original transmission 75.3%	100	Not recorded	84.3	81.0	81.2	80.3	75.0	68.6
II—Original transmission 75.5%	100	Not recorded	74.4	69.4	67.4	63.6	36.6	33.2

I—Good quality sheet made with GP resins and surfacing tissue.
II—Poor quality sheet made with special non fire-retardant resins but no tissue.

intermediate grades of material have been introduced and their weathering qualities also are expected to be intermediate between best and worst.

For opaque sheeting and panels, a well cured gel coat that is well bonded to the glass fibre reinforcement is likely to give good protection against breakdown. A pigmented panel is likely to be more durable than translucent

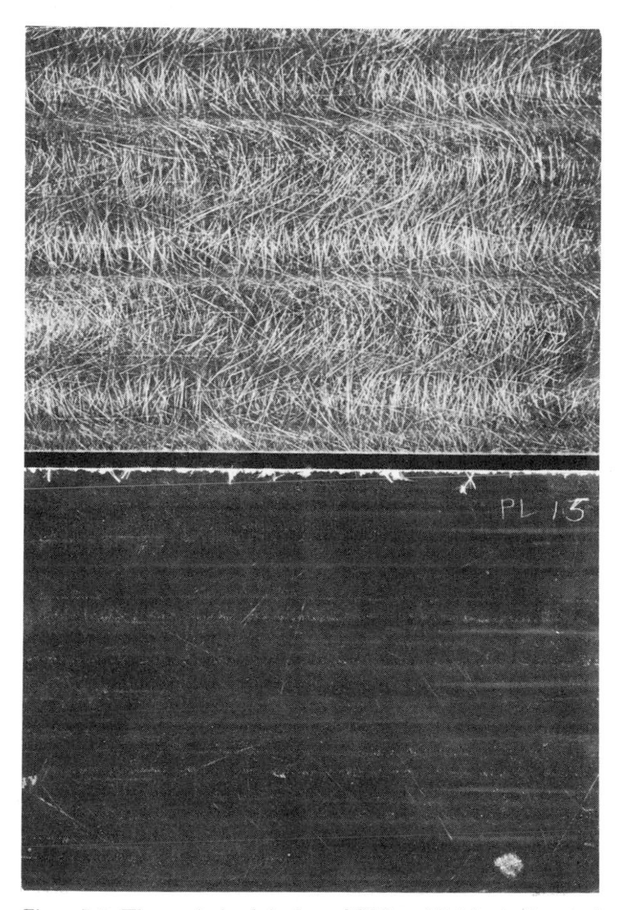

Figure 7.5. The weathering behaviour of GRP roof lighting is illustrated by this comparison between weathered and unweathered samples from the same sheet (Crown copyright)

sheeting, because the pigment excludes ultraviolet light from the resin. The appearance of a deliberately uneven finish is likely to remain acceptable for much longer than that of an even surface, which rapidly shows up defects. It has been estimated[10] that a properly made panel is likely to remain structurally sound for considerably more than 30 years. Suitable raw materials and closely controlled manufacture probably hold the key to long-term satisfactory performance.

4. COMPARISON WITH ALTERNATIVE MATERIALS

Corrugated plastics roof lighting was first introduced as an alternative to glass. For reasons of economy, the latter tended to be used in continuous overhead runs, but required suitable framework and was installed by glaziers. Corrugated translucent plastics sheet could be installed by a roofer, without the need for specialist labour. Transparent acrylic sheeting was the first to be used in this way but is heavy, expensive, requires care in handling and presents a considerable fire risk. It has good weathering qualities, having an estimated life of 40 years or more but has been used mainly in farm buildings where regulations are less restrictive. When corrugated GRP sheets were introduced, they were used mainly in industrial and factory buildings, a market which they still hold. Another alternative material, corrugated transparent or translucent PVC, with or without an incorporated wire mesh, was introduced more recently. This has been used largely in the 'do-it-yourself' market because of its cheapness, but is also used in some industrial buildings. It is available in a range of qualities and some grades are covered by BS 4203: 1967[11]. Its weathering properties also cover a range and though it is cheap and easy to handle, it is not as strong as reinforced polyester and after a period of weathering becomes brittle. A life of about 10 years[10] may be as much as can be expected for typical commercial corrugated translucent PVC sheeting. Opaque PVC will last for longer, but dark sheets may distort from solar heating on weathering.

5. REGULATIONS FOR USE IN BUILDING

Reference has been made to the need to incorporate fire-retardant additives in sheet for roofing purposes. Until the Building Regulations[12] for England and Wales came into force in 1966, local authorities required sheet for use in most sloping roofs to have an EXT. S AA grade when tested according to BS 476 Part 3[9] and to be used only in chequerboard pattern. Regulation E15 of the Building Regulations still imposes limitations on some grades of sheeting, but a material which achieves an A rating in the 'Fire Penetration' test of this Standard, and a C rating or better for 'Spread of Flame' may now be used without restriction in sloping roofs. The A rating for 'Fire Penetration' is usually obtained because of the incorporation of glass fibre mat; 'Spread of Flame' characteristics are controlled by incorporation of suitable additives in the resin.

Fire precautions for the external cladding of buildings are defined by Building Regulations E5 to E7. In general, reinforced plastics cannot confer any significant degree of fire resistance by themselves, in terms of BS 476 Part 1. They must be used in conjunction with more fire resistant materials, combined in panels, if these are to reach even the half hour or one hour fire resistance required by the Regulations for many applications. Failing their attainment of the level specified for a particular part of an external wall, they must be classified as 'unprotected areas' and their use is limited as specified in Schedule 9 of the Regulations.

Furthermore, reinforced plastics cannot be classed as non-combustible or achieve Class O as defined in the Regulations, and only exceptionally can they achieve even Class I in the 'Surface Spread of Flame' test in BS 476 Part 1[13]. Although they may still be used in a very large proportion of buildings, the Regulations do restrict their use because of these limitations in their fire properties. Except by special dispensation, GRP cannot be used as complete cladding for buildings over 50 ft (15.24 m) in height and are limited for large buildings in some of the purpose groups defined in the Regulations.

Despite these limitations, GRP has been used within the terms of the Regulations to clad many buildings in the U.K., some of not inconsiderable size. It is unlikely that it will ever displace other cladding materials, but there is no doubt that it has earned a position of wide acceptance as an attractive and adaptable cladding material.

REFERENCES

1. *British Standard* 4154: 1967, 'Corrugated Plastics Translucent Sheets made from Thermo-setting Polyester Resins (Glass Fibre Reinforced)'.
2. C. Joseph, *2nd British Plastics Fed. Reinf. Plast. Conf.* London, 1960, Section 7.
3. Libby-Owens-Ford Glass Corpn. *Brit. Pat.* 760 530 (1956).
4. Montecatini Co., *Brit., Pat.* 820 762 (1959).
5. Filon Corp., *Brit. Pat.* 826412 (1960).
6. Ashdowns Co., *Brit. Pat.* 780 667 (1957).
7. *Brit. Plast.* **31**, (9) (1958) 372.
8. Crowder, J. R., *4th Int. Reinf. Plast. Conf.,* London, 1964, Paper 27.
9. *British Standard* 476: *Part* 3; 1958, 'External Fire Exposure Roof Tests'.
10. 'Applications and Durability of Plastics', Building Research Station Digest 69 (Second Series)', HMSO., 1966.
11. *British Standard* 4203: 1967, 'Extruded Rigid PVC Corrugated Sheeting'.
12. *The Building Regulations* 1965, HMSO 1965.
13. *British Standard* 476: *Part* 1: 1953, 'Fire Tests on Building Materials and Structures'.

8

Aerospace Applications

M. A. DENNEY

1. INTRODUCTION

Much of the early impetus to the development of glass reinforced plastics was derived from the needs of military aircraft during the Second World War. With the introduction of airborne radar there was an urgent requirement for a structural material which was also transparent to radar. The needs of military aircraft were soon supplemented by those of the civil aircraft operators who introduced radar and radio navigational aids in the post-war commercial airline boom.

From these requirements techniques were developed for the fabrication of fairly large structures, having a high strength/weight ratio and closely controlled electrical properties. These production techniques were closely aligned to the small scale requirements of the aircraft industry for particular components. Many people believe that the adoption of these techniques by general industry may have retarded growth of reinforced plastics where high production rates and numbers were a prerequisite.

In the 1945–55 period, production relied almost entirely on wet lay-up and low consolidation pressures. With the shrinking numbers of any one type of aircraft produced in the U.K. the wet lay-up technique has persisted for many aircraft components up to the present day. Since the reliability and closely guaranteed properties achieved with metallic materials cannot be obtained using wet lay-up techniques the growth rate of reinforced plastics in the aircraft industry has not matched that achieved in other industries. However, there are a number of specialised products developed for aircraft which are worthy of a closer study.

2. RADOMES AND DIELECTRIC PANELS

The most critical requirement for a radome is that the electrical thickness shall be closely controlled and shall match the wavelength of the radar equipment with which it is to be used. In general, there are two main methods which satisfy these requirements:

2.1. SINGLE SKIN RADOMES

For a limited type of application where structural considerations are small it is possible to make a thin laminate, particularly for long wavelength equipment where the effect of the thin plastics diaphragm can be ignored.

Where short wavelength radar equipment is utilised and where structural loads are more significant it is still possible to design an adequate single skin radome. However, in order to meet the more critical electrical requirements a thick laminate, closely matched to the radar equipment, is essential. Two techniques have been used to match the radome to electrical requirements. Either the laminate is made slightly thicker than required and then ground back, or, alternatively, a laminate, made slightly thinner than calculated, is sprayed on the inside with resin to correct it. In both of these operations it has been found essential to monitor the electrical properties of the laminate continuously while adjusting the thickness. For certain applications, where there are no critical electrical requirements but structural properties are of prime importance, single skin epoxy resin/glass laminates have been used. To achieve the necessary stiffness 'top hat' or angle stiffeners have been moulded into the component.

2.2. SANDWICH RADOMES

It is not often that a thick single skin radome is acceptable to a designer, since, although it may give adequate radar performance and strength, a considerable weight penalty is involved. A much more attractive design is a sandwich construction usually two thin skins stabilised with a low density filler. In other large structures it has been found necessary to utilize a double sandwich construction, i.e. two layers of filler separated by one skin and stabilised by two further thin skins.

Most of the early radomes were produced using a preformed foam filler of an expanded nitrile-ebonite, 'Hycar' to which were bonded polyester/glass-cloth skins. Subsequent developments have incorporated resin impregnated glasscloth honeycomb as core with polyester or epoxy resin/glasscloth skins.

Alternative constructions which have been developed include (*a*) *in situ* foamed polyurethane cores with thin polyester/glasscloth skins which were developed by the Lockheed Aircraft Corporation. (*b*) Fluted Core radomes which were developed by Bristol Aircraft Ltd. and used on certain Hawker Hunters and subsequently on Bristol Britannia aircraft. (*c*) A number of companies are seriously investigating the use of 'Nomex', polyamide fibre, honeycomb for a core material mainly in conjunction with epoxy resin/glass-cloth skins.

A diagrammatic outline of typical radome and dielectric panel constructions is given in *Figure 8.1*.

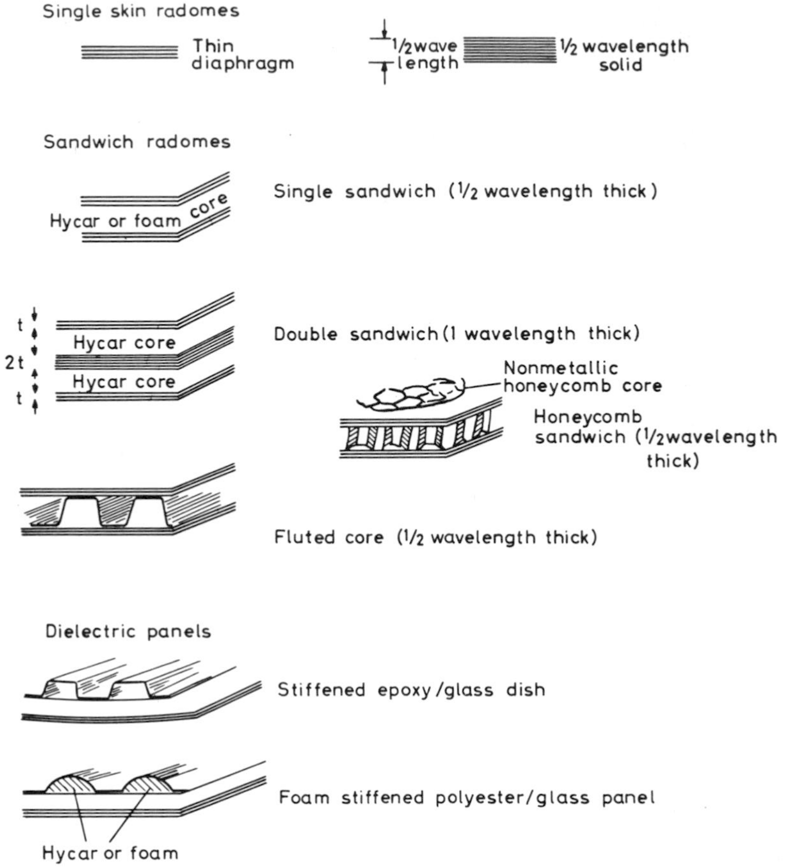

Figure 8.1. Construction of typical radomes and dielectric panels

2.3. PRODUCTION TECHNIQUES FOR RADOMES

Single Wall Radomes—Single wall radomes have been built in three ways using: a male mould, a female mould or with matching male and female tools.

The use of a male mould has been developed by two techniques. Where a cold curing polyester resin system has been employed a plaster mould, often built-up in segments from a wooden master, is coated with paraffin wax or another suitable parting agent. The pre-cut cloths are laid up dry, carefully staggering all joints, and the catalysed resin poured slowly over the top of the radome and carefully worked into the cloth by hand. A thin gel coat of resin is then applied over the outer surface. An alternative technique using either cold curing or hot curing resin formulations relies on impregnating each layer of glasscloth either before or immediately after lay-up. For hot curing it is, of course, necessary to use a temperature resistant tool, usually an aluminium fabrication.

Where the external profile of the radome is of primary importance it is usual to employ a cast aluminium female tool. The glasscloth is layed up and impregnated layer by layer and either the excess resin screened out by hand or a vacuum blanket is used to consolidate the laminate.

The third technique, using matched tools was developed for missile radomes and is now used for the manufacture of the Concorde nose radome. Briefly, closely matching tools are made from aluminium or plastic and coated with a suitable release agent. A pre-determined number of layers of glasscloth is assembled on to the male tool and the female tool is then almost closed on to it. Vacuum is applied to the apex of the inverted mould and resin is injected around the lower periphery of the tool. Assisted by the vacuum and by gravity, this resin is slowly drawn up through the cloth. When impregnation is complete, the tool is finally closed to ensure the correct consolidation, and curing is carried out.

In early radomes the glasscloth was tailored to the correct shape and then stitched on the mould. The next development was the use of a loosely knitted glass shape which stretched to fit the mould contours. More favoured on present radomes are contoured shapes which are woven to fit the tool accurately. The weaving process is controlled with a fully automatic punch card system. Depending on the complexity of the shape, it may be necessary to weave a number of 'socks' each with slightly differing dimensions to compensate for the change in thickness as lay-up proceeds. It is also possible to introduce changes in the thickness of any shape to meet a particular design or manufacturing requirements.

The Concorde radome shown with its tool in *Figures 8.2* and *8.3*, is an excellent example of the standard which can be achieved using the matched mould resin impregnation technique. A number of structural components with non-critical electrical requirements have been made from pre-impregnated glasscloth. Typical of these is the V.O.R. Aerial cover for the Vickers VC 10 which is manufactured from epoxy resin pre-impregnated glasscloth moulded in a steam heated autoclave at 150°C and 475 kN/m^2 using a female tool and rubber vacuum bag technique. This is a highly-loaded dish shaped component stiffened with mould-in 'top hat' section longitudinal members.

Sandwich construction radomes—A number of slightly different techniques have been utilized for Hycar sandwich radomes. The Hycar is cut to shape and then moulded under heat and pressure to the correct contour. Either before or after the moulding operation, the surface of the Hycar is treated to promote adhesion to the resin/glasscloth laminate. The Hycar is scrubbed to remove all chalk, treated with benzoyl peroxide in trichlorethylene and then abraided and coated with resin or hycar primer. Alternatively, the surface of the Hycar is removed using a wood planer.

As for single skin radomes, Hycar sandwich structures have been made on male and female moulds. The Avro Shackleton radome was manufactured in a male mould using a 10 mm thick Hycar core with cold cured polyester resin/ glasscloth skins. Vickers Vanguard, VC 10 and B.A.C. 1-11 radomes are of similar construction but made in a female mould.

Figure 8.2. Concorde radome. Contour woven glasscloth shapes being drawn on male tool (Courtesy of B.A.C. Ltd (G.W. Division) Stevenage)

Figure 8.3. Foreground, a finished Concorde radome. Background, radome tool showing resin-injection unit
(Courtesy of B.A.C. Ltd (G.W. Division) Stevenage)

Two excellent examples of double sandwich radomes are the Vickers Valiant and the even larger Avro Vulcan radomes. Both of these were made in female moulds using a hot curing polyester resin system.

Although no *in situ* foam filled sandwich radomes have been made in this country, the manufacturing techniques developed for radomes by the Lockheed Corporation, have been used on other structures. Briefly, polyester or epoxy resin glasscloth laminates for the inner and outer skins are cured in the normal way and assembled together on their respective moulds to give an accurately controlled gap between the skins. The mixed alkyd resin/isocyanate blend is then injected into the interspace and allowed to cure. The density of the final foam can be controlled by adjusting the resin blend or the temperature of the moulds.

Another interesting sandwich construction is the fluted core developed by the Bristol Aeroplane Company but never really exploited. The technique is similar to the resin injection system described for single skin radomes in that layers of glasscloth are placed with wax mandrels into a female tool. Further cloths are placed on to the male mould which is finally closed into position and catalysed resin pumped into the cloth until it is fully impregnated. The lay-up is allowed to cure and the component is then removed from the mould for post-curing in a suitable oven. During the post-cure the wax mandrels melt and drain away to form the fluted core.

3. DUCTING

Although radomes are best known, probably the widest use of GRP in aircraft is for ducting. Apart from protective ducting for electrical cable runs, most ducting is employed in air conditioning systems. There are several considerations which lead to their general acceptance for this application.

(i) Excellent resistance to vibration compared with the fatigue prone thin metal ducts which they replaced.

(ii) Ease of fabrication into the many unusual shapes which are typical in this type of aircraft system.

(iii) Weight advantage over any metal ducting which might be thick enough to give adequate service life.

The most commonly used manufacturing technique for ducting is tape winding either on a lathe, for straight runs, or by hand for more complex structures. Although wooden split moulds may be employed on simple shapes it is more usual to prepare break-up plaster formers. More complex ducting may be made in sections and finally assembled by joining the cured sections with overlapping tapes.

During several years of successful operation of polyester resin/glasscloth air-conditioning ducting one major problem has arisen. In order to make maximum weight savings, the thickness of much of the low pressure air distribution ducting is defined by the loadings which might occur during assembly into the aircraft. Once fitted, these ducts appear substantial and unfortunately they are often used as 'stepping stones' during servicing. Inevitably the resin cracks and leakages occur. In order to overcome this problem, a new concept of flexible ducting was developed at B.A.C. Weybridge. Whilst retaining the production 'know-how' and advantages of plaster moulds and tape winding a flexible polyurethane/isocyanate laminating resin system with excellent fire retardant properties was developed. The only real modification to the polyester laminating technique was enforced by the small amount of solvent contained in the polyurethane system. This necessitated a dwell between successive layers of tape/resin to allow the solvents to flash off.

Special attention was also required to ensure that a basic two layers of 0.08 mm tape would give a non-porous duct. Co-operation with the glasscloth supplier was needed to control the tension in the selvedge of the tapes. The proof of this successful co-operation can be seen in *Figures 8.4* and *8.5*—an extremely flexible and non-porous, non-flammable lightweight ducting.

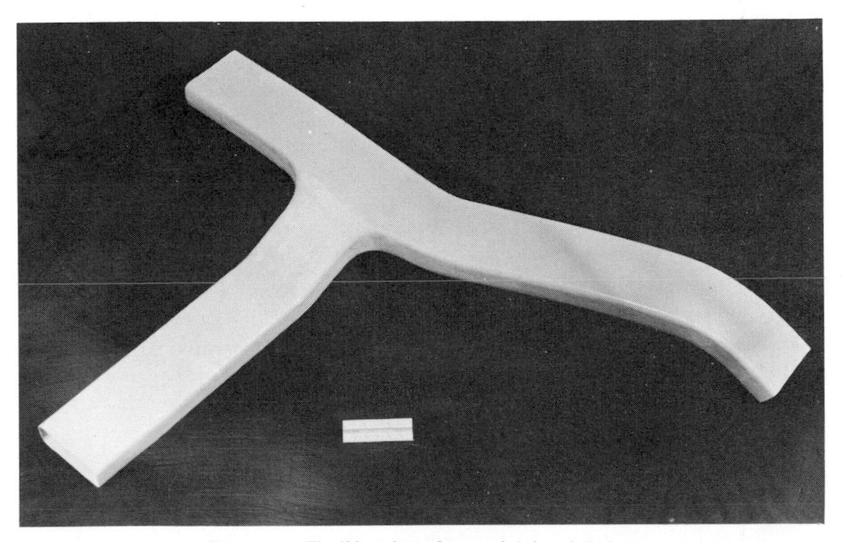

Figure 8.4. Flexible polyurethane resin/glasscloth duct

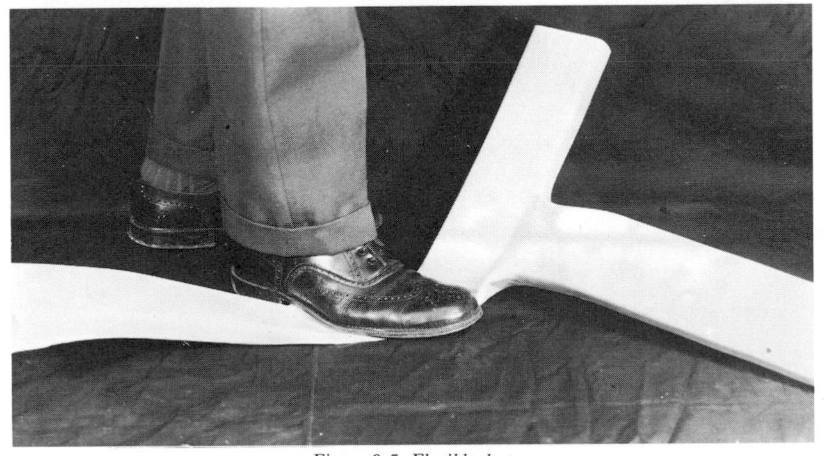

Figure 8.5. Flexible duct

4. SECONDARY STRUCTURES

A large variety of secondary components are made by wet lay-up of polyester resin and glasscloth using the techniques developed for radome or duct manufacture. In general, these are made using a cold curing resin system, either without pressure or at best with a rubber vacuum bag technique, and usually with a subsequent post-cure.

Conventional stiffeners may be moulded in to provide the requisite stiffness or, alternatively, a sandwich construction with Hycar, polyurethane or P.V.C. foam may be used.

4.1. FAIRINGS ETC.

Typical applications include the Beagle Miles M 218 and M 117 light aircraft where 60 per cent of the surface area utilised glasscloth and glass mat with polyester resin. The nose sections and general fairings of several gliders manufactured by Slingsby Sailplanes Ltd are made of GRP. In addition the wing and empennage tips of several aircraft and gliders are made by wet lay-up of polyester resin/glasscloth.

4.2. ELECTRICAL DISTRIBUTION PANELS

In these types of application fire resistance is an essential requirement and although chlorinated resins are offered, the more usual approach of the aircraft industry is to add a paraffin wax and antimony trioxide to the resin. Although this gives increased viscosity and a less satisfactory appearance operators soon develop adequate expertise and with a small addition of pigment, satisfactory standards are achieved.

However, due regard must be taken by the designer of the maximum operating temperature since the fire retardant properties of these resin mixes may deteriorate above 70°C.

4.3. PROPELLER SPINNERS

Following joint developments between de Havilland Propellers Ltd. and Ciba (A.R.L.) Ltd. a number of aircraft have been successfully fitted with epoxy resin/glasscloth spinners. These are usually prepared by wet lay-up followed by an elevated temperature cure under vacuum pressure.

4.4. TANK PLATINGS

Thin aluminium alloy skins, which support and separate rubber bag fuel tanks, have been found to fail in fatigue due to resonance in certain conditions of flight. By applying a wet lay-up of a thin polyester resin/glasscloth skin to the tank platings the fatigue problems have been almost entirely eliminated.

5. FURNISHINGS

Although not receiving the publicity of radomes and ducting, a great deal of polyester resin/glasscloth is utilized within the furnishings of the aircraft passenger cabin. As with electrical panelling, a primary requirement is low flammability and the addition of antimony trioxide and chlorinated paraffin wax is usual. Hat racks often have moulded-in foam stiffeners and bulkheads utilize thin polyester resin/glasscloth skins bonded with a cold curing epoxy resin to paper honeycomb or expanded PVC core to form a stiff lightweight sandwich construction.

5.1 TRIM PANELS

These often incorporate cabin distribution ducting and fall into three categories:

GRP panels—covered with vinyl fabrics were used widely in such aircraft as the Viscount but uneven adhesion of the fabrics led to adverse comments. The backs of these panels usually form one wall of air distribution ducts.

Rigid PVC decorative panels—offer the maximum variability in colours and patterns. In many current aircraft these are backed with polyester resin/glasscloth ducts and replaced by polyester resin/glasscloth panels in areas not visible to passengers, for example, freight bays.

A very interesting variation in trim panels is offered by textured finish laminates developed by C. F. Taylor (Plastics) Ltd. in conjunction with Hawker Siddeley Aviation Ltd. Although only allowing limited variation in decor schemes, these panels are extremely hard wearing and most suitable for trooping and similar applications.

For Concorde—a very lightweight rigid PVC type of trim panel is used and polyester resin/glasscloth has been laminated on the backs of these PVC panels to improve stiffness.

5.2. FREIGHT BAY LINING

Special problems are presented in the lining of aircraft freight bays. Heavy sharp cornered packages give unusual impact and tear problems, while fire containment requirements give flammability and heat distortion problems. As always in aircraft, weight saving is another consideration. One answer to these problems is a fire resistant epoxy resin formulation used in conjunction with non-woven glass developed in the U.S.A. This sheeting is similar to high strength epoxy laminates developed for the electrical industry but differing in that it incorporates a fire resistant resin and is available in very thin sheets.

6. MOULDINGS

When dough moulding compounds were first developed, the aircraft industry examined them but did not accept their use. This was mainly due to the expense of tooling for small numbers of components. With the introduction of 'pre-preg' mat materials suitable for low pressure mouldings, a much more attractive position arose. It was found possible to produce cheap aluminium tools which were more than adequate for the small numbers of any one component required in the aircraft industry. As with furnishings, it was necessary to achieve an adequate standard of fire resistance. Once a suitable specification had been agreed with the suppliers, 'pre-preg' mat material found increasing

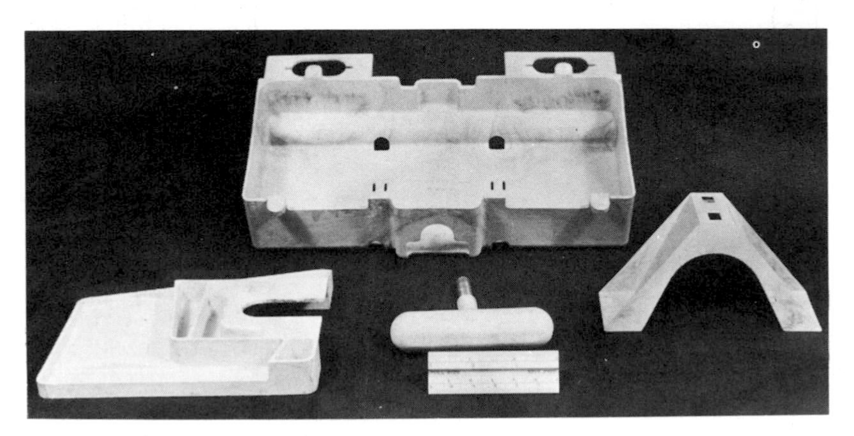

Figure 8.6. Typical pre-preg. mat mouldings

usage in secondary structural brackets, cover plates and numerous small mouldings. As designers have become more aware of this material, the utilisation has increased and some typical complex mouldings are shown in *Figure* 8.6.

Although thermoplastic mouldings have been used to some extent on aircraft, the designer has been very wary of the low strength and creep characteristics of these materials, particularly at elevated temperatures.

With the recent developments in glass reinforced thermoplastics these materials, particularly glass/nylon, are finding new applications, replacing phenolic mouldings, polyester resin/glass and in some cases aluminium alloys.

7. TOOLING

Polyester resins are not widely used for tooling purposes due to their shrinkage during cure. However, the use of glass reinforced epoxy resin tools is well developed for many aspects of aircraft construction. Although not as widely used as in the U.S.A., the increase in epoxy formulating companies and epoxy tooling specialists is a clear indication that this is a healthy part of the industry. The replacement of traditional materials by GRP for tooling has been readily accepted, particularly in areas where size and complexity of shape gives an obvious advantage to the plastics material. Another advantage of particular importance to the aircraft industry is that minor modifications are easily incorporated.

Similar tools have largely replaced metal dies for vacuum forming of thermoplastics sheeting. As might be expected, many GRP mouldings are now manufactured from GRP tooling. Large radome tools are typical examples which offer many attractive advantages including the ability to build in heating elements. On these moulds it is usual to introduce fillers such as aluminium wool or powder to provide better strength retention at elevated temperature and dimensional stability. This type of tooling is also spreading into the adhesive bonding field. In this field adhesives curing at 175°C have previously precluded the use of reinforced plastic tooling. However, recent advances in epoxy resins and the introduction of a family of structural adhesives curing at 120° have opened a new field to plastics tooling.

Another widely used tooling application is for assembly and drill jigs where light, stiff fixtures usually backed with a resin/glasscloth tubular framework are now generally accepted and immediately visible in any aircraft production line.

Part 3

Materials

9

Resin Systems

A. A. K. WHITEHOUSE

This chapter is concerned mainly with unsaturated polyester resins (UP) as they form the matrix of most GRP. Epoxide resins are used to a more limited extent and these are treated briefly.

1. HISTORY

In the context of reinforced plastics technology the term unsaturated polyester resin is used to describe an unsaturated polyester cross-linked by means of a vinyl monomer; or since the final cross-linked structure is produced by a free-radical copolymerisation, it is perhaps more logical to say that it is a vinyl polymer cross-linked by means of an unsaturated polyester. This latter view is supported by the fact that the copolymer chains are typically an order of magnitude longer than the polyester chains.

As is so often the case with technological development, UP resins were first made by purely empirical work in the 1920s, with publication in the 1930s by Carleton Ellis[1] and by Dykstra[2]. However, this first work was soon put on a rational basis and further stimulated by Staudinger's concept of macromolecules[3], Kienle's theory of functionality[4] and Carothers theory of polycondensation[5].

Carothers in 1929 prepared linear polyesters[5] from ethylene glycol and phthalic and maleic anhydrides and fumaric acid essentially as we do today. Bradley in 1937 showed that unsaturated polyesters could be cured to infusible and insoluble products, whilst Ellis in 1936 showed that the rate of cure was increased at least thirtyfold in the presence of styrene, and he further advocated the use of benzoyl peroxide to promote cure. The commercial development of UP resins was now possible and in fact began in the U.S.A. in 1941.

However, it required another 10 years before the reinforced plastics industry really became established. During this decade several further key developments occurred: Muskat[8] in 1943 showed the advantages of using mixtures of phthalic and maleic anhydride in obtaining reduced crystallinity in the base polyester resin and better compatibility with styrene. Hurdis[9] in 1944 applied for a patent on the use of tertiary aromatic amines as promoters, accelerating the catalytic action of benzoyl peroxide so that room temperature cure became possible. This was published in 1949 but the system tends to give yellowish products and short pot lives. The use of cobalt naphthenate as a promoter with other peroxide catalysts gives more satisfactory results and its use with cyclohexanone peroxide and with methylethylketone peroxide were the subject of patent applications by Fraser[10] and by Simons[11] in 1947, published respectively in 1950 and 1951.

All these systems of cure give a tacky surface because of oxygen inhibition. Parkyn and Bader [12] showed in 1951 that this could be avoided by including a minute amount of paraffin wax in the resin. These various developments in resin technology together with the introduction in 1951 of relatively low cost glass fibre reinforcement in the form of chopped strand glass mat finally established the RP industry on a commercial basis.

World production of reinforced polyesters probably reached 50 000 tons a year by 1955 and 500 000 tons a year by 1968 and the rate of growth is not yet falling off.

Whereas unsaturated polyester resins were first developed in America, epoxide resins originated in Europe. W. H. Moss[13] reacted glycerin dichlorhydrin with diphenylolpropane and caustic soda and made the first epoxide resin in 1937. This was followed by the work of Castan[14, 15] in Switzerland first published in 1940. They consist of linear polycondensates of epichlorhydrin and diphenylolpropane, which may then be cross-linked in a variety of ways. The use of amines for cross-linking epoxide resins was proposed by Castan [16] and independently by Greenlee[17] in America, so that by 1950 commercial production of epoxide resins was possible. With the availability of epichlorhydrin in bulk, commercial production followed quickly and had become established on both sides of the Atlantic by 1955.

The history of the unsaturated polyester and epoxide resins has been outlined because it is the development of these two families of resins which launched the GRP industry because of their ability to cure without heat or pressure and hence to be used in contact moulding processes. Other resin systems, which have since been reinforced with glass fibres, such as phenolics, silicones, nylon and polypropylene, were already in existence for different purposes and hence it is not thought appropriate to outline their history here.

2. COMPOSITION AND MANUFACTURE

General purpose polyester resins when cured are copolymers of styrene and unsaturated polyesters. Before cure they consist of solutions of linear unsaturated polyesters in styrene. These base polyesters are made by condensing phthalic and maleic anhydrides with propylene glycol and some diethylene glycol. The polyesterification is a typical polycondensation reaction between roughly stoichiometric proportions of the anhydrides and the glycols with usually a slight excess of the glycol components as these tend to be lost more easily during the reaction.

Since the glycols and anhydrides are both difunctional, each able to form two ester links, the esterification reactions continue step by step building up a linear chain molecule and splitting off a molecule of water as each link is made.

$$\mathrm{HO{\cdot}OC{\cdot}R{\cdot}CO{\cdot}OH + HO{\cdot}R'{\cdot}OH \rightarrow}$$
$$\mathrm{HO{\cdot}OC{\cdot}R{\cdot}CO{\cdot}O{\cdot}R'{\cdot}OH + H_2O}$$
$$\downarrow \mathrm{+ HO{\cdot}OC{\cdot}R{\cdot}CO{\cdot}OH}$$
$$\mathrm{HO{\cdot}OC{\cdot}R{\cdot}CO{\cdot}O{\cdot}R'{\cdot}O{\cdot}OC{\cdot}R{\cdot}CO{\cdot}OH + H_2O}$$
$$\downarrow \mathrm{+ HO{\cdot}R'{\cdot}OH}$$
$$\mathrm{HO{\cdot}R'{\cdot}O{\cdot}OC{\cdot}R{\cdot}CO{\cdot}O{\cdot}R'{\cdot}O{\cdot}OC{\cdot}R{\cdot}CO{\cdot}OH + H_2O \rightarrow etc.}$$

Theoretically chains of indefinite length can be built up but in practice the reaction is only continued until some 10 to 30 links have been made giving polyesters with a molecular weight of around 1000–2000. The polycondensation is usually carried out without catalyst at about 200°C under an inert atmosphere, the reaction being continued for 10 or more hours until an acid value of 30–40 mg KOH/g is achieved. Alternatively, a small percentage of an azeotrope such as xylene may be added, facilitating the removal of the formed water and allowing rather lower reaction temperatures to be used. The azeotrope is distilled off under vacuum when the required acid value has been reached.

The speed of the reaction falls as it proceeds and if carried on for too long a period, side reactions become more important and there is a risk of gelation. In some cases it is advantageous to carry out the polycondensation in two stages, condensing first the saturated acid and then the unsaturated acid.

The unsaturated polyester prepared as above is stabilised by small additions of phenols and is then dissolved in styrene to give the viscous liquid which the fabricator knows as polyester resin. *Figure 9.1.* shows the plant arrangement used for this manufacture.

The main alternatives to the general purpose raw materials for UP resins are the use of other glycols such as ethylene glycol or neopentyl glycol, or the more complex hydroxypropyl diether of diphenylolpropane (often called bisphenol 'A'): the most usual alternative acids are fumaric, isophthalic, tetrachlorophthalic and hexachloro-endomethylene-tetrahydrophthalic (Het acid); some vinyl-monomers other than styrene may also be used particularly α-methylstyrene and methyl methacrylate. These are used to provide particular properties, for example, fumaric and isophthalic acids for heat resistance, bisphenol 'A' for chemical resistance, chlorinated compounds for fire-resistance. For information on relationships between structure and properties Boenig's[18] 1964 book is particularly useful.

These raw materials for UP resins are derived ultimately from petroleum or to a much lesser extent from coal. For reference their molecular structures are given in *Figure 9.2*. The symbolism is that there are carbon atoms with attached hydrogen atoms at both ends of the long straight lines, whilst the shorter lines show links to oxygen or chlorine atoms.

During the polyesterification there are several side reactions which may affect the quality of the resin.

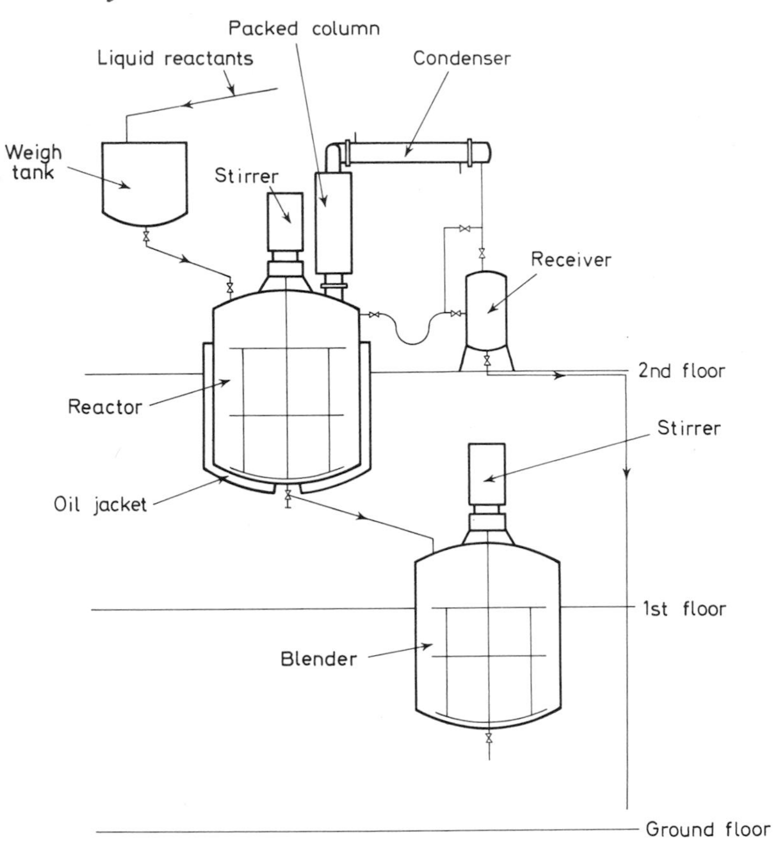

Figure 9.1. Diagram of plant

The unsaturated acids may decarboxylate losing carbon dioxide and forming acrylic esters.

$$RO{\cdot}OC{\cdot}CH{:}CH{\cdot}COOH \rightarrow RO{\cdot}OC{\cdot}CH{:}CH_2 + CO_2$$

This reaction is promoted by high temperatures, and catalysed by impurities. It blocks the chain growth and may lead to premature gelation. Gelation can also be caused by free radical chain reactions initiated by oxygen, or traces of peroxides, removing hydrogen from α-methylene groups and so causing cross-linking.

The glycols may decompose splitting off water to form aldehydes or ethers.

$$CH_2OH{\cdot}CH_2OH \rightarrow CH_3{\cdot}CHO + H_2O$$

$$2CH_2OH{\cdot}CH_2OH \rightarrow HO{\cdot}CH_2CH_2O{\cdot}CH_2CH_2{\cdot}OH + H_2O$$

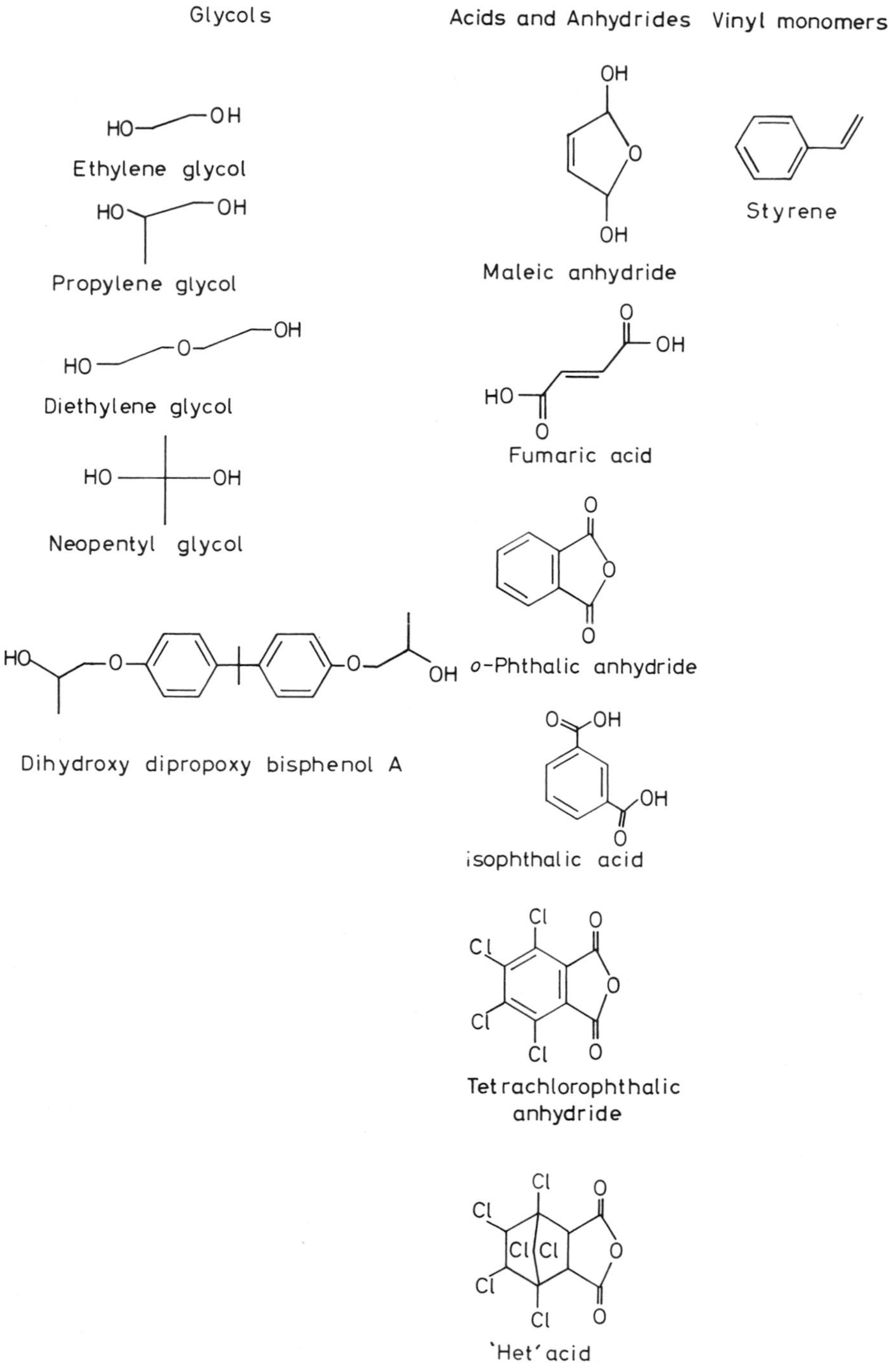

Glycols

Acids and Anhydrides

Vinyl monomers

HO—OH
Ethylene glycol

HO—OH
Propylene glycol

HO—O—OH
Diethylene glycol

HO—OH
Neopentyl glycol

HO—O—O—OH
Dihydroxy dipropoxy bisphenol A

OH O OH
Maleic anhydride

O OH HO O
Fumaric acid

O O O
o-Phthalic anhydride

O OH OH O
isophthalic acid

Cl O Cl O Cl Cl O
Tetrachlorophthalic anhydride

Cl O Cl Cl Cl O Cl Cl O
'Het' acid

Styrene

Figure 9.2. Molecular structures

Branched chains and ring structures can be formed by addition reactions between the unsaturated acids and the glycol, for example,

$$\text{—CH=CHCOOH + HOROH} \rightarrow \text{—}\underset{\displaystyle\text{OROH}}{\underset{|}{\text{CH}}}\text{—CH}_2\text{COOH}$$

with possible further reaction to form lactones.

An important secondary reaction when maleic anhydride is used as the unsaturated acid is a *cis-trans* isomerisation. This reaction has been particularly studied in Hungary by Vancso-Szmercsanyi[19] who showed that the amount of isomerisation was not only dependent on the esterification time but also on the glycols used. (*Figure 9.3.*) Thus she showed that with propylene

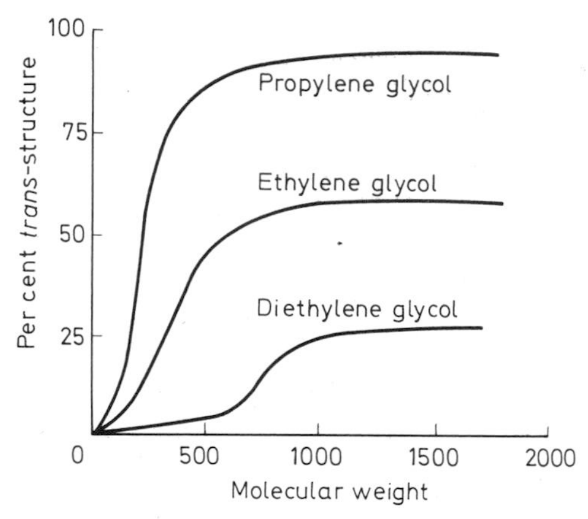

Figure 9.3. Showing percentage isomerisation in glycol maleate polyesters

glycol a mainly *trans*-structure soon developed whereas with diethylene glycol a mainly *cis*-structure remained even after long periods of condensation.

In practice the amount of *cis-trans* isomerisation tends to be greater than shown in *Figure 9.3.* because it is promoted by aromatic dicarboxylic acids such as phthalic acid.

The techniques of polarography, infra-red spectroscopy and nuclear magnetic resonance have all been used to study this isomerisation. Curtis[20] has shown by n.m.r. that with a general purpose propylene glycol, maleic, phthalic system over 95 per cent isomerisation occurs, and that increasing condensation temperatures promote this change. The reaction is important because during copolymerisation fumarate unsaturation has a much greater reactivity and so participates more in the cross-linking than does maleate unsaturation. Lewis and Mayo[21] measured the reactivity ratios of various maleate and fumarate esters with styrene. The diethyl esters at 60°C gave the following ratios of reactivity with their own species to copolymerisation:

	Ester radical	*Styrene radical*
Maleate	< 0.01	6.5
Fumarate	0.07	0.3

These differences will markedly affect the number and length of the styrene bridges between the polyester chains. These effects have been discussed by Funke[22] and by Park[23]. Typically, there are about two styrene residues between linked polyester chains and the structure is as shown in *Figure 9.4*.

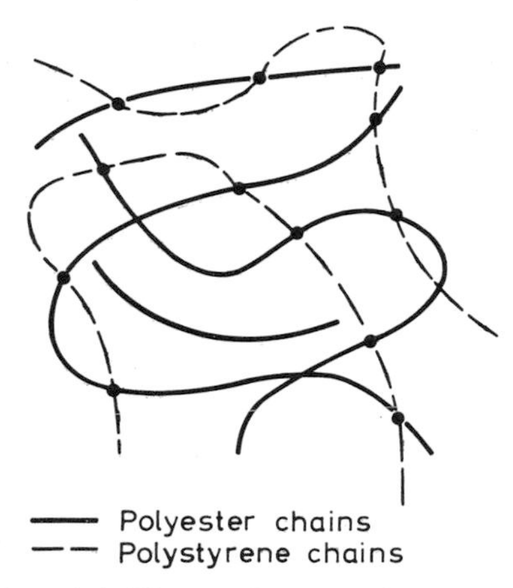

Figure 9.4. Diagram of structure of cured resin

Epoxide resins are made with similar equipment to that used for polyesters, by condensing diphenylolpropane with a large molar excess of epichlorhydrin in the presence of ethanol as a diluent and sodium hydroxide as a catalyst. The reaction is continued for some hours at about 60°C, after which excess epichlorhydrin and the alcohol are distilled off and the sodium hydroxide is removed by washing with water. This process gives a low molecular weight linear polycondensate.

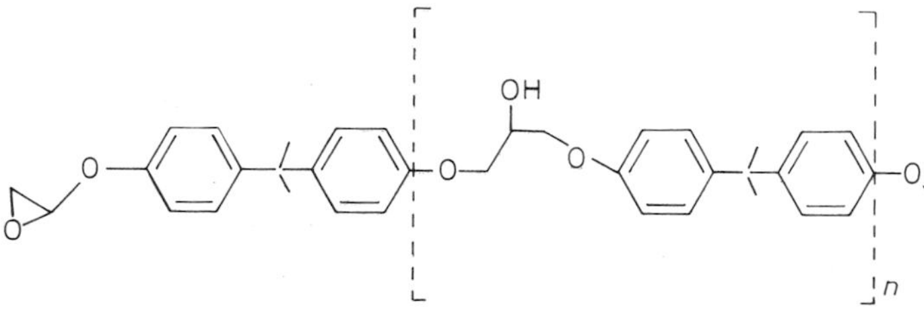

Whereas polyester resins are self-curing, requiring only a free-radical initiator to assist the reaction, the epoxide resins require an added hardener which will react with the epoxy or hydroxyl groups forming cross-links which are an essential part of the overall structure of the hardened resin. The most usual hardeners are aliphatic polyamines and acid anhydrides, and these must be used in substantial quantities. Examples are the use of 10–15 per cent of diethylenetriamine ($NH_2CH_2CH_2NHCH_2CH_2NH_2$), triethylenetetramine or tetraethylenepentamine or 40–50 per cent of phthalic anhydride.

3. CATALYSTS AND CURING

Polyester resins are cured by a free-radical copolymerisation reaction, and if this is initiated by a catalyst system which will give free radicals at room temperature, then the curing or setting process can take place at room temperature. This is the factor more than any other which gives polyester resins their unique suitability for fabricating very large structures as monolithic units.

The catalyst systems consist of organic peroxides, activated by substances known as accelerators or promoters. The organic peroxides are usually known as catalysts, even though they do not comply with the traditional usage of being a substance which speeds up a chemical reaction whilst itself remaining unchanged. In fact they must necessarily decompose in order to initiate the reaction. Their special virtue is that when the peroxide link ruptures, two free radicals are formed. This link like many chemical bonds is made with a shared pair of electrons, but whereas many bonds when broken leave the electron pair on one fragment so that the two fragments are oppositely charged electrically, the two fragments from these peroxide links each retain one electron. These free radicals with an odd electron are uncharged but impatient to revert to a paired state by capturing another electron from somewhere. This they do by abstracting one from the vinyl monomer so forming a new free radical and initiating the rapid chain reaction whereby the cross-linked structure is built up.

$$RO{:}OR \rightarrow 2\,RO\cdot$$

$$RO\cdot + R'CH\vdots CH_2 \rightarrow ROCHR'\underset{\bullet}{C}H_2$$

$$ROCHR'\underset{\bullet}{C}H_2 + R''CH\vdots CHR''' \rightarrow ROCHR'CH_2CHR''\underset{\bullet}{C}HR'''$$

etc.

Now the initial rupture of the peroxide link may take place under the action of heat, and benzoyl peroxide alone is an effective catalyst for hot curing polyester resins above 70°C. Its breakdown may be accelerated by tertiary aromatic amines such as dimethylaniline or diethylaniline. The reaction will then take place at room temperature and this was the earliest type of room temperature catalyst used for polyesters during the late 1940s. However, although very active, this system has the disadvantage of causing some yellowing. It should be mentioned here that the direct reaction of peroxides and accelerators can be explosive. They must therefore never be mixed together but one must be mixed into the resin before the other is added.

The yellowing can be avoided by using aliphatic peroxides such as cyclohexanone peroxide and methylethylketone peroxide. These alone only break down to free-radicals at around 80–100°C, but they may be activated at room temperature by various transition metals. The most widely used accelerator is cobalt usually as the octoate or naphthenate. The amount required to give an acceptable gelation and setting rate is around 0.01 per cent of cobalt depending on temperature as against around 1 per cent of the peroxide.

The peroxides are always supplied diluted with an inert substance because of their hazardous nature, for example, benzoyl peroxide may be mixed with an inert powder, cyclohexanone peroxide may be dispersed in a plasticiser to give a thick paste, whilst methylethylketone peroxide may be dissolved in a plasticiser to give a colourless liquid. The accelerators are used as dilute solutions in styrene because the amounts required are so small that they would otherwise be difficult to mix in uniformly. In many cases resins are supplied pre-accelerated so that only catalyst has to be added before use.

The curing reaction takes place in several stages. Firstly there is an induction period when nothing seems to be happening. In fact the free radicals being formed are immediately reacting with the inhibitor which was added to the resin to give it adequate shelf life. This induction period is, of course, very

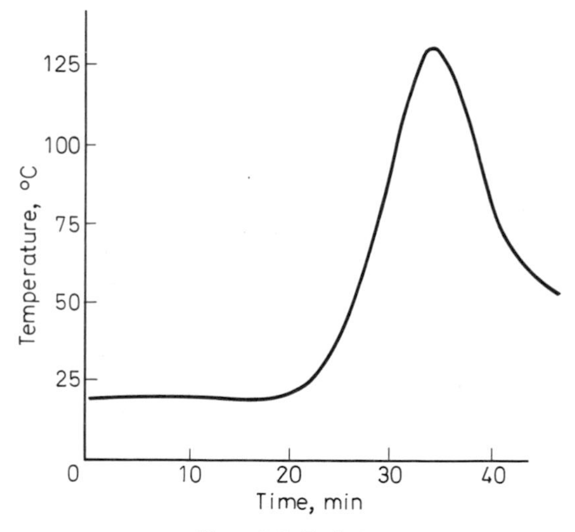

Figure 9.5. Typical exotherm

useful in enabling the glass reinforcement to be impregnated. Then rather quickly gelation occurs, as the cross-linked network structure of *Figure 9.4.* begins to be formed. Substantial amounts of heat are evolved and this exotherm builds up the temperature as shown in *Figure 9.5.*

The actual temperature rise depends on the thickness of the resin layer and the nature of the mould surface. In thin sections the rate of heat loss will prevent much temperature rise.

Air inhibits cure resulting in a tacky surface because oxygen reacts preferentially with the free radical so blocking the chain reaction by forming less reactive radicals. This can be prevented by a plastic film or by dissolving not more than 0.1 per cent of paraffin wax in the resin[12]. As the resin cures, this wax is thrown out of solution forming a protective film on the surface. The tendency for a tacky surface is less pronounced when cobalt accelerators are used, perhaps because they act also as driers. If a tacky surface is formed, it can always be removed by wiping with a solvent.

After the exotherm, final hardening of the resin continues for a long time of days or weeks. This hardening can be speeded up and made more complete by post-curing at say 80°C, for a few hours. This, however, is clearly not possible when making large objects which offer unique possibilities for reinforced polyesters. When one monitors different properties during the cure of polyester resins, one finds that some continue to change after others have ceased to change.

The problem of how to establish that cure is complete has received much study but no simple answer. A full picture can only be obtained by a comprehensive examination of a wide variety of properties such as solvent extraction, residual styrene, indentation hardness, static and dynamic mechanical tests, sonic tests, electrical tests, infra-red absorption, chemical resistance. A useful review and study of possible tests has been carried out by Pritchard[24]. He finds the dielectric loss peak and the torsional pendulum as particularly worthy of further study.

In practice the Barcol hardness is the most useful control test. It is not very precise, but is extremely simple and quick to carry out and is non-destructive. A dozen or more readings can be taken in a minute or two and an average of these provides a useful check for a particular resin system that effective cure has occurred.

4. ANALYSIS

Many of the newer instrumental techniques of analysis may be used both for the qualitative detection of polyester resins and for the quantitative determination of each type of resin. Infra-red spectroscopy is of particular value in that it can be applied to the cured resin. For further information the resin must be pyrolysed or broken down by hydrolysis, methanolysis or aminolysis to its original components or to other small molecular fragments. Infra-red spectroscopy, gas chromatography, paper chromatography and polarography may then be applied to identify and estimate the various components. However, it is often necessary to modify these components further to make them more amenable to the proposed technique. Thus the polyhydric alcohols are acetylated, and the carboxylic acids are methylated to obtain derivatives with adequate volatility for gas chromatography. The dibasic acids may be separated in a phenolic solution by paper chromatography or they may be determined by polarography[25].

5. ADDITIVES

The polyester resins which we have described are activated and mixed with reinforcing glass fibres. Full details of this and all aspects of their use are given in textbooks such as those by Selden[26] and by Parkyn[27]. One further point must, however, be mentioned here, and this is that besides the essential ingredients already mentioned, various ancillary materials are often added. These must all be selected not only to achieve their particular purpose, but also for freedom from interference with the catalyst-accelerator system to be used, and as far as possible, the storage, flow and weathering properties.

Up to 25 per cent of finely powdered mineral fillers such as calcium carbonate, asbestos or kaolin may be added to reduce shrinkage and costs and to improve compressive strength. Mixtures of paraffin wax and chlorinated paraffin are added to give fire retardant properties such that flame spread does not occur, any combustion ceasing on removal of the source of flame. Thixotropic additives are added to prevent drainage of the resin as it is applied to vertical surfaces. The most widely used of these is silica aerogel which is highly effective in additions of a few per cent. Ultraviolet absorbers may be added to improve the durability of polyesters used out of doors. Amounts under 1 per cent of various hydroxybenzophenone and triazole derivatives are highly effective in this respect. Another type of additive is magnesium oxide to give high viscosity for pre-impregnated glass mats. Yet another is the whole range of pigments and dyestuffs which are needed to produce resins of every colour; these must be carefully selected for inertness to resin and catalyst system and for light stability in the resin matrix.

REFERENCES

1. Ellis, C. U.S. Pat. 1 897 977 (1922–1933).
2. Dykstra, H. U.S. Pat 1 945 307 (1930–1934).
3. Staudinger, H. *Helv. chim. Acta*. **5**, 785 (1922).
4. Kienle, R. *Ind. Engng. Chem*. **22**, 590 (1930).
5. Carothers, W. H. *J. Am. chem. Soc*. **51**, 2548, 2560 (1929).
6. Bradley, T. F. *et al. Ind. Engng. Chem*. **29**, 440, 579, 1270 (1937).
7. Ellis, C. *U.S. Pat*. 2 195 362 (1936–1940).
8. Muskat, I. E. *U.S. Pat*. 2 423 042 (1943–1947).
9. Hurdis, E. C. *U.S. Pat*. 2 480 928 (1944–1949).
10. Fraser, G. L. *U.S. Pat*. 2 516 309 (1947–1950).
11. Simons, W. G. *U.S. Pat*. 2 537 375 (1947–1951).
12. Parkyn, B. S. and Bader, E. *Br. Pat*. 713 332 (1951–1954).
13. Moss, W. H. *Brit. Pat*. 506 999 (1937).
14. Castan, P. *Swiss Pat*. 211 116 (1938–1940).
15. Castan, P. *U.S. Pat*. 2 324 483 (1942–1943).
16. Castan, P. *U.S. Pat*. 2 444 333 (1944–1948).
17. Greenlee, S. O. *U.S. Pat*. 2 494 295 (1946–1950).
18. Boenig, H. V. *Unsaturated Polyesters: Structure and Properties,* Elsevier Publishing Co. (1964).
19. Vancso-Szmercsanyi, J. *et al. J. Polym. Sci*. **53**, 241 (1961).
20. Curtis, L. G. *et al. Ind. Eng. Chem. Prod. Res. Dev*. **3**, 218 (1964).
21. Lewis, F. M. and Mayo, F. R. *J. Am. chem. Soc*., **70**, 1533, (1948).
22. Funke, W. *et al. Makromolek. Chem*. **50**, 188 (1961); **57**, 192 (1962) 120

23. Park, R. E. *et al. S.P.E. Jl* **17**, 1088 (1961)
24. Pritchard, G. 'Cross-Linking of Polyesters'. *Ph.D. Thesis* University of Aston (1968).
25. Traxton, M. E. *Chemy Ind.* **39**, 1613 (1966).
26. Selden, P. H. *Glasfaserverstärkte Kunststoffe,* Springer (1967).
27. Parkyn, B. *Polyesters,* Iliffe (1967).

10

Reinforcement

L. A. R. WARING

1. INTRODUCTION

It has been known for many centuries that molten glass could be drawn into glass threads. In ancient times it was usual to colour the glass and use coloured threads for decorative purposes. It is also known, that during the eighteenth century coarse glass mono-filaments were produced and these were woven into fabrics and used as heat insulating materials, but apart from this end use they did not find many industrial applications.

During the late 1930s Owens–Corning Fiberglas Corporation of the U.S.A. developed continuous filament glass fibre strand. They discovered that certain types of molten glass could be drawn from orifices in the base of an electrically heated platinum crucible, attenuated to get continuous filaments of around 10 μm in diameter, gathered together into a multifibre strand, sized and wound on a high-speed winding head. It was found that this type of sized glass strand could be processed into yarn on conventional textile machinery and the yarn woven into fabric. The main outlet for the yarn and fabric at that time was their use as a reinforcement for electrical insulation systems. A glass composition which gave good electrical properties was developed to meet this condition; this was a calcium-alumina boro-silicate glass with an alkali content of less than 1 per cent and which was capable of being drawn, as already described, into fine filaments; this glass became commonly known as E glass.

It was soon realised that the unique properties of E glass fibre, high strength coupled with a high Young's modulus and also its good resistance to weathering, could be used as reinforcement for thermosetting resins for industrial applications outside the electrical industry. Early attempts to use glass fibre as a reinforcement for resins for structural use were disappointing because the brittle nature of the material did not lend itself to high moulding pressures. However, with the advent of polyester type resins in the U.S.A. in 1942, which could be cured using a very low moulding pressure, or even at atmospheric pressure, and without producing by-products, the use of glass fibre as a reinforcement material became a reality and its use in reinforced plastics started to grow rapidly throughout the world.

2. TYPES OF GLASS

There are now other types of glass, as well as E glass, which are capable of being formed into continuous filament glass fibres. A calcium-alumina boro-silicate glass with an alkali content of up to 5 per cent is capable of producing glass fibres which have physical properties equal to those of E glass, but it is inferior electrically. Soda lime glass with an alkali content varying from 10 to 15 per cent commonly known as 'A' glass can also be used, but it is inferior in both physical and electrical properties to E glass. 'A' glass fibre does, however, because of its cheapness, find some applications in reinforced plastics, where high physical strength and good electrical properties are not of paramount importance and the end product is not submitted to severe attack by moisture, but its resistance to attack by acids is superior to E glass.

More recently, a new fiberizable glass has been developed by Owens–Corning Fiberglas Corporation[1]. It consists of silica, alumina and magnesia combined in proportions to provide a glass composition having the necessary viscosity-temperature relationship and miscibility and is known as 'S' glass. It is claimed to produce continuous filament glass fibres 40 per cent stronger and its resistance to temperature 100°C higher than for E glass. It is, however, much more expensive to produce than E glass and its outlets are mainly in the missile and space capsule applications in the U.S.A. E glass, on the other hand, because of its versatility and unique all-round properties, such as good textile processing, its suitability for the electrical industry and as a fibre reinforcement in both fabric and strand form, has become the major type of glass fibre used in the reinforced plastics industry. The properties of all the end products outlined in *Figure 10.3* are based on this type of glass.

3. MANUFACTURE OF GLASS FIBRES

E glass is produced by first blending a mixture of inorganic components, of which silica accounts for a little over 50 per cent and transferring them into a hopper which is used to feed a high temperature melting furnace. As the metal oxides interact and pass through the furnace the temperature rises to around 1600°C, at which temperature the dissolved gases are removed and homogeneity of the glass composition is obtained. In the direct melt system the refined glass is fed directly to filament forming platinum crucibles (termed 'bushing'), *Figure 10.1*, but in the marble process the refined glass is first made into marbles which are annealed; these are subsequently re-melted in a platinum bushing, *Figure 10.2a*.

In both processes an electrically heated platinum bushing is used. The base of the bushing contains between 100 and 800 nipples of special design. The diameter of the orifice in the nipple can also be varied, depending on the type of end product required. The beads of molten glass at the base of the orifices are attenuated mechanically into continuous filaments at a speed of approximately 50 m/s. They are gathered together into a continuous strand at a sizing

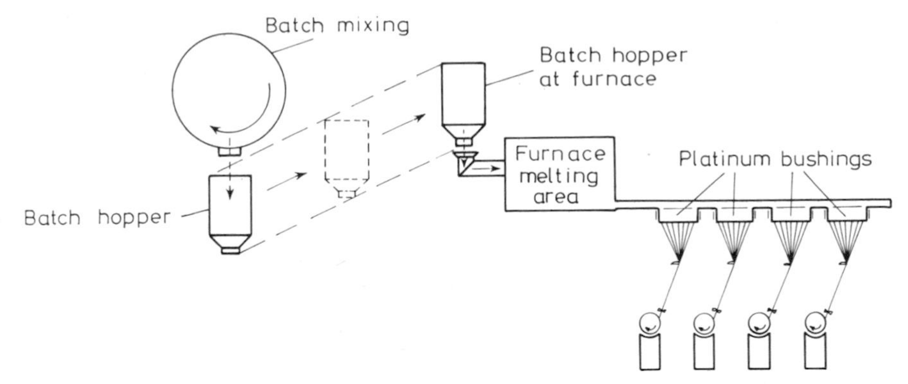

Figure 10.1. Flow diagram to illustrate 'Direct Melt System' of manufacture of E Glass continuous filament fibres

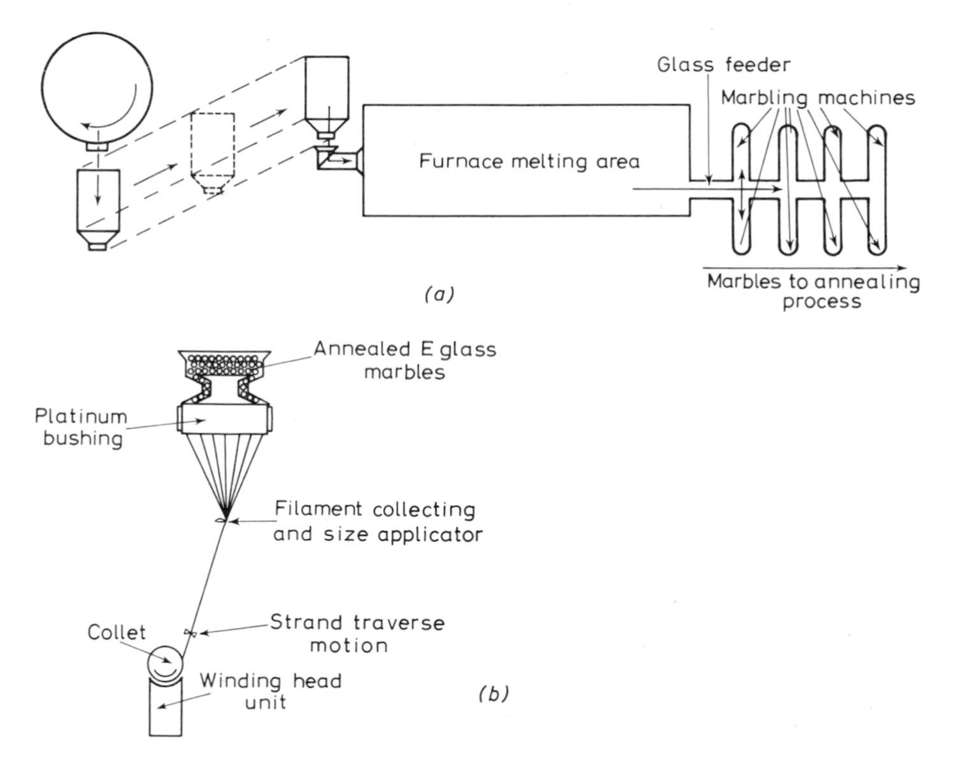

Figure 10.2. Flow diagram to illustrate 'Marble System' of filament forming for manufacture of E glass continuous filament fibres

application pad and wound on a high speed collet, *Figure 10.2b*. The diameter of the fibres is controlled by the size of the hole in the bushing orifice, the viscosity of the glass and the rate of attenuation; this means that the count of the glass strand may be varied without necessarily changing the bushing.

4. SIZES

Single glass filaments are easily damaged mechanically and are also subject to self-abrasion. In order to minimise these effects a textile or plastics size (also known as 'dressing') is added, depending on the end product requirements, at the collection pad (*see Figures 10.1* and *10.2*b). The addition of a size at this early stage of processing provides protection to the virgin filaments and also binds them into a strand, but does not bond adjacent strands together, thus allowing the strands to be easily unwound from the cake at later processing stages such as yarn preparation and weaving on textile machinery and the conversion to roving, chopped strand and chopped strand mat.

The type of size applied to the glass filaments during the drawing operation is extremely important. The 'textile size' consists of a dextrinised starch/emulsified vegetable oil system, which allows the strand to be twisted, doubled and woven into fabrics without undue mechanical damage to the glass fibres. This type of size is non-compatible with polyester and epoxide resins and it has to be either partially or completely removed from the fabric depending on the end product and in many applications a resin coupling agent is added.

The second type of size is known as 'plastics size' and is based on polyvinyl acetate, suitably plasticised and containing a resin coupling agent. This type of size is compatible with polyester resin and also with epoxide and phenolic resins, depending on the type of coupling agent used, and therefore does not require to be removed after the processing has been completed. The incorporation of a resin coupling agent assists in improving the bond between the laminating resin and the sized glass fibre and improves the mechanical and electrical properties of the reinforced laminate. The chemistry of the resin coupling agents is dealt with later in the chapter.

A more recent development in sizing is the use of a 'plastics size', containing a resin coupling agent, which is compatible with polyester resins and also has improved resistance to abrasion. Rovings with this type of size can be woven on suitable looms into fabric without excessive fibre damage. This has resulted in the manufacture of a range of woven roving fabrics, which do not need the desizing and finishing processes required for yarn based woven fabrics.

5. PROPERTIES OF E GLASS FILAMENTS

Much experimental work has been carried out to determine the effect of production conditions on the tensile strength of E glass fibres. Most of the earlier work by Griffith[2], Smedal[3], Murgatroyd[4], and others, had shown that the tensile strength of E glass filaments increased with decrease in the diameter of the fibre. Recent work carried out by Thomas[5] showed that the breaking strength of E glass fibres was independent of the fibre diameter and around 345×10^4 kN/m^2, provided the temperature of the molten glass, from which the fibre was drawn, was sufficiently high to permit a fibre of

Table 10.1

PROPERTIES OF E GLASS FIBRES

Properties	*Values*
Physical and Mechanical	
Specific gravity	2.55
Tenacity	62 g/tex
Tensile strength (nominal)	175×10^4 kN/m^2
Young's modulus	7×10^7 kN/m^2
Extension at break (nominal)	2.5%
Poisson's ratio	0.2
Hysteresis	none
Creep	none
Thermal	
Coefficient of thermal expansion	4.7×10^{-6}/K
Coefficient of thermal conductivity	1.05 W/mK
Electrical	
Dielectric constant	6.43 at 10^2 Hz 6.11 at 10^{10} Hz
Optical	
Refractive index	1.548

Table 10.2

TYPE AND DIAMETER OF BASIC E GLASS FILAMENTS

Filament designation	*Diameter of basic filament* mm $\times 10^{-5}$
Beta	<457
D	508–632
E	635–759
G	889–1013
J	1143–1267
K	1270–1395
L	1397–1521

uniform diameter to be produced and that the fibre was very carefully mounted prior to testing. However, under normal production conditions the breaking strength of E glass fibres is almost halved due to mechanical damage during processing and is in the range of 140–200 kN/m^2 $\times 10^4$.

Single E glass filaments have the properties shown in Table 10.1 and Table 10.2 gives the type and diameter of basic E glass filaments.

Amory[6] has shown in a recent paper, Table 10.3, that the bending radius at rupture of different types of glass filament increased with increase in filament diameter, confirming the tendency to use finer filaments for textile processing and coarser filaments for strand reinforcements. The coarser filaments, which are cheaper to produce, stand up readily to the degree of processing required for strand reinforcements.

Recent improvements in manufacturing techniques by Owens–Corning Fiberglas Corporation have resulted in the commercial production of Beta glass filaments[7] and, as can be seen from the Table 10.3, these glass filaments

Table 10.3

Filament designation	*Fibre diameter* mm × 10^{-5}	*Bending radius to rupture* mm × 10^{-5}
Beta	368	3810
DE	640	6604
G	902	9398

are of extreme fineness and bend much more sharply before they rupture—an important property for filaments for textile processing.

The production and the major properties of E glass fibres have now been covered. It can be seen that E glass fibres have several advantages over other materials when used as a reinforcement for plastics. These are:

(1) Production from readily available raw materials
(2) High tensile strength
(3) High Young's modulus
(4) Low extensibility
(5) Good dimensional stability
(6) Do not exhibit creep
(7) Resistant to temperatures up to 550°C
(8) Complete fire-proofness
(9) Immunity to microbiological attack
(10) Good resistance to most forms of chemical and solvent attack
(11) Good weathering properties
(12) Do not absorb moisture
(13) Good electrical properties
(14) Availability in a variety of forms to suit particular end uses in the reinforced plastics industry.

6. TYPES OF GLASS FIBRE PRODUCTS FOR REINFORCEMENT

Having described very briefly how continuous filament E glass is made and examined its physical, mechanical, thermal and optical properties, and the advantages it has over other reinforcing materials, it is now necessary to consider the various end products outlined in *Figure 10.3*. These can be divided into two distinct categories, those which require to be processed on textile machinery into yarn and fabric and those which can be used in strand form.

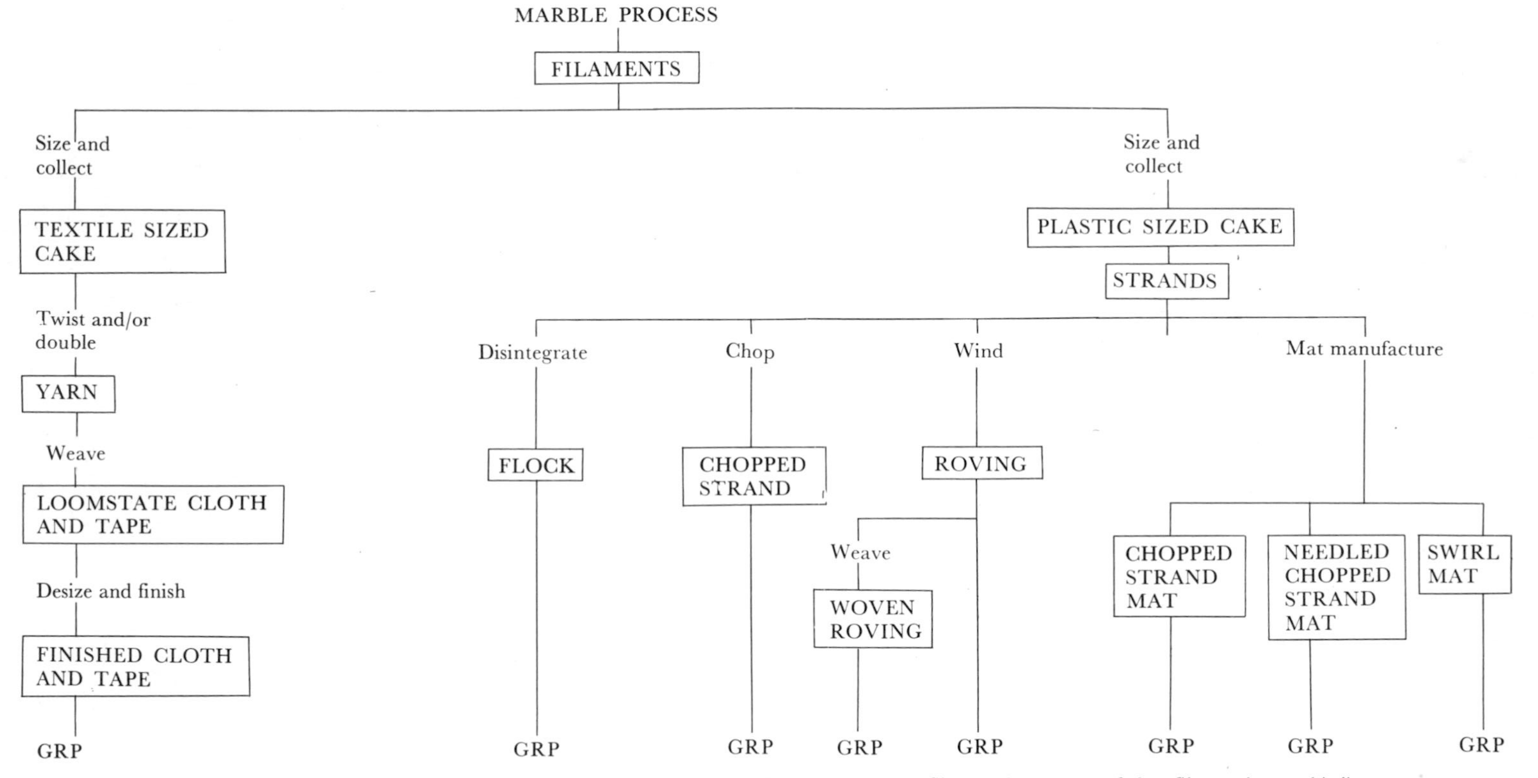

Fig 10.3. *This chart shows the processes involved in the conversion of E glass continuous filaments into a range of glass fibre products and indicates the products used in the manufacture of glass reinforced plastics*

7. YARN

Yarn is produced from sized cake by first twisting the strand, followed by plying a number of the twisted strands together to form a doubled balanced yarn. An important property which is used to designate the type of yarn is called the 'Count'. The direct system is now used in which the yarn count is expressed in terms of weight per unit length—this is known as the Tex system and is the weight in grammes of 1 km of strand. In the first stage of twisting a 33 Tex strand, 1.8 turns/cm are generally put into the strand using Z twist. It will be obvious that in order to finish with a balanced yarn the doubling twist will be in the opposite direction and this is generally 1.6 turns/cm and is known as S twist.

The plies of yarn are designated by the digits following the count number. The first shows the number of strands twisted together and the second is the number of these strands doubled together. Using the following symbols:

E—Electrical glass
C—Continuous filament
G—Average filament diameter (*see* Table 10.2)
T—Basic strand count in tex
A—Original strands twisted
B—Twisted strands doubled together.
The yarn can be designated as ECG.T–A × B, for example ECG. 33–1 × 2.

Table 10.4

TYPICAL CONTINUOUS FILAMENT SINGLE AND DOUBLED GLASS FIBRE YARNS

Strand count in Tex	*No. of filaments new strand*	*Yarn construction*	*Approximate* m/kg	*Minimum breaking strength in* N
2·75	51	2.75–1 × 0	355 600	1.33
2.75	51	2.75–1 × 2	177 800	2.67
5.5	102	5.5 1 × 0	177 800	2.67
5·5	102	5.5- 1 × 2	88 900	5.34
11.0	204	11·0 × 1 × 0	88 900	5.78
11.0	204	11.0–1 × 2	44 450	11.57
22.0	204	22.0–1 × 0	44 450	10.68
22.0	204	22.0–1 × 2	22 230	21.35
33.0	408	33.0–1 × 0	29 630	15.57
33.0	204	33.0–1 × 0	29 630	14.23
33.0	204	33.0–1 × 2	14 820	31.14
66.0	204	66.0 1 × 0	14 820	25.35
132.0	204	132.0–1 × 0	7 310	34.70

In certain cases the basic strands are put through the twisting operation only. Such yarns have a lower twist, generally 0.4 turns/cm. Using the same symbols they are designated, e.g. ECG. 33–1 × 0.

The main use for yarns in the reinforced plastics field is for the manufacture of woven fabrics, but they also find outlets for plastic reinforcement in the electrical industry and winding applications. Table 10.4 gives details of the yarns most commonly used.

8. FABRICS AND TAPES

Fabrics and tapes may be considered together, since the latter may be regarded as a narrow fabric. Many types of E glass fabrics and tapes are available, ranging from 0.03 mm to 0.56 mm in thickness and up to 547 g/m^2 in weight; these are listed in the British Standards Institution specification for fabrics[8] and tapes[9]. The weave of a fabric is of considerable importance to the designer of reinforced plastics. The tensile strength of the fabric can be balanced in the warp and weft directions or concentrated in either direction according to the weave chosen.

The commonly used weave constructions are as follows:

Plain weave—A fabric in which each warp and weft yarn is passed under one yarn and over the next. By weaving the fabric tightly, yarn slippage is

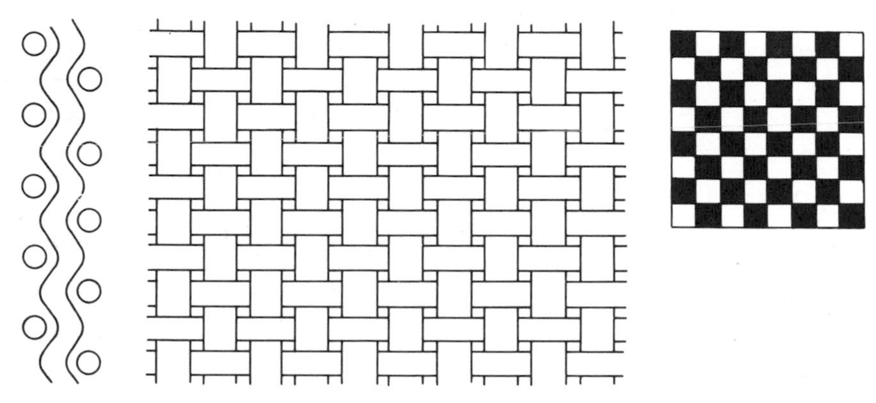

Figure 10.4. Plain weave

reduced to a minimum, but it loses its drape. Loosely woven fabrics, known as 'scrim' can also be made in plain weave (*Figure 10.4*).

Square weave—This is a plain weave fabric, which has identical count yarns and an equal number of ends and picks in the warp and weft directions.

Twill weave—A fabric woven in such a manner that a characteristic pattern of diagonal lines is produced on the surface of the fabric (*see Figure 10.5*).

Satin weave—A fabric in which the surface consists almost entirely of either warp or weft ends. In the repeat of the weave each yarn of the weft passes over all but one end of the warp, generally 3 or 7 yarns, and the weaves are known as 4 and 8 shaft satin weaves (*Figure 10.6*).

Unidirectional fabric—A fabric in which the strength is very high in one direction compared with the other. This is achieved by using heavy yarns in the warp and relatively light yarns, opened spaced, in the weft direction.

The physical properties of some typical cloth constructions are given in Table 10.5.

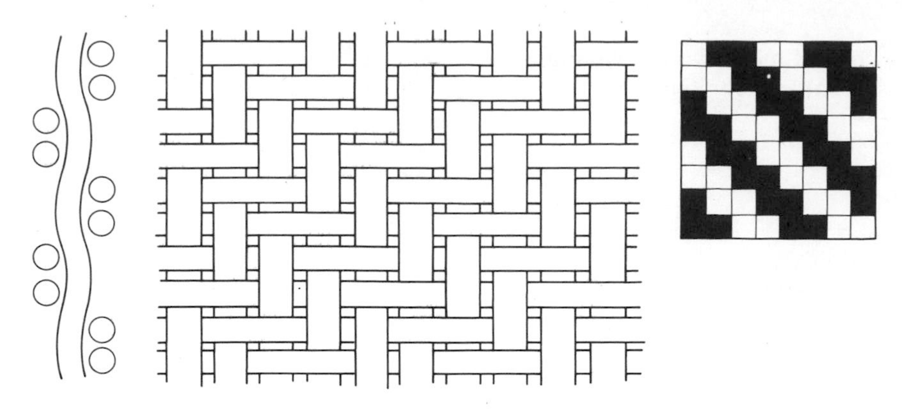

Figure 10.5. 2 × 2 twill

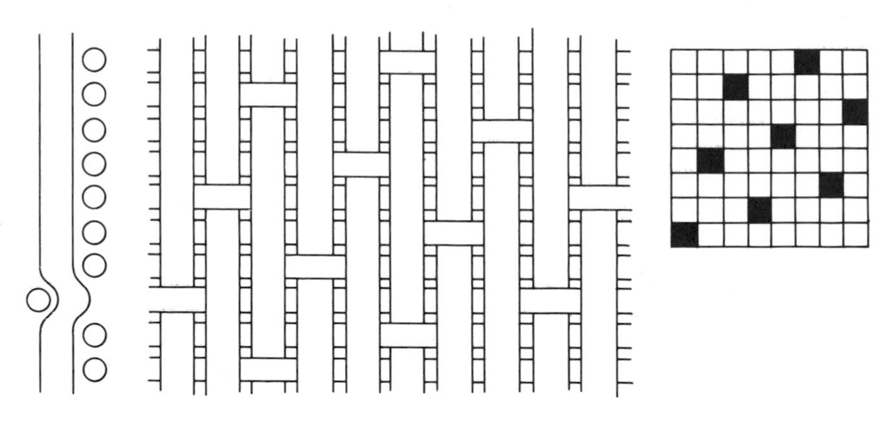

Figure 10.6. 8-shaft satin

Table 10.5

Weave		*Plain*		*2 × 2 Twill*	*8-shaft satin*
B.S. designation		P2/11	P6/33	T3/22	S2/22
Nominal thickness (mm)		0.08	0.15	0.28	0.20
Average weight (g/m^2) ±10%		99.4	188	331	306
Ends/10 cm ±2½%		252	142	118	224
Picks/10 cm ±2½%		173	126	118	213
Nominal count of yarn:					
warp—tex		11–1 × 2	33–1 × 2	22–2 × 3	22–1 × 3
weft—tex		11–1 × 2	33–1 × 2	22–2 × 3	22–1 × 3
Minimum average breaking strength (kN/m)	warp	19.0	26.5	50.4	47.9
	weft	13.1	23.5	50.4	45.4

8.1. EFFECT OF CLOTH STRUCTURE ON LAMINATE PROPERTIES

It can be seen from the diagrams on the various types of cloth weave that no yarn remains in one plane, but passes over and under other yarns. This deviation of length in relation to the straight path of the yarn is known as crimp. The degree of crimp, which is usually expressed as a percentage, will vary with the type of weave; the greatest crimp is in a plain weave and smallest crimp is in a satin weave.

The fabric shown in *Figure 10.5* is a 2 × 2 twill; other types of twills are woven by different groupings of the warp and weft. Twill fabrics have better draping properties, reduced crimp and higher tensile strength in laminate form than plain cloth of equal weight. The individual ends are more easily distorted and moulding finishes are not so good as with plain and satin weaves.

The satin type of weave almost eliminates crimp in the fabric (*see Figure 10.6*). The fabric has, therefore, good laminate tensile strength and flexural modulus. The fabric has excellent drape and good surfaces on mouldings are also achieved.

Haythornthwaite[10] found that the breaking strength of fabrics made with equivalent weights of yarn increased with the fineness of the basic filament used. For example, a laminate made from fabric with 11-tex yarn gave a stronger laminate than one made using the same weight of fabric, but with 33-tex yarn. He also found that in the case of satin weave it was important to make the laminate with similar fabric faces uppermost, if maximum laminate properties were to be obtained. It was also found that the impact strength of laminates increases with increase in fabric thickness.

The choice of fabric must, therefore, be based on design requirements. Square weave fabrics should be used where equal strength is required in both directions and satin weave where good drape properties coupled with minimum distortion of yarns are required. If maximum strength in one direction is the main criterion, then a unidirectional fabric should be used.

8.2. DESIZING

It has already been stated that all yarns for weaving are coated with about 1.5–2.0 per cent of textile size and that this type of size is not compatible with either polyester or epoxide resins. It is generally necessary to remove the size completely, but for some applications removal of the volatile content will suffice. This can be carried out by the following methods:

(*a*) *Carmelization*—This consists of a short heat treatment which drives off the volatile components of the size and carbonises the starch and gives the fabric a golden brown colour. The residual size content is about 0.6 per cent. This treatment gives good laminates with epoxide resins.

(*b*) *Aqueous desizing*—This consists of passing the fabric through a scouring bath and removing as much of the organic matter as possible. The resulting fabric is white and has a residual size content of below 0.3 per cent. It is a satisfactory method of desizing fabric, if high physical properties are not required in the laminate.

(*c*) *Heat treatment*—When maximum physical, mechanical and electrical properties are required in a laminate, the heat treatment process should be used. There are two methods of heat cleaning in which the residual size content is less than 0.1 per cent. The first is the batch method, where rolls of fabric are put into an oven and gradually heated up to a temperature of 350°C and kept at this temperature for 30 h. The second process is a continuous process where the cloth is first aqueous desized and then passed continuously through an oven at a temperature of 300°C–320°C.

8.3. FINISHES

After the organic size has been removed from the glass fabric, finishes which improve the bonding between the laminating resin and the fabric can be

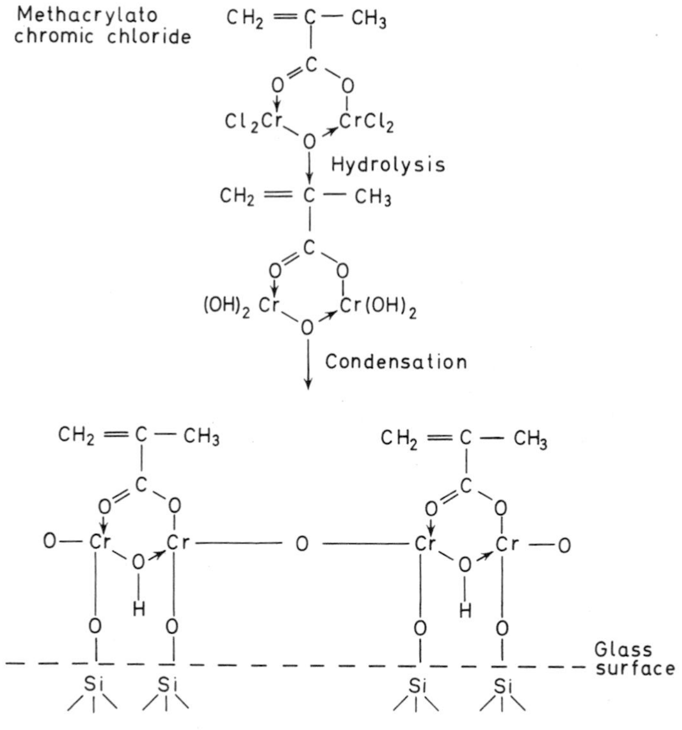

Figure 10.7. Reaction of methacrylato chromic chloride with the glass surface

applied. The use of a finish improves both the mechanical and electrical properties of the laminate and also its resistance to moisture.

One of the first finishes to be used in the treatment of glass fabrics was methacrylato chromic chloride, commonly known as 'chrome finish', with trade names—Volan (U.S.A.) and Vitrafix (U.K.). It was developed for use

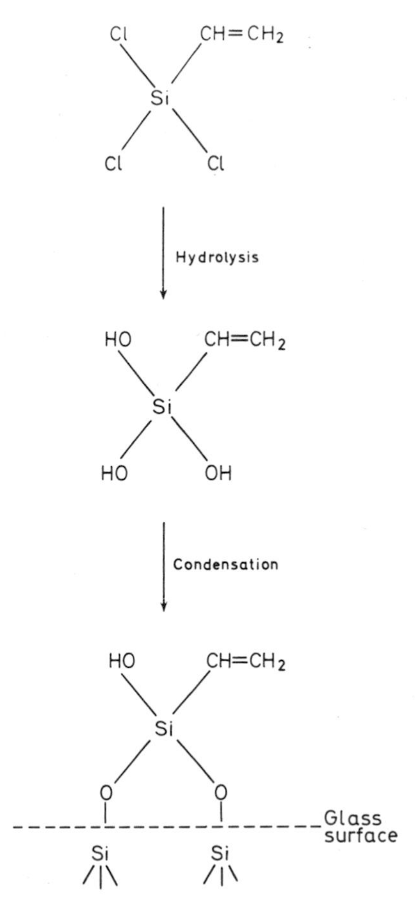

Figure 10.8. Reaction of vinyltrichlorosilane with the glass surface

with polyester resins, but it also improves the laminate properties with both epoxide and phenolic resins. When glass fabric is treated with this compound, the chromium attaches itself through the silanol groups on the glass surface leaving the methacrylato group to link up with the laminating resin. *Figure 10.7.*

With the advent of silane chemicals a number of new glass fibre/resin coupling agents was developed. The first in this range was vinyltrichlorosilane. This compound is very unstable and it hydrolyses immediately it comes into contact with moisture in the atmosphere liberating hydrogen chloride fumes; this makes it most obnoxious as a finishing agent under production conditions, *Figure 10.8.* It was, therefore, soon superseded by vinyltriethoxysilane, commonly known as the Garan finish[11]. Since that time considerable development work has been done on these silane coupling agents. The finish now most commonly used with polyester resins is gamma-methacryloxypropyl-trimethoxy-silane, commonly known as methacrylato-silane. This finish gives glass fibre/polyester resin laminates higher physical properties, particularly flexural strength, and the resistance of the laminate to moisture is also improved. *Figure 10.9* shows the reaction of this compound with glass fibre.

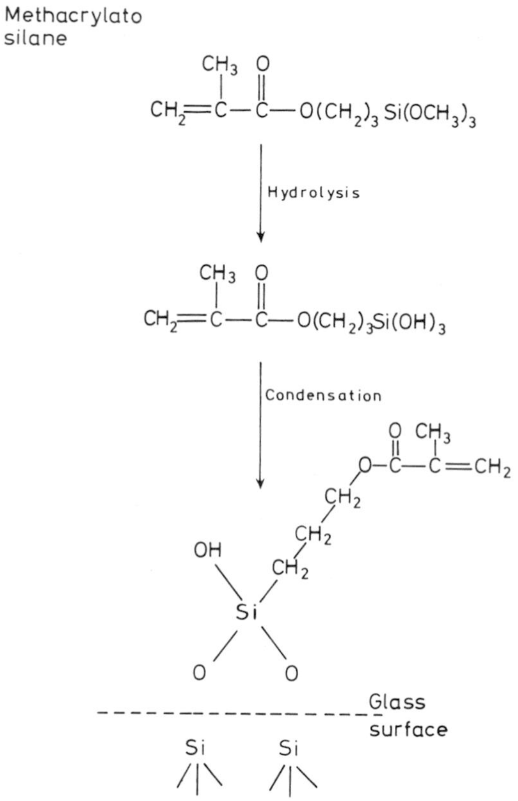

Methacrylato silane

$$CH_2{=}C(CH_3){-}C({=}O){-}O(CH_2)_3Si(OCH_3)_3$$

Hydrolysis

$$CH_2{=}C(CH_3){-}C({=}O){-}O(CH_2)_3Si(OH)_3$$

Condensation

$$CH_2{=}C(CH_3){-}C({=}O){-}O{-}CH_2{-}CH_2{-}CH_2{-}Si(OH)(O{-})_2$$

Glass surface

Si Si

Figure 10.9: Reaction of gamma-methacryloxypropyl-trimethoxy-silane with the glass surface

It is thought that the presence of the long chain group, in linking with the resin allows some degree of flexibility in the interface bond, thus improving the transfer of strain and enhancing the properties of the laminate.

Silane coupling agents have also been developed for use with epoxide resins. One of these commonly used is a similar type of coupling agent to that previously described, except that the methacrylato group has been replaced by an amino group and the trimethoxy group replaced by a triethoxy group. *Figure 10.10* shows the reaction of gamma-aminopropyltriethoxy-silane[12], commonly known as amino-silane, with the glass surface.

The effectiveness of a finish is assessed in terms of laminate strength, in accordance with the relevant British Standard specification[13]. Minimum

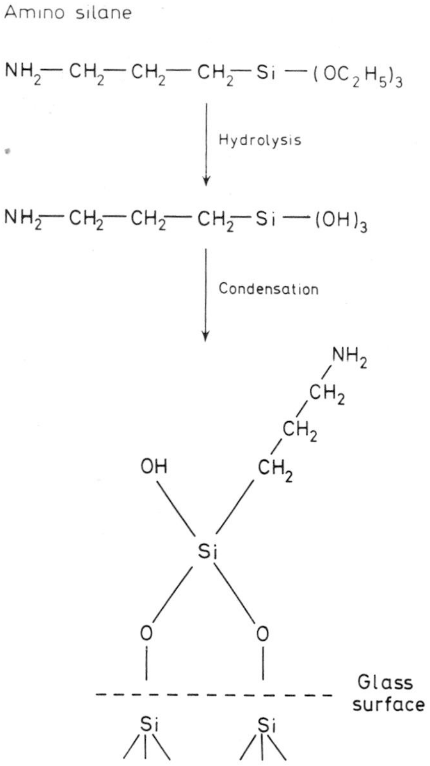

Amino silane

$NH_2—CH_2—CH_2—CH_2—Si—(OC_2H_5)_3$

↓ Hydrolysis

$NH_2—CH_2—CH_2—CH_2—Si—(OH)_3$

↓ Condensation

NH_2
CH_2
CH_2
OH CH_2
Si
O O
Glass surface
Si Si

Figure 10.10. Reaction of gamma-aminopropyltriethoxy-silane with the glass surface

Table 10.6

B.S. designation	Silane and chrome finishes			
	Minimum flexural strength—dry kN/m^2 × 10^4		*% Wet strength retention minimum*	
	Warp	*Weft*	*Silane*	*Chrome*
P2/11	39	36	85	75
P6/33	35	34	85	75
T3/22	36	33	85	75
S2/22	41	41	85	75

values are given for finished glass fabrics/polyester resin laminates[18] in Table 10.6.

A wet strength retention of over 90 per cent would be obtained with methacrylato silane finish.

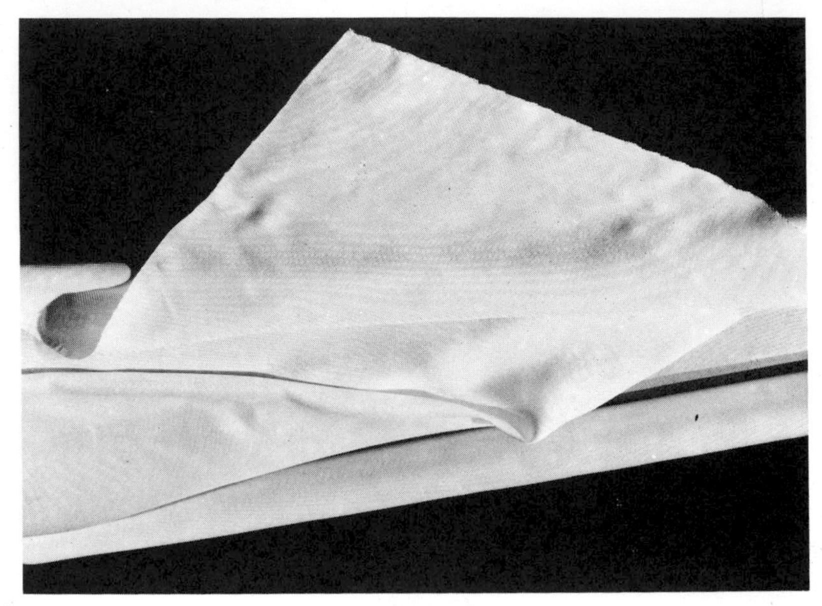

Figure 10.11. Glass fabric

9. ROVINGS

A roving is a number of strands wound together into a cheese or cone. The use of roving in the reinforced plastics industry has increased considerably during the last few years, the effect being that rovings have been developed for specific end uses. They can be sub-divided into the following categories:

(*a*) Depositor roving
(*b*) Preform roving
(*c*) Translucent sheeting roving
(*d*) Filament winding and rod stock roving
(*e*) Weaving grade roving.

The number of ends in a roving can vary from two for weaving grade roving up to 120 ends for a chopping grade roving.

The type and amount of plastics size on the strand and the resin coupling agent used are extremely important and range considerably with the type of roving.

(*a*) The use of depositor roving using the spray gun technique, at the expense of hand lay-up using glass mat, has been increasing rapidly over the last few years. The main reasons are lower raw material costs; faster production with a reduction in waste and reduced labour costs. Special rovings have been developed for use with spray guns. There are now several spray guns on the market, their function is to chop the roving into specified lengths, usually around 20–40 mm, add the resin and catalyst at the gun and spray on the mould a mixture of chopped strand and catalysed polyester resin. A roving suitable for use with a spray gun must have the following characteristics:

Figure 10.12. Roving

Ease of chopping—This is a measure of the ability of the roving to be chopped consistently and break up into individual strands without adjustment to the chopping unit. The principal fault with many depositor rovings is their tendency to tape, i.e. not break up into individual strands.

No build up of static charge—Build up of static causes the chopped fibre to stick to and cause blockage at the spray gun.

Ease of consolidation—This characteristic covers the rapid wetting out of the fibres with the resin and the removal of air.

(*b*) If the moulder wishes to use matched metal die moulding and to make his own preforms, then rovings which are suitable for depositor work are also suitable for preforming.

(*c*) There has been a tendency to move away from glass mat to roving for the manufacture of translucent sheeting. Again, utilisation of roving instead of mat results in considerable cost saving on raw material and its use also lends itself to automation of the process. A roving suitable for the manufacture of translucent sheeting should have the following characteristics:

Ease of chopping and freedom from taping.

Fast wetting out with the polyester resin.

Freedom from variation in size content, which, if present, causes fibre pattern in the finished sheet.

Special size to achieve good laminate translucency by making the refractive index of the glass fibre/size a near match to that for the polyester resin.

Good weathering properties.

(*d*) The main characteristic of a roving suitable for filament winding is that it is wound under even tension and has the ability to stay in a taped form, but wet-out quickly as it passes through the resin bath. As creels are normally used to supply the roving to the filament winding machine, good abrasion characteristics are also required.

(*e*) Roving suitable for weaving should have the following characteristics:

The ability to stay in a taped form from package to loom, but break up into individual strands during the weaving operation. Good resistance to abrasion at a low size content level is also a significant factor.

Table 10.7 gives typical properties of depositor and translucent grade rovings. The values given in Table 10.7 are for laminates with 30 per cent glass content. The standard B.S. test for a roving laminate is the cross-breaking

Table 10.7

TYPICAL PROPERTIES OF SPRAYED-UP GLASS FIBRE/POLYESTER RESIN LAMINATES USING DEPOSITOR TYPE ROVINGS AND TRANSLUCENT SHEETING ROVINGS

	Type of roving		
	Depositor		*Translucent*
	Hard silane roving	*Hard chrome roving*	*Silane roving low size content*
Specific gravity	1.42	1.46	1.42
Tensile strength $kN/m^2 \times 10^4$	9.8	9.9	10.0
Tensile modulus $kN/m^2 \times 10^7$	0.66	0.69	0.72
Flexural strength $kN/m^2 \times 10^4$	18.8	22.6	23.8
Flexural modulus $kN/m^2 \times 10^7$	0.55	0.55	0.55
Flexural strength retention after 2 h boil in water	90%	85%	95%
Glass fibre content	30%	30%	30%
tex	2280	2280	2280

strength of a polyester rod containing 60 ± 5 per cent glass content[14], but the data given in Table 10.7 is considered to be more useful to a design engineer.

10. WOVEN ROVING FABRICS

Woven roving fabrics are made from weaving grade roving. As the twisting processes are eliminated, coarser glass fibre filaments can be used and also a resin compatible size can be applied at the bushing. The rovings, therefore, contain practically no twist so both ends and picks lie flat in the fabric; this has also the effect of reducing crimp. Woven roving fabrics have excellent draping properties. The silane resin coupling agent is in the size so the fabric imparts great strength to a laminate at low cost, as it does not require the heat cleaning and finishing processes for yarn based fabrics.

The fabrics are generally in a heavier weight/m^2 range than conventional fabrics, the weight/m^2 varying from 200 g to 850 g and the nominal thickness from 0.20 mm to 0.90 mm. They are manufactured in both plain and twill

Figure 10.13. Woven roving

Table 10.8

TYPICAL WOVEN ROVING FABRICS

Weave	*Plain*			2 × 1 *Twill*
B.S. designation	RP6	RP10	RP25	RT25
Nominal thickness (mm)	0.20	0.33	0.84	0.76
Average weight (g/m^2) ±10%	203	363	827	827
Ends/10 cm ±2½%	67	61	20	20
Picks/10 cm ±2½%	63	55	16	16
Roving structure (Nominal count of yarn)				
warp—tex	152	305	2290	2290
weft—tex	152	305	2290	2290

weaves[15]. They find many applications in composite laminates, where woven roving is used in conjunction with random chopped strand mat to give a thinner laminate with improved mechanical properties. Typical woven roving fabrics are given in Table 10.8 and the properties of laminates in Table 10.9.

The standard B.S. test for woven roving laminate is the cross-breaking strength of a polyester rod containing 60 ± 5 per cent glass content. All the

Table 10.9

TYPICAL PROPERTIES OF WOVEN ROVING/POLYESTER RESIN LAMINATES

B.S. designation		RP6	RP10	RP25	RT25*
Tensile strength	$kN/m^2 \times 10^4$	30.3	36.5	27.6	35.2
Tensile modulus	$kN/m^2 \times 10^7$	1.7	1.9	1.6	1.5
Flexural strength	$kN/m^2 \times 10^4$	29.7	22.1	37.9	31.0
Flexural modulus	$kN/m^2 \times 10^7$	1.5	1.5	1.4	1.7
% Flexural strength retention after 2 h water boil		92	93	93	92
% Glass fibre content		52	52	51	58

* (By permission of Fothergill & Harvey Ltd.)

qualities in Table 10.9 comply with this test, but the data given in the table is considered to be more useful to a design engineer.

11. CHOPPED STRANDS

Chopped strands are available in lengths of 6 mm, 12.5 mm, 25 mm and 50 mm. The major outlets of chopped strands are in the production of 'doughs' and 'gunks'. It can be shown that the strength of glass reinforced mouldings falls rapidly as the strand length is reduced below 16 mm. This is of importance in the preparation of doughs—the incorporation of the filler and glass fibres into the resin is a difficult process, and much care must be exercised to prevent the strand length being broken down below the critical limit. The degree of filamentisation also has a great bearing on the strength of dough mouldings, and strands with a high 'strand integrity' should be used to counteract the tendency for the filaments to break apart during the mixing process. More recently an improved H.S.I. (high strand integrity) grade containing a less soluble binder has become available.

Figure 10.14 shows the effect of strand length on the strength of laminates, and *Figure 10.15* shows fall of impact strength with increasing mixing time.

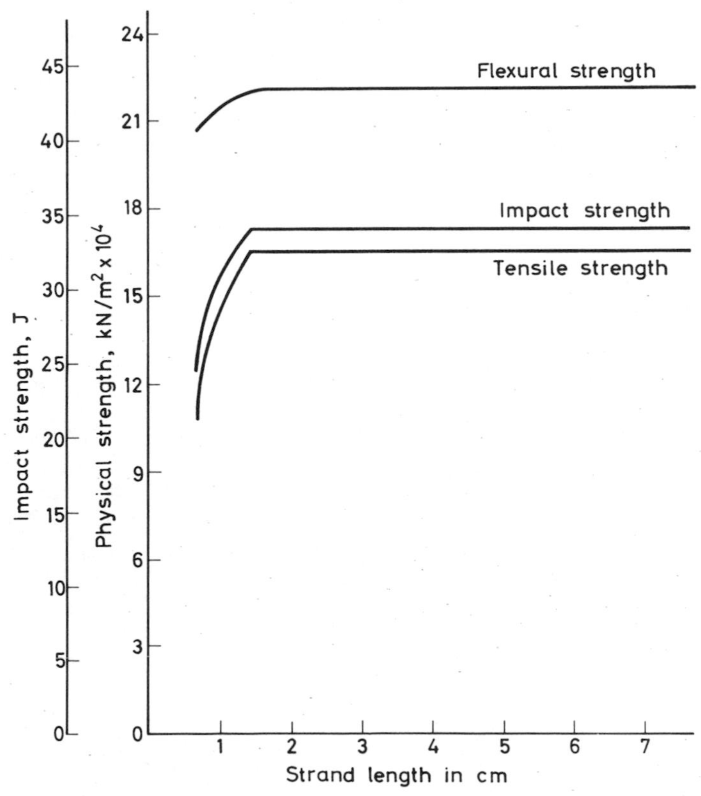

Figure 10.14. Physical strength of chopped strand reinforced mouldings at 25 per cent glass by volume

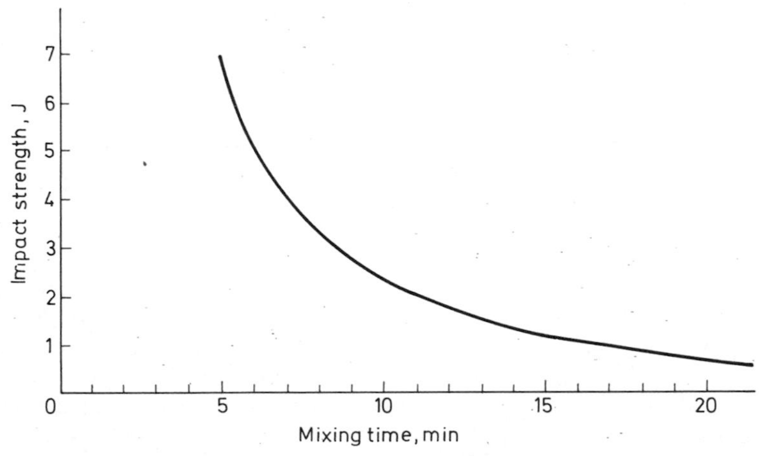

Figure 10.15. Effect of mixing time on impact strength of polyester dough moulding compound
(By permission of B.I.P. Chemicals Ltd.)

12. CHOPPED STRAND MAT

In many applications in the reinforced plastics industry, glass fabrics and woven rovings give mechanical properties in the laminate which are much greater than that required, apart from stiffness and for such applications a cheaper material is adequate. For these reasons chopped strand mat is ideal and is the most widely used sheet-like material.

There are three basic types of chopped strand mat. All are non-woven and are made from chopped strands distributed in a random pattern to ensure uniformity of strength in all directions. They are made in the following

Figure 10.16. Chopped strand mat

nominal weights, 300, 450 and 600 g/m^2, and are available to the relevant British Standard specification[16]. These grades of mat are known as follows:

(*a*) Emulsion bound mat
(*b*) Powder bound mat
(*c*) Mechanically bound mat.

(*a*) Emulsion bound mat is ideal for hand lay-up applications using a general purpose polyester resin. It has fast wetting-out properties. The dry mat has sufficient strength to allow it to be cut to the required shape and to be capable of easy impregnation, even in complex moulds. Any entrapped air must be easily removed during the rolling process. The binding emulsion can be varied to present the widest possible choice to moulders.

(*b*) Powder bound mat is made from chopped strand bonded with a powdered polyester resin. It is a much stronger mat than the emulsion bound type and is suitable for matched die moulding, translucent sheeting (if made from mat), electrical laminates and pre-preg moulding materials.

(*c*) Mechanically bound mat, also known as needled mat, is made by needling chopped strand to glass tissue or cotton scrim backing. It is designed for pressure moulding as the needled strands are not displaced during the pressing cycle. It has also excellent drape characteristics.

13. OTHER TYPES OF MAT

Continuous strand mat—This is a mat also known as swirl mat developed by Owens–Corning Fiberglas Corporation[17]. It is made from multiple layers of continuous filament in a swirl pattern. The strands are held together in mat form with a resinous binder. This type of mat is recommended for matched die mouldings where relatively deep and complex contours require maximum 'draw' characteristics.

Weave mat—This is a combination of chopped strand bonded to woven roving fabric. The glass fibre strands can be either bound together mechanically or chemically by the use of a powdered polyester resin. This type of

Table 10.10

B.S. designation	*Emulsion bound mat*	*Powder bound mat*	*Continuous* strand mat*
Tensile strength, $kN/m^2 \times 10^4$	15.2	15.2	15.2
Tensile modulus, $kN/m^2 \times 10^7$	1.0	1.0	1.1
Flexural strength, $kN/m^2 \times 10^4$	23.5	24.8	26.2
Flexural modulus, $kN/m^2 \times 10^7$	0.9	0.9	1.0
% Flexural strength retention after 2 h water boil	90	90	80
% Glass fibre content	40	40	40

(By permission of Chemical Trading Co. Ltd.)

mat is used where high mechanical properties are required and also for extrusion processes, which require minimum movement of the fibre during extrusion.

Surfacing mat—This is a thin tissue of glass fibre bonded together with a soluble binder and is about 0.30 mm thick and around 27 g/m^2 in weight. It is used to give a smoother finish to hand lay-up and mouldings.

Table 10.10 gives typical physical properties of laminates with chopped and continuous strand mats and polyester resin.

14. QUALITY CONTROL

The importance of modern methods of quality control is fully realised and most glass fibre reinforcement products are manufactured to British Standard specifications. Production is geared to the relevant specification and all glass fibre manufacturers in the U.K. are Government approved Companies. *Figure 10.17* shows glass fabric being tested for tensile strength under constant

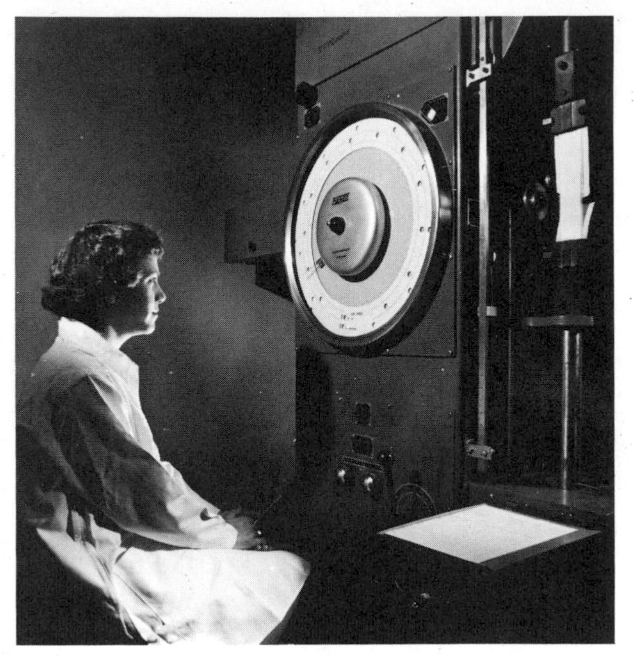

Figure 10.17. Testing of glass fabric for tensile strength (By permission of Turner Brothers Asbestos Co. Ltd.)

rate of loading conditions; this is only one of the large number of control tests carried out to ensure that the quality of the reinforcement is correct throughout the manufacturing stages and that the range of reinforcement products shown in *Figure 10.3* comply with the British Standard Specifications[8,9,13–16].

REFERENCES

1. *Brit. Pat. 979 196*—Owens-Corning Fiberglas Corporation (1965).
2. Griffith, A. A. *Trans. R. Soc.* **A221**, 163 (1920).
3. Smedal, A. *J. Soc. Glass Technol.* **20**, 432 (1936).
4. Murgatroyd, J. B. *J. Soc. Glass Technol.* **28**, 368 (1944).
5. Thomas, W. F. *Physics Chem. of Glasses* **1**, 4 (1960).
6. Amory, G. *Studies in Modern Yarn Production* **4**, 48. Annual Conference of the Textile Institute (1968).
7. *Brit. Pat.* 1031273—Owens-Corning Fiberglas Corporation (1966).
8. B.S. 3396—*Woven Glass Fibre Fabrics for Plastics Reinforcement. Part 1, Loom-State Fabrics* (1966); *Part 2, Desized Fabrics* (1966).
9. B.S. 3779—*Glass Fibre Woven Tapes* (1964).
10. Haythornthwaite, E. *The Effect of Fabric Construction and Properties on the Physical Properties of*

Glass Fabric-Resin Laminates. British Plastics Federation Reinforced Plastics Technical Conference (1958).

11. *Brit. Pat. 745374*—Johns Manville Inc. (1956).
12. *Brit. Pat. 978042*—Union Carbide Corporation.
13. B.S. 3396—*Woven Glass Fibre Fabrics for Plastics Reinforcement. Part 3, Finished Fabrics for use with Polyester Resin Systems* (1970).
14. B.S. 3691—*Glass Fibre Roving for the Reinforcement of Polyester Resin Systems* (1969).
15. B.S. 3749—*Woven Glass Fibre Roving Fabrics for the Reinforcement of Polyester Resin Systems* (1964).
16. B.S. 3496—*Glass Fibre Chopped Strand Mat for the Reinforcement of Polyester Resin Systems* (1962).
17. *Brit. Pat. 978042*; *Brit. Pat. 1035069*: Owens-Corning Fiberglas Corporation (1964 and 1966).
18. B.S. 3532—*Unsaturated Polyester Resin Systems for Low Pressure Fibre Reinforced Plastics* (1962).

11

Specifications

K. PARVIN

1. INTRODUCTION

A *specification* has been defined as 'a concise statement of the requirements for a material, process, method, procedure or service including, whenever possible, the exact procedure by which it can be determined that the conditions are met within the tolerances specified in the statement: a specification does not have to cover specifically recurring subjects or objects of wide use, or even existing objects.'[1]

The same body, the American Standards Association or A.S.A. (now the United States of America Standards Institute or U.S.A.S.I.) has defined a *standard* as 'a specification accepted by recognised authority as the most practical and appropriate current solution of a recurring problem.'

Specifications can take several forms as will be seen later, but why do we need specifications? Standards have been known since the advent of man when his need to communicate with other men led to the use of standard sounds, i.e. a language. This gradually led to a standard set of symbols for written purposes and hence an alphabet. As civilisation progressed there came the need for standard units of length and weight. One of the earliest units was the Egyptian cubit which was the distance from the average man's fingertip to his elbow. This seems very inexact today and it was not until the sixteenth century that the length of a certain bronze bar was decreed to be the British standard yard.

Until the Industrial Revolution articles were produced by individual craftsmen but with the advent of mass production it became essential to produce interchangeable parts, each manufactured to material and dimensional standards. In 1841 Joseph Whitworth advocated standardisation to achieve interchangeability of threaded parts in the engineering industry, but it was not until 1901 that the Engineering Standards Committee was formed. This was the forerunner of the British Standards Institution which was formed in 1931. Today all the major countries of the world have standards organisations and 40 are listed by Houghton[2].

2. TYPES OF SPECIFICATIONS

There are five main sub-divisions:

1. Dimensional standards which specify the dimensions needed to achieve interchangeability.
2. Standards of performance or quality which ensure that a product is adequate for its designed purpose.
3. Standard methods of test enabling materials and products to be compared.
4. Standard terminology or abbreviations to enable technologists to communicate.
5. Codes of practice which lay down methods of accomplishing an operation.

So far few performance standards have been published in the reinforced plastics field for, as mentioned by Everard[3], reinforced plastics are still at the stage when knowledge of performance is being accumulated. Indeed the standards which are published are too often written down to the lowest possible levels in order to satisfy all the parties involved in writing the specifications. It will take a further period before there is sufficient information on material properties to allow the materials to be used with confidence and low safety margins.

3. WHO ISSUES SPECIFICATIONS?

An apparently bewildering array of bodies issue specifications, from companies through industries, trade organisations, technical societies, government departments and national bodies to international organisations. Company specifications are often based on national standards with modifications, where necessary, to suit the needs of the company. They normally include raw material specifications, product specifications and manufacturing specifications together with standard test methods. Many of these are not published for reasons of commercial security.

Semi-official bodies publishing specifications covering reinforced plastics include Lloyd's, the Ship and Boat Builders National Federation (S.B.B.N.F.), the American Society for Testing and Materials (A.S.T.M.), Society of Plastics Industry (S.P.I.), National Electrical Manufacturers Association (N.E.M.A.), and The Society of Automotive Engineers (A.M.S.). These vary from dimensional standards to codes of practice and the addresses of these organisations and others are given by Yescombe[4].

British government specifications include D.E.F. Specifications and D.T.D. Specifications available from H.M.S.O., D.Eng.R.D. Specifications issued by the Ministry of Technology and Defence Standards, C.S. and T.S. specifications available to contractors from the Chemical Inspectorate, Woolwich.

There is also a series of Approved Firm's Schedules (AFS series) which are specific to a product of one manufacturer. These documents are confidential to the firms named to whom applications should be made for their trade literature.

In the U.S.A. there are a whole series of Military Specifications (MIL Series) and Federal Specifications (FED Series).

National standards include British Standards Institute specifications (B.S. Series), the U.S.A. Standards Institute Specifications (ASA Series) and the Deutsche Normenausschuss (DIN Series).

Internationally there is the International Organisation for Standardisation (ISO) which is an agency of the United Nations Organisation devoted wholly to standardisation. This body issues ISO recommendations which are available in the Library of the British Standards Institution.

4. GLASS FIBRE SPECIFICATIONS

There are four current B.S. Standards and two D.T.D. specifications which cover the whole range of glass fibre products. These are as follows:

B.S. 3396 Woven Glass Fibre Fabrics for Plastics Reinforcement
Part 1: 1966: Loom State Fabrics
Part 2: 1966: Desized Fabrics
Part 3: 1966: Finished Fabrics for use with Polyester Resin systems

(This specification replaced DTD 5518 and DTD 5519 which were withdrawn.)

B.S. 3496: 1962: Glass fibre chopped strand mat for the reinforcement of polyester resin systems
B.S. 3691: 1963: Glass fibre rovings for the reinforcement of polyester resin systems
B.S. 3749: 1964: Woven glass fibre rovings fabric for the reinforcement of polyester resin systems
D.T.D. 5546: 1959: Woven glass fibre tape and webbing
D.T.D. 5559A: 1959: Contour woven glass fibre fabric

4.1. GLASS FIBRE FABRICS

B.S. 3396 is currently under revision and the following points are worth noting:

(*a*) It only covers fabric for use with unsaturated polyester resins.
(*b*) 'A' glass fabric could be omitted as it is not used.
(*c*) Part 3 refers to chrome and silane finished fabrics without attempting to specify the type of silane used.
(*d*) The resin for making the test laminate is not specified.

The corresponding A.S.T.M. specifications are more precise and cover a wider field, namely:

A.S.T.M. 2408—67 Woven glass fabric (specifies an amino-silane finish for use with epoxy or phenolic resins)
A.S.T.M. 2409—67 Woven glass fabric (specifies a silane (A172 or Garan) finish for use with polyester resins)
A.S.T.M. 2410—67 Woven glass fabric (specifies a chrome complex finish for use with epoxy, phenolic or polyester resins)

A.S.T.M. 2660—67 T Woven glass fabric (specifies an acrylic-silane for use with polyester resins)

These specifications can be used to highlight some of the difficulties in writing specifications which are unnecessary limitations and insufficient specification.

Taking the B.S. Specification a new finish which was neither chrome or silane based could not comply even though the performance requirements of the finished fabric were met. Secondly, differing polyester resins can give different results obtained on various fabrics unless a standard resin is used.

Other American specifications covering glass fabrics are:

MIL–C–9084 Finished glass cloth for polyester resin laminates
MIL–F–9118 Finish for glass fabric (Chemical composition not stated)
MIL–F–12298 (MV) Woven glass fabric (chrome finished for use with phenolic resins)
A.M.S. 3825 Glass fabric (181) (Chrome treated)
A.M.S. 3835 Glass fabric (181) (Decorative grade)

In Germany there is also a series of standards in the 8,4500 series on glass fibre fabric of 'E' glass issued by Normenstelle Luftfahrt.

4.2. GLASS FIBRE MATS

As regards mats B.S. 3496 again has some drawbacks. The first is the fact that the filament diameter of the fibres used is specified: this is unnecessary and undesirable as filament diameters are tending to increase. Secondly, the specification is directed towards material for use with polyester resins but the resin type is not specified.

On the other hand, the American MIL–M–15617 does not specify the fibre to be used or require a performance in laminate form but specifies six different types of binder. MIL–M–43248 (MR) lists more types of mat but has no performance requirements at all.

A.S.T.M. D 1529–58T covered four types of mat but has now been discontinued.

4.3. GLASS FIBRE ROVINGS

B.S. 3691 covers both 'A' and 'E' glass rovings for use with polyester resins. The type of finish is loosely stated as methacrylatochromic chloride and/or a silane. MIL–R–60346 (MV) on the other hand, covers rovings for use with epoxy resins for filament winding. It is the only specification covering 'S' glass.

B.S. 3749 includes both 'A' and 'E' glass roving fabrics and specifies a silane or modified silane keying agent. The performance of rovings taken from the warp and the weft and tested as for rovings to B.S. 3691 is specified. In contrast MIL-C-19663 (Navy) does not specify the keying agent but requires minimum strength figures in laminate form. A.S.T.M. D 2150–63T is very similar.

New Zealand has issued B.S. 3691 and B.S. 3749 as N.Z.S.S. 1875/1964 and N.Z.S.S. 1972/1965 respectively.

4.4. OTHER GLASS FIBRE PRODUCTS

Other miscellaneous glass specifications are D.T.D. 5546 covering woven glass fibre and webbing which stipulates a residual size content plus a minimum warpway breaking strength. D.T.D. 5559 A covers contour woven glass fibre fabric and MIL-Y-1140 covers yarn, cord, sleeving, cloth and tape glass and includes staple fibre yarn, cordage, cloth and tape.

In Germany DIN 60850/1960 covers glass fibre yarns whilst A.S.T.M. D2343–67 gives a test for tensile properties of glass fibre strands, yarns and rovings used in reinforced plastics.

5. POLYESTER RESIN SPECIFICATIONS

B.S. 3532 covers five types of polyester resins namely general purpose, improved general purpose, heat resistant, low flammability and very low flammability. D.T.D. 5537/D.T.D. 5549 can be considered complementary, extending the range to resins for use up to 150°C, 200°C, and 260°C. Both of these specifications cover liquid resin properties and have performance requirements in laminate form. Whilst some of the test methods laid down in the specifications agree others are different. The D.T.D. specifications also call for electrical, chemical resistance and corrosive ingredients testing.

Two points that are worth noting with these specifications are:
(*a*) The requirement for the gelation time, after storage for a minimum of 3 months, to remain within 30 per cent of the original is difficult to meet especially in the case of pre-accelerated resins which have become more popular. (*b*) The non-specification of a particular finish for the glass fabric to be used in making test laminates can lead to misunderstandings. This is similar to the position already mentioned in the case of B.S. 3396 on glass fabric.

The Americans have several specifications covering polyester resins, namely, MIL-R-7575 which is similar to D.T.D. 5537, MIL–R–25042 high temperature resistant resins which approximates to D.T.D. 5537 Class F and MIL–P–43038 (Ord) which also covers high temperature resistant resins.

Other than these the only countries having specifications for polyester resins are New Zealand (N.Z.S. 1964, a copy of B.S. 3532) and Israel (SI 541).

6. OTHER RESIN SPECIFICATIONS

Although polyesters are used in the overwhelming majority of GRP applications specifications exist for other low pressure laminating resins.

In the epoxy field there is B.S. 3534 which is split in two parts: Part 1 dealing with wet lay-up systems and Part II covering pre-impregnating systems. Part 1 is similar in many respects to B.S. 3532 but also includes electrical tests. Once again the finish to be used on the glass fabric for making test laminates is not specified.

The corresponding American Specifications are MIL–R–9300 and MIL–P–43398 (GL). Both have performance requirements in laminate form whereas A.S.T.M. D1763–67 only covers liquid epoxy resins. U.S. specifications also exist for phenolic resins, MIL–R–3745 and MIL–R–9299 (ASG), and silicone resins MIL–R–25506 (ASG).

Other foreign specifications are limited to the Russian GOST 10587/1963 for uncured epoxy resins.

7. SPECIFICATIONS FOR GLASS REINFORCED PLASTICS

The use by the moulder of consistent materials to certain specifications is only the first step in the case of GRP. Specifications must be laid down for the mouldings and it is in this area that we are sadly lacking. It is possible to specify materials, shape, thickness, etc., but the quality of the final laminate will depend on the skill of the moulder and most people in the industry have seen very poor laminates made with the best resins and glass reinforcement available. One solution to this situation is to formulate a code of practice. This is done by the Ship and Boat Builders National Federation. Their 'Code of Recommended Practices for Boats up to 20 ft in Length' gives the type of resin, glass reinforcements and working standards using B.S. where applicable. No tests are given for the final product.

On the other hand, the performance of the moulding can be specified but firstly it is difficult to test mouldings non-destructively and secondly, the versatility of GRP to give laminates with a wide range of mechanical, electrical and chemical resistant properties could lead to a large number of specifications.

The fact that corrugated roof sheeting is the largest individual market for GRP is reflected in the number of specifications from all over the world covering this usage. In Great Britain there is B.S. 4154 'Corrugated Plastics Translucent Sheets made from Thermosetting Polyester Resins' whilst there are no less than three American specifications:

LP–505 Plastic Sheet, Corrugated, Translucent, Glazing
C.S. 214–57 Glass Fibre Reinforced Polyester Corrugated Structural Plastic Panels
A.S.T.M. D1919-65T Glass Fibre Reinforced Thermoset Corrugated Structural Plastic Panels

Similar foreign specifications are:

France P38–301 and 302 Glass Fibre Reinforced Corrugated Polyester Sheet
Israel SI 567 Reinforced Polyester Corrugated Sheet
Australia A 66 Glass Fibre Reinforced Polyester Plastic Building Sheets
Canada 41–GP–6a Sheets: Translucent Thermosetting Plastics, Glass Fibre Reinforced

The general laminate specifications can best be shown in the Table 11.1.

In addition to these there are a number of American specifications written for particular end uses:

MIL–W–10505 (Ord) Water Tank, Plastic Laminated
MIL–B–13592 (CE) Boats, Assault Plastic 16 ft
MIL–R–19907 Repair Kit, Glass Reinforced, Plastic Laminate (Epoxy Resin)
MIL–T–25363 Tank, Pneumatic Pressure, Aircraft, Glass Fibre
MIL–G–46095 (MV) Grommets, Plastic Filament Wound
MIL–R–81202 (WP) Receiver Ass, Nitrogen, Glass Filament Wound Construction

Table 11.1

Country and Spec. No.	Laminating resin						Comments
	Polyester	Epoxy	Melamine	Silicone	Phenolic	PTFE	
U.K.							
B.S. 3953	*	*	*	*	*		
U.S.							
L.P. 383	*						
L.P. 509	*	*	*	*	*		
MIL-P-997				*			
MIL-P-8013	*						Superseded by L.P. 383
MIL-T-10652	*						
MIL-P-13949	*	*				*	Copper clad
MIL-P-15037			*				
MIL-P-17549	*						
MIL-P-18177		*					
MIL-P-19161						*	
MIL-G-21729		*					
MIL-P-22245		*					
MIL-P-22693	*	*	*		*		
MIL-P-25395	*						
MIL-P-25421		*					
MIL-P-25515					*		
MIL-P-25518				*			
MIL-P-46040					*		
ASTM-D-1532-67T	*						
ASTM-D-709-67		*	*	*	*		
ASTM-D-1867-64T		*					Copper clad
AMS 3598						*	Copper clad
AMS 3601		*					Copper clad
AMS 3642		Not specified					
AMS 3666						*	
Germany							
51400							Acrylic
52000	*						
52001	*						
52002	*						
52003	*						
52005	*						
52006	*						
52007	*						
52400				*			
52404				*			
52405				*			
52406				*			
52407				*			
53104			*				
53200		*					
53204		*					
53205		*					
53206		*					
53207		*					
DIN 7735							
Japan							
K 6905		*					
U.S.S.R.							
GOST 10292							

C.S. 221–59 Gel Coated Glass Fibre Reinforced Polyester Resin Bathtubs
C.S. 222–59 Gel Coated Glass Fibre Reinforced Polyester Resin Shower Receptacles
A.S.T.M. C 582–68 Reinforced Plastic Laminates for Self-Supporting Structures for use in a Chemical Environment
A.S.T.M. D 2310–64T Reinforced Thermosetting Plastic Pipe
A.S.T.M. D2517–67 Reinforced Thermosetting Plastic Gas Pressure Pipe and Fittings

In Great Britain a draft specification is being considered for Light Duty Vessels, Storage Tanks and Associated Equipment in Reinforced Plastics. This covers design, construction and erection, testing and control of materials and inspection.

Dough moulding compounds are covered by:

B.S. 3840: 1965 Polyester Dough Moulding Compounds
A.S.T.M. D10201–62 Polyester Moulding Compounds
DIN 16911 Polyester Moulding Compounds

There are no British Specifications for pre-impregnated materials but two American specifications exist for 'B' stage materials together with an American and a draft German specification for pre-impregnated glass mats. These are as follows:

MIL–P–23944 (Wep)	*Tape, Pre-impregnated Bias, Non-woven Continuous Glass Filament Reinforced Phenolic Resin*
NEMA VF10	*Glass Polyester 'B' Stage Materials*
MIL–P–43127 (MR)	*Plastic Pre-impregnated Glass Mat, Polyester Resin*
DIN 16913 (Draft)	*Plastics: Types of Prepregs impregnated with polyester resins*

Some American Specifications also exist for sandwich constructions, namely:

MIL–C–8073	*Core Material, Plastic Honeycomb, Laminated Glass Fabric Base*
MIL–S–9014	*Sandwich Construction: Plastic Resin, Glass Fabric Base, Laminated Facings and Honeycomb Core*
MIL–S–25392	*Sandwich Construction, Plastic Resin, Glass Fabric Base, Laminated Facings and Polyurethane Foamed-in-Place Core*
A.M.S. 3710	*Sandwich Structures, Glass Fabric Resin, Low Pressure Moulded, Heat Resistant*

One point that should be borne in mind by the writers of specifications and codes of practice is the undesirability of specifying a particular product, e.g. Blogg's Polyester 999 or Smith's Glass Mat 123. Over the years resin and glass suppliers' products change and such materials are often supplanted by improved versions. If batches of mouldings are required perhaps only every 2 years, the moulder finds that the original raw materials have become obsolete by improved products and the small amounts needed of the original would be more expensive as they have to be specially made. The addition of the words '. . . or equivalent product' makes the difference as the improved resin or glass product can usually be shown to have the same performance.

8. STANDARD TEST METHODS

Many of the specifications already mentioned contain standard methods of test but there are some other specifications which are solely methods of test. B.S. 2782:1965 and amendments is entitled 'Methods of Testing Plastics' and covers all plastic materials. Federal Specification LP 406 is the corresponding American specification. Other specialised methods of test have been published and amongst those for reinforced plastics are:

A.S.T.M. D 1494–60 Diffuse Light Transmission of Reinforced Plastics Panels
A.S.T.M. D 1502–60 Transverse load of Corrugated Reinforced Plastic Panels
A.S.T.M. D 1602–60 Bearing Load of Corrugated Plastics Panels
A.S.T.M. D 2105–67 Longitudinal Tensile Properties of Reinforced Thermosetting Plastic Pipe and Tube
A.S.T.M. D 2143–63T Cyclic Pressure Strength of Reinforced Thermosetting Plastic Pipe
A.S.T.M. D 2563–66T Classifying Visual Defects in Glass Reinforced Laminates
A.S.T.M. D 2733–68T Interlaminar Shear
A.S.T.M. D 2734–68T Void Content

The B.S.I. are currently considering a draft standard on test requirements for reinforced plastic mouldings. No performance specification is given but it covers certain properties of mouldings and methods for their determination.

Some of the tests vary considerably and it is worth considering these in more detail.

8.1. GELATION TIME

B.S. 3532 calls for the measurement of gelation time at 25°C and at 82°C. The former is made using a Techne gelation timer and, whilst it works satisfactorily with normal polyester resins, it is not suitable for highly thixotropic gel coats and the gel times of these are often measured manually.

The 82°C gelation time is measured manually and no account is taken of the peak exotherm. By comparison the S.P.I. Standard 180°F exotherm curve[6] gives more information obtained only by temperature/time measurements. Ten gramme samples of resin catalysed with 1 per cent benzoyl peroxide are placed in a bath at 180°F with a thermocouple in the resin. The temperature of the resin sample is recorded and a curve of the type shown below is obtained (*Figure 11.1*).

The time from 150°F to 190°F is taken as the gelation time and the time from 150°F to peak exotherm is taken as the cure time. The peak exotherm is noted.

This method can be adapted to other temperatures which may mean the use of alternative curing systems. For example DIN 16945[7] uses either 3 per cent benzoyl peroxide paste + 1 per cent of a 10 per cent solution of dimethylaniline or 2 per cent ethyl ketone peroxide (50%) + 0.3 per cent cobalt accelerator (1% cobalt) at 25°C and the time intervals between 25°C and 35°C and 25°C and peak exotherm are noted. Certain polyester manufacturers use the S.P.I. test as a standard control procedure and its wider use would be

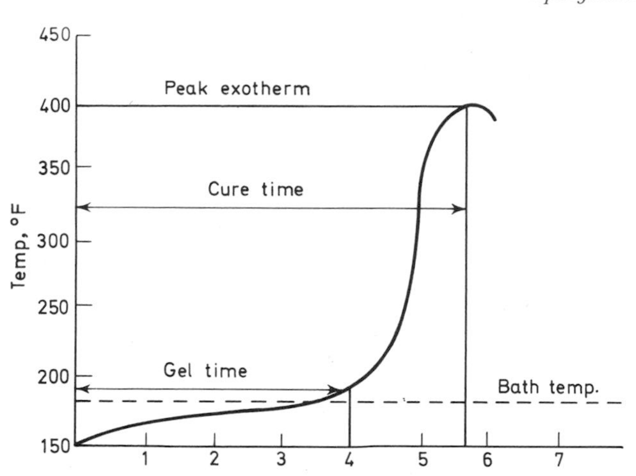

Figure 11.1. Resin sample temperature curve

an advantage as it can be used to compare the effects of various curing systems and inhibitors.

8.2. FLAMMABILITY TESTS

Probably more confusion exists in this field than in any other section of reinforced plastics testing. Scott in a recent review[8] lists over 40 different tests for the flammability of plastics.

As far as reinforced plastics are concerned, it is convenient to divide the tests into screening tests which can be made in the laboratory and design tests which can only usually be undertaken by few organisations having the necessary large scale facilities. Unfortunately, there appears to be no correlation between the two types of test.

Of the screening tests B.S. 476 Part 5: 1968 describes 'Ignitability Tests for Materials' which classifies materials as easily ignitable or not easily ignitable. It is to be used in conjunction with B.S. 476: Part 6:1968 which describes a 'Fire Propagation Test for Materials,' from which an Index of Performance is calculated. Experience has so far shown that consistent results are difficult to obtain on the second test in the case of plastics.

Most of the screening tests are burning tests, the two main tests used being B.S. 2782, Method 508 A and A.S.T.M. D 635–63.

These tests are usually carried out on laminates as is also the case with DTD 5537/5549 and British Railways Spec. No. 602 but B.S. 3532 Appendix B calls for resin casts.

The American HLT–15 test which requires the laminated test piece to be supported in the vertical plane is a much more stringent test.

However, none of these tests are as searching as full scale tests like A.S.T.M. E 84–68 which requires a specimen 21 in wide and 25 ft long, *B.S. 476: Part*

1: 1953 'Surface spread of Flame Test' which requires six specimens 3 ft long and 9 in wide or *B.S. 476 Part 3: 1958 'External Fire Exposure Roof Tests'* which requires six specimens 33 inches square. The trouble with these tests is that they require large areas for testing and few organisations have such equipment. An attempt was made to introduce a laboratory scale version of the *B.S. 476 'Surface Spread of Flame Test'* but results on the apparatus, described in the Appendix to B.S. 476 Part 1, do not correlate with the results obtained on the full scale apparatus in the case of reinforced plastics.

We are left in the position of having a large number of tests, each measuring a different property and none measuring one property that can be of importance, namely the evolution of toxic gases and smoke.

A comparison of the flammability of standard 'self-extinguishing' polyesters when tested by various methods would be valuable.

8.3. IMPACT TESTS

A word of warning must be introduced about the results of impact tests as misleading figures are sometimes quoted for reinforced plastics.

When a pendulum type of impact tester is used, the energy absorbed to break the specimen is higher than that to damage the specimen to such an extent that any similar structure could be useless. This is due to the fact that, in addition to the energy required to crack the resin, energy is also required to tear the glass fibres apart or out of the resin. In the case of glass fabric or mat reinforced polyesters this can be considerably more than the energy required to crack the resin and make the structure permeable to water or other environment.

8.4. CHEMICAL RESISTANCE

Chemical resistance is another area where the results of various tests are difficult to interpret in practice. B.S. 3532, Appendix M gives a test method for determining the chemical resistance of resin casts which involves changes in dimensions and weight, but unreinforced polyesters are rarely used. *A.S.T.M. D 543–67 'Test for Resistance of Plastics to Chemical Reagents'* covers the same test on laminates 3 in × 1 in × $\frac{1}{8}$ in.

A.S.T.M. 'C581–68 Chemical Resistance of Thermosetting Resins used in Glass Fibre Reinforced Structures' is more realistic using mat laminate specimens 4 in × 5 in × $\frac{1}{8}$ in. After exposure these are cut into three test specimens 4 in × 1 in which are used for flexural strength tests.

In practice most applications are in the form of tanks and three main differences between the conditions of test and practical conditions become evident: (*a*) Cut edges are not normally present in practice. (*b*) Under test conditions attack is on both faces of the specimen. (*c*) Surface to volume ratios vary considerably.

Some tests have been made using laminates as the ends of cylindrical glass cells containing the chemical and this gives far more realistic results. After exposure the laminates can be cut into test specimens and certain mechanical tests can be made. Probably the most useful information is the change in flexural modulus after exposure.

8.5. TOXICITY

Toxicity can be considered as the converse of chemical resistance, that is what possible toxic constituents can be leached from GRP in contact with foodstuffs? The whole question of the use of polyesters in contact with foodstuffs is complex and a B.P.F. sub-committee is currently considering the position. In other countries, notably Germany and the U.S.A., similar works to codes of practice have been published[9]. Some of these include extraction tests on the finished mouldings with extractants simulating common foodstuffs.

8.6. COLOUR

Many resins are sold today in coloured form and the accuracy of colour matching from batch to batch becomes important especially in the case of gel coats. For many years colour matching has been done by skilled colour matchers. However, the matching of colours depends on the light under which they are viewed and the eye of the colour matcher. People have therefore looked at scientific methods of comparing colours and instrumental methods in which a value is given to the colour difference between a sample and a standard are now available. The main difficulty to be overcome is how to set tolerances between batches and further work is required before instrumental methods replace the human eye.

9. METRICATION

It is now the intention that the U.K. will convert to a metric system by 1975. This will mean that we shall all have to rethink in metric units and many of our day-to-day expressions such as a flexural strength of 22.4×10^3 p.s.i. will become obsolete.

Perhaps as technologists we shall be less troubled than the layman, but the proposed use of International System (SI) units for technical purposes is likely to cause some confusion.

This system was adopted by a resolution of the 10th Conférence Général des Poids et Mesures in 1954 and endorsed by the International Organisation for Standardisation (ISO).

There are six basic SI units as follows:

Length	metre	m
Mass	kilogramme	kg
Time	second	s
Electric current	ampere	A

Thermodynamic temperature	degree kelvin	K
Luminous intensity	candela	cd

From these a number of other units are derived, some having special names. Thus the unit of force becomes a Newton, which equals a kilogramme metre per second squared. Dynamic viscosity is expressed in Newton seconds per metre squared and kinematic viscosity in metres squared per second. For example 22.4 × 10^3 p.s.i. should become 154 400 Newtons/m^2 or 154.4 kN/m^2 as the multiples and submultiples of units have standard names and symbols. In other words to convert lb/in^2 to kN/m^2 multiply by the factor 6.89476.

More details are given in B.S. 3763:1964 and other publications[10, 11].

APPENDIX

Some of the points that have been considered earlier, namely that certain items are not specified closely enough and that the limits stipulated are sometimes lower than those commonly attainable can be illustrated by comparing specifications. B.S. 3396, B.S. 3532 and D.T.D. 5537 all call for glass fabric

Table 11.2

	B.S. 3396	*B.S. 3532*	*D.T.D. 5537*
Flexural strength MN/m^2			
Dry	414	345	379
2 h boil	352	276	310
Resin content (%)	35–42	35–40	35–40

laminates to be made using finished S/225/E glass fabric. However, as shown in Table 11.2 the requirements for wet and dry flexural strengths are not the same.

From the figures it can be seen that a resin according to B.S. 3532 or even D.T.D. 5537 would not necessarily be suitable for use in making test laminates to B.S. 3396. On the other hand, if resins to B.S. 3532 and D.T.D. 5537 easily give laminates with the required mechanical properties for B.S. 3396 then the limits set in the former specifications are too low.

REFERENCES

1. *Standardisation* 41, Jan., 10 (1960).
2. Houghton, B. *Technical Information Sources,* Clive Bingley, London (1967).
3. Everard, H. R. *6th Int. Reinforced Plastics Conf.,* B.P.F. Paper 7 (1966).
4. Yescombe, E. R. *Sources of Information on Rubber, Plastic, and Allied Industries,* Pergamon Press, London (1968).
5. Beach, N. E. *Plastic Laminate Materials,* Forster Publishing Co., California (1967).

6. Oleesky, S. S. and Mohr, J. G. *Handbook of Reinforced Plastics of the S.P.I. Inc.* p. 51, Reinhold, New York (1964).
7. Selden, P. H. *Glasfaserverstarkte Kunststoffe*, Springer Verlag, Berlin (1967).
8. Scott, K. A. *R.A.P.R.A. Technical Review* No. 27 (1965).
9. Lefaux, R. *Practical Toxicology of Plastics,* Iliffe, London (1968).
10. *The Use of S.I. Units* (*PD 5686*), British Standards Institution (1967).
11. Anderson and Brigg. National Physical Laboratory. *Changing to the Metric System*, H.M.S.O. (1966).

BIBLIOGRAPHY

Beach, N. E. *Guide to Test Methods for Plastics and Related Materials, Plastec Note 17*, U.S. Dept. Defence, Plastics Technical Evaluation Centre, Springfield, Va. (1967).

Lever, A. E. and Rhys, J. A. *Properties and Testing of Plastics Materials,* Temple Press, London (1968).

1968 Book of A.S.T.M. Standard Methods, Parts 26 and 27, A.S.T.M. Philadelphia (1968).

Part 4
Engineering Design

12

Nature of Composites

R. W. CAHN and B. HARRIS

1. INTRODUCTION

The three most important requirements for materials which are to be used in highly demanding load-bearing applications at normal temperatures are a high *resistance to plastic deformation,* a high *resistance to fracture,* and *elastic stiffness*. A material which combines the first two properties is said to be *tough*. If it is stiff as well, it is a paragon among structural materials.

The high resistance to plastic deformation has traditionally been achieved in two quite distinct ways. Either a fairly soft metal, such as copper or iron, is alloyed so that the passage of the crystalline defects which 'carry' deformation is blocked by the alloying additions. This is how aluminium is strengthened for use in airframe construction. Alternatively, one can use materials like glass or porcelain, which are intrinsically incapable of plasticity, because the aforementioned defects cannot move at all in such substances; these defects might as well not be there. Such materials can have a very large intrinsic strength, but this strength is useless, because the resistance to fracture is poor. This in turn is due to the fact that such intrinsically strong, brittle materials contain minute, invisible cracks at the surface; the cracks magnify any applied load to such a degree that they break easily, just as a piece of cloth or paper tears easily if—and only if—a deep narrow cut is first made in it. The ubiquitous minute cracks are known as 'Griffith cracks', after A. A. Griffith, a distinguished Rolls-Royce engineer, who first postulated their existence and deduced their consequences. Their universality is now established beyond doubt.

Materials strengthened by dispersing small hard particles in an intrinsically soft matrix, thus raising the resistance to plastic deformation, have a long history of development, and this line of advance has neared its limits. The great benefit of this strategy is that the soft matrix neutralises the Griffith cracks as soon as a load is applied, because the roots of the cracks, which are the sensitive sites, are blunted by localised plastic distortion before the crack can begin to spread. (A good analogy is to make a 'starter cut' in a sheet of paper and then punch a round hole at the tip of the cut. The cut will then be found to have lost its effectiveness as a tear-starter.) Dispersion-hardened matrices are currently also the only reliable way of achieving good strength at high temperatures, for instance for gas-turbine blades (as opposed to compressor blades, which operate cool). Aluminium alloys for aircraft frames and all the more elaborate structural and special steels fall in this category.

2. CERAMICS AND WHISKERS

One strategy to increase the strength of constructional materials is to use those which are brittle (i.e. incapable of plastic deformation) and have a high intrinsic resistance to fracture—that is, that in the absence of Griffith microcracks, would be exceedingly strong—and to use these materials under conditions which minimise the dangers from the microcracks which are inevitably present. (There is no point in striving to avoid microcracks altogether, since even a brush with the fingertips will generate them!) Thus glass, which is currently undergoing a renaissance, is now available in forms which are enormously resistant to fracture in compression (hence its imminent use for submersibles), and by putting the surface layer in compression, as in toughened windscreens, the influence of surface microcracks can be reduced and glass can be made very attractive for some uses. However, designing with strong brittle materials—*ceramics* is their family name—is as yet a very tentative skill: In December 1968 the Institution of Mechanical Engineers held a meeting on this topic, and judging from published reports the outlook in this field is not currently promsing, though there are enthusiastic supporters of particular materials for particular uses, especially silicon nitride, which has been impressively developed by the Admiralty Materials Laboratory. For one thing, the statistical scatter in the properties of apparently identical pieces of ceramic is a major difficulty.

The spectacular *intrinsic* strength of some kinds of ceramic was first widely recognised when people started growing *whiskers*, that is, fine filamentary crystals of various crystalline solids which are free of defects and of Griffith cracks. These were at first purely a scientific curiosity, and the main interest was in understanding how to grow them, or in some instances (as in tin foil condensers) how to stop them growing and short-circuiting the components. Then people began to recognise that so long as they are handled with care and their surfaces not touched, whiskers are by far the strongest materials known—even a pure metal, in filamentary form, can be stronger than a bar of high-tensile steel, in terms of fracture stress. Plastic deformation cannot happen because the necessary *internal* defects are entirely absent. It was soon realised that the very highest intrinsic resistance to fracture was restricted to a small group of substances containing elements such as carbon, boron, silicon, aluminium, oxygen and nitrogen, which form very strong interatomic chemical bonds and have small interatomic distances. For a while, a few years ago, there was great excitement at the prospect of using the great strength of whiskers for constructional purposes. It was immediately realised that this was feasible only by putting these whiskers into a soft matrix. This was necessary for several reasons: to enable load to be conveniently transferred to the whiskers, to protect their surfaces from crack-formation, to balance the statistical fluctuation in whisker strength and (because the matrix was to be soft and deformable) to assure toughness, so that fracture would not spread from an accidental cut or notch in the composite. All these principles were well-established in connection with glass fibre, but the innovation now was that the best whiskers were considerably stronger than glass fibre. Wood is a natural material which owes its admirable toughness to the fact that it is a natural fibre composite.

Much work has been done, mostly in the U.K. and U.S.A. on the production of whiskers, their incorporation in composites and the mechanics of the composites themselves. In Great Britain, Government laboratories, especially E.R.D.E. and more recently A.W.R.E., have been very active, and there was a very lively group of investigators at Cambridge University, whose leader, Dr. A. Kelly, is now a Deputy Director at N.P.L. Rolls-Royce at Derby were active in the field at an early stage and it was because of this hard-won expertise that they were so quick to seize on the potential of carbon fibre when this was first perfected.

3. CARBON FIBRES

It was realised after a while that the production costs of ceramic whiskers (silicon carbide, silicon nitride and aluminium oxide were much favoured) would never make whisker-reinforcement economically feasible. This is because the standard methods of growing whiskers are intrinsically slow, batch processing cannot be avoided, whiskers are normally at most a few inches long and they grow in tangled masses which are very hard to separate and align without serious damage. However, a novel breakthrough is always possible, and it is significant that very recently a U.S. Air Force laboratory demonstrated sapphire whiskers 30 m long. The race was then on to find other ways of making long fibres stronger than glass and cheaper than whiskers. They might be somewhat less strong than the single-crystal whiskers and yet be very useful. The materials worth considering were limited by the aforementioned chemical bonding considerations, and pure elements were most likely to be easy to prepare. The Americans chose boron largely because it is easily deposited by the thermal decomposition of vapours (the same principle as is used in the long-life quartz-iodine lamps in which evaporated tungsten forms a volatile iodine compound which at once deposits tungsten at the hot filament); in fact, boron fibres are grown on fine tungsten wires. R.A.E., which had an existing expertise in graphite, chose carbon. The notion of breaking down plastic precursors to form carbon has a respectable ancestry, but the R.A.E. team were the first to bring it to perfection. (A corresponding technique is to produce large pieces of glassy or 'vitreous' carbon by firing pieces of plastic chemically rather similar to the Courtelle fibres used for making carbon fibres. Vitreous carbon is a remarkable material, pioneered in this country by Plessey, and now manufactured by a subsidiary, Vitreous Carbons Limited, of Evenwood, Bishop Auckland, Co. Durham. It is also made in U.S.A., France and Japan. This material, until recently unfamiliar, is rapidly finding applications: many are summarised in *Design Engineering* for March 1970.

A property mentioned at the beginning of this summary was *elastic stiffness*. This is as important as strength in most structural applications, especially in aircraft manufacture. It has been pointed out that if normal aircraft materials could be stressed without fracture to the level that whiskers can sustain, then wings of normal construction would be bent almost into semicircles! R.A.E.

Table 12.1

MECHANICAL PROPERTIES OF REINFORCING FIBRES

(Materials arranged in order of increasing specific stiffness)

Material	*Specific gravity* ρ	*Tensile strength* σ $kN/m^2 \times 10^4$	*Young's modulus* E $kN/m^2 \times 10^7$	*Specific strength* σ/ρ $kN/m^2 \times 10^4$	*Specific modulus* E/ρ $kN/m^2 \times 10^7$
Drawn Nylon 66	1.1	83	0.5	75	0.5
Tungsten, cold drawn wire 0.001 in dia.	19.3	310	41	16	2.1
High tensile steel wire	7.87	130	21	16	2.7
E glass fibre	2.54	170	7	67	2.8
S glass fibre	2.54	260	8	104	3.2
Asbestos fibre	2.5	580	19	232	7.6
Boron fibre (0.003 in dia.)	2.65	350	34	134	12.8
Carbon fibre, type 2	1.74	300	23	174	13.2
Carbon fibre, type 1	2.0	200	40	100	20
Sapphire whisker (6μ dia.) (maximum values)	3.96	4300	206	1090	52

Note that the order would be slightly different on a specific strength basis.

for instance has for many years sought to produce materials for aircraft construction with particularly high elastic stiffness, and thus they were well placed to recognise the vital importance of this parameter. Fortunately, the chemical bonding characteristics that make for high resistance to fracture normally also make for high stiffness, though (as the existence of Types I and II carbon fibres show) to some extent the fibre can be designed either for optimum stiffness or for optimum strength. These both vary with the geometrical arrangement of the minute carbon crystallites in the fibre, and it has been the skill of R.A.E., Rolls-Royce and other teams concerned to optimise this arrangement in various ways. Glass fibre can be made as strong as carbon fibre, but not nearly so stiff.

Even at the present price in the U.K., carbon fibre is a good deal cheaper than any available grade of whiskers, and the understandable pessimism which many expressed as to the economics of using whiskers is less apposite to carbon fibre reinforced plastics. If the price could be lowered ten- or twentyfold, the situation would be transformed. Awareness of economics is very acute, and comparisons have to be done in a sophisticated way: for instance, a recent paper from I.C.I. (who are much concerned in the matter because they, among other firms, make the matrix plastics) calculates the cost-effectiveness of composites using various fibres of different stiffness and cost, in varying concentration in the matrix, on the criterion that under a given load a panel should deflect by not more than a specified amount. However, such a stiffness criterion is only one among several: strength comparisons, impact resistance (important in the turbine compressor blade), cost of fabrication, ease of repair, corrosion-resistance, can all be of importance.

Table 12.2

MECHANICAL PROPERTIES OF COMPOSITE MATERIALS

(Materials arranged in order of increasing specific stiffness)

Material	*Specific gravity* ρ	*Tensile strength* σ $kN/m^2 \times 10^7$	*Young's modulus* E $kN/m^2 \times 10^7$	*Specific strength* σ/ρ $kN/m^2 \times 10^4$	*Specific modulus* E/ρ $kN/m^2 \times 10^7$
Nylon 66 reinforced with 20 vol.% Courtaulds type A carbon fibre	1.22	8.1	1.2	6.6	1
Epoxy resin with 70 vol% E glass	2.1	107	4.8	51	2.3
Epoxy with 14 vol.% Al_2O_3 whiskers	1.64	78	4.1	48	2.5
Titanium alloy T/AM (Rolls-Royce)	4.5	45	11.7	10	2.6
Aluminium alloy A/FLS (Rolls-Royce)	2.67	31	7.2	12	2.7
Epoxy with 70 vol.% S glass	2.11	207	6.1	98	2.9
Epoxy with 40 vol.% type I carbon fibre (Courtaulds)	1.54	66	16	43	10
Epoxy with 50 vol.% type I carbon (Morganite)	1.92	90	21	47	11
Epoxy with carbon fibre. Rolls-Royce unspecified composition	1.68	62	28	37	17

Some of the data in this table were taken from Courtaulds and Morganite advertising literature and some from a Rolls-Royce paper in *Nature*. Others are from the book 'Modern Composite Materials' Ed. by Broutman and Krock, Addison-Wesley, 1968.

It is probable that insufficient attention has as yet been paid to *precise* methods of assessing the cost-effectiveness of carbon fibre reinforced plastics as a function of the costs of fibre and of fabrication, and it is important that this should be done by Government laboratories or Research Associations, so that the calculations should be widely available. Everything depends on price: if the price of fibre and of fabrication methods (both are equally important) can be brought down sufficiently by economies of scale and technical improvements, then one day plastics reinforced with carbon fibres may become industry's 'man-made timber', with all the virtues and none of the drawbacks of natural timber, used as widely as timber is now but capable of supporting enormously greater loads. After all, both materials largely consist of carbon.

13

Predicting Mechanical Properties of Fibre Composites

P. H. H. BISHOP

1. INTRODUCTION

Fibre reinforced materials have great potential in reducing the weight of engineering structures, but unfortunately this is not always exploited in practice. The capacity to vary fibre orientation, resin type etc. should enable high efficiencies to be attained because a special material can be formulated for each task, but, in fact, the existence of these variables confuses the engineer and increases the difficulty of building up reliable design data. High safety factors tend to be incorporated which compromise the weight savings of which composite materials are capable.

Theoretical expressions can account for the effects of some of the variables, in particular fibre and resin volume fractions, and fibre orientation.

Predictions will not always agree with observation, but gross discrepancies will usually serve as a warning light, so that one can run through the arithmetic used in analysing observations and thus detect an error. If there is no error, but the discrepancy remains, the observed laminate must disagree with theoretical assumptions in some fundamental way, for example the fibres may be kinked whereas theory assumes them straight.

To what extent theory should be used 'blind', i.e. to what extent one is justified in taking a bold and costly step not easily experimentally verified until done, depends on circumstances. Theoretical expressions can often be adjusted semi-empirically to fit observation, and then extrapolated safely to make accurate quantitative predictions.

In this chapter expressions will be derived to do this, and it is hoped they will be of practical use.

2. RELATIONS BETWEEN BASIC LAMINATE PARAMETERS

2.1. GENERAL

Quantities such as density, weight fraction and volume fraction play a fundamental role in measurement and prediction. Their accurate measurement and calculation are as important as the accurate measurement of strength and stiffness, and should not be thought of as subsidiary. Only the formal algebraic relations are considered here; if high accuracies are required the practical measurements involved, for example, determination of fibre and resin densities on which calculations depend, is not simple.

A fibre composite material usually consists of the following components:

(*a*) fibre reinforcement in the form of plies of cloth, mat or layer of filaments. Fibre may carry a surface finish and contain voids to which the laminating resin cannot penetrate. Let the apparent density of the fibre, including surface finish and internal voids, be ρ_f*. After preparation of the laminate let each reinforcement ply have a laminated thickness t cm and a weight u g/cm^2; t is influenced by resin surface tension effects and is reduced by external pressure; u is a function of filament size, density and spacing.

(*b*) *resin* to which are added curing agents. The density of the cured resin, including any solid catalyst, but excluding voids or additional filler, is ρ_{re}.

Composites may contain also

(*c*) *inert filler* to modify resin properties or to extend the resin. The density ρ_q used in prediction is the intrinsic density of the filler (averaged to include internal voids, not reached by the resin). It is not the bulk density of the form in which the filler is added to the resin.

The term 'matrix' will be used to denote the cured mixture of resin and filler in position in the laminate, but not including fibre reinforcement or laminate voids.

2.2. WEIGHT AND VOLUME FRACTIONS

In the prediction of mechanical properties, volume fractions are used. These cannot usually be measured directly and are estimated from weight fractions. Let 1 g of matrix contain R g of resin and Q g of filler. It has volume $(R/\rho_{re} + Q/\rho_q)$ where ρ_{re} and ρ_q are densities of the resin and filler respectively. Density of the matrix is given by

$$\rho_m = (R/\rho_{re} + Q/\rho_q)^{-1} \tag{1}$$

When the matrix is applied to the laminate, voids may be formed or entrapped. The volume of fibre per unit area of one ply is u/ρ_f. The volume of voids is tv_v where v_v is the volume fraction of voids in the laminate. The volume V of matrix is given by

$$V = t - u/\rho_f - tv_v \tag{2}$$

The weight of resin per ply is $RV(\rho_m)$ and the weight of filler per ply is $QV(\rho_m)$. The total weight of 1 ply may be expressed as $u + V(\rho_m)$ or, alternatively, as $t\rho_L$ where ρ_L is the density of the laminate. In general, the above variables are known and values of laminate density, weight fractions w, and volume fractions v may be estimated from the following equations, where

* Symbol list, pages 187 and 188

suffices f, re, q and m denote fibre, resin, filler and matrix respectively:

$$\rho_L = (u + V\rho_m)/t \tag{3}$$

$$w_f = u/t\rho_L \tag{4}$$

$$w_{re} = RV\rho_m/t\rho_L \tag{5}$$

$$w_q = QV\rho_m/t\rho_L \tag{6}$$

$$w_m = V\rho_m/t\rho_L \tag{7}$$

$$w_v = 0 \tag{8}$$

$$w_m = 1 - w_f \tag{9}$$

$$v_f = u/\rho_f t \tag{10}$$

$$v_{re} = RV\rho_m/\rho_{re}t \tag{11}$$

$$v_q = QV\rho_m/\rho_q t \tag{12}$$

$$v_m = 1 - u/\rho_f t - v_v \tag{13}$$

$$v_v = 1 - (v_f + v_{re} + v_q) \tag{14}$$

$$v_v = 1 - (v_f + v_m). \tag{15}$$

If details of the reinforcement are not known, the relation between weight fractions, volume fractions and densities can be obtained by eliminating u and t from the above equations. This can be shown to give:

$$w_f = 1\Bigg/\left[1 + \frac{\rho_m}{\rho_f}\{1 - v_v)/v_f - 1\}\right] \tag{16}$$

$$v_f = (1 - v_v)\Bigg/\left\{1 + \frac{\rho_f}{\rho_m}\left(\frac{1}{w_f} - 1\right)\right\} \tag{17}$$

$$w_f = \left\{1 - \frac{\rho_m}{\rho_L}(1 - v_v)\right\}\Bigg/\left(1 - \frac{\rho_m}{\rho_f}\right) \tag{18}$$

$$v_f = \{\rho_L - \rho_m(1 - v_v)\}/(\rho_f - \rho_m) \tag{19}$$

Equation (17) allows fibre content by volume to be calculated from fibre content by weight; equation (19) allows fibre content by volume to be calculated from laminate density. ρ_m is calculated from equation (1), and it is necessary to make an arbitrary assumption concerning void volume fraction v_v, for example that it is zero. If, however, both laminate density ρ_L and fibre weight fraction w_f are known, the void volume fraction v_v can be calculated by re-arranging equation (18):

$$v_v = \rho_L\left\{w_f\left(\frac{1}{\rho_m} - \frac{1}{\rho_f}\right) - \left(\frac{1}{\rho_m} - \frac{1}{\rho_L}\right)\right\} \tag{20}$$

If v_v is unknown, or one is not able to estimate it accurately, it is preferable to assume a nominal value in calculation, say 0.05. To set v_v to zero is to assume a perfect laminate with no voids at all, which is unlikely.

An alternative form to equation (1) for the calculation of ρ_m is

$$\rho_m = (w_{re} + w_q)/\{w_{re}/\rho_{re} + w_q/\rho_q\} \tag{21}$$

If the total weight of the reinforcement is known, and none is lost during laminating and cleaning-up then obviously w_f is simply the ratio of reinforcement weight to specimen weight.

Otherwise, weight fractions may be determined by observing change in weight of the laminate after heating in air (ashing) at an appropriate temperature. The following expressions may be derived:

$$w_f = (r_L - r_{re}R - r_qQ)/(r_f - r_{re}R - r_qQ) \tag{22}$$

$$w_{re} = R(r_f - r_L)/(r_f - r_{re}R - r_qQ) \tag{23}$$

$$w_q = Q(r_f - r_L)/(r_f - r_{re}R - r_qQ) \tag{24}$$

$$w_m = (r_f - r_L)/(r_f - r_{re}R - r_qQ) \tag{25}$$

where r_f, r_{re}, r_q and r_L are the residues after ashing of 1 g of fibre, cured resin, filler and laminate respectively.

It is difficult to estimate weight and volume fractions for a laminate containing constituents in unknown proportions, if the resin contains an unknown amount of filler, or if the various components react chemically during ashing. In such cases the ashing must be followed by chemical analysis of the residues. If P and Q are known, a typical procedure would be to determine laminate density ρ_L and then weight fractions using equations (22) to (25).

In principle these equations are also applicable to laminates containing components such as silicone resin, which neither remain completely stable during ashing nor volatilise totally. To be on perfectly safe ground, even where it is known that chemical interaction will not occur, one must be sure that the physical proximity of the components in the ashed composite does not affect the values of r_f, r_{re} and r_q. This can be proved by taking a known weight of reinforcement in a tared ashing crucible, adding a reasonable proportion of resin/hardener/filler mixture, curing and weighing, thus preparing a model laminate of known proportions in the crucible itself. Subsequent ashing will give a weight for the combined residue which should agree with that calculated from the residues of the individual ashings using equations (22) to (25).

If the component densities are known, ρ_m may be calculated from equation (21) then v_v from equation (20). Fine algebraic differences may yield a negative value of v_v, in which case it is taken as zero. v_f may be calculated from equation (19) and v_m by re-arranging equation (15).

Work is sometimes simplified by thinking of the resin/filler as a single entity with effective density ρ_m, residue r_{re} etc. and setting $P = 1$, $Q = 0$ in the above equations. (Filler must not be confused with reinforcement anywhere in this chapter.)

2.3. SOME USES OF WEIGHT AND VOLUME FRACTIONS

Volume fractions are important in estimating mechanical properties, as will be seen in the remainder of the paper; with weight fractions they are useful also in estimating thermal properties. Specific heat can be estimated with reasonable accuracy on the basis of a simple mixture law, i.e. effective specific heat $= \Sigma$ (wc_p for each component). Thermal conductivity parallel to a unidirectional laminate can be estimated with sufficient accuracy to give 'feel' for a problem, on the basis of effective $k = \Sigma vk$. Conductivity normal to any laminate, or across the fibres of a unidirectional laminate, can be estimated as effective $k = (\Sigma v/k)^{-1}$.

The prime cost of unit weight of laminated material will be Σwc, where the quantities c are component costs per unit weight.

3. MECHANICAL PROPERTIES

3.1. GENERAL

In principle one could take a variety of reinforcements and glue them together with any arbitrary resin. For example, we often see random chopped mat faced with one or two layers of satin fabric to improve appearance. Limiting the discussion to the planar case, such an arbitrary laminate can be loaded in tension, compression or shear, separately or simultaneously. The laminate responds in terms of direct strains ε_x, ε_y and a shear strain ε_{xy}.

As loads increase, failure may occur because of excessive fibre or resin strain. Usually the resin fails first, giving a cracked or cloudy appearance, but the laminate continues to take a useful load. The fibre/resin interface is also a source of weakness.

3.2. FITTING THEORY TO OBSERVATION

Most theories of fibre composite materials attempt to predict laminate properties from the known properties of the separate fibre and resin. This usually gives satisfactory answers, but it is not always possible to take account of all the factors that affect laminate properties (e.g. quality of the current batch of resin) in which case it pays to adjust the fibre and resin modulus constants in the theory to fit observation. This procedure has been shown to have a sound basis[1]. It makes the adjusted theory quantitatively correct and gives much safer predictions.

3.3. 'LATERAL' FIBRES

Theories of composites are based either on the netting analysis[2,3], which assumes that only the fibre bears load, or by the continuum analysis[4–6] which bases prediction on the measured properties of a single unidirectional ply or reinforcement. The writer prefers the netting analysis for engineering design because it is easier to apply and can be modified as follows to give good accuracy.

Usually in practical applications of the netting analysis, the resin is imagined as an isotropic medium which pervades the net of fibre; its contributions are represented by additional terms to the netting analysis. Unfortunately with this approach, predictions can be very inaccurate, and even adjustment of constants cannot put the matter right.

For example, if we consider a unidirectional laminate, simple netting analysis with additional resin terms will predict:

$$E_L = v_f E_f + v_m E_m \tag{26}$$

where E_L is Young's modulus of the laminate parallel to the fibre.

For transverse properties it will predict:

$$(E_L)_T = v_m E_m \tag{27}$$

and for shear modulus

$$G_L = \tfrac{1}{2} v_m E_m/(1 + \nu) \tag{28}$$

In the last two cases, only resin terms are present; this results from the assumption implicit in the netting analysis that fibres contribute lengthwise but not transversely. In reality the fibres can make a very substantial transverse contribution; the writer has seen a published value of 3.77×10^6 lb/in^2 for the transverse modulus of a well-packed unidirectional glass-fibre sheet, which the netting analysis-with-resin would have allowed approximately 0.1×10^6 lb/in^2.

To improve prediction without introducing excessive complication, the writer has suggested the concept of 'lateral fibres'[8]. If we imagine every real fibre of modulus E_f to have two hypothetical lateral fibres of modulus E_l

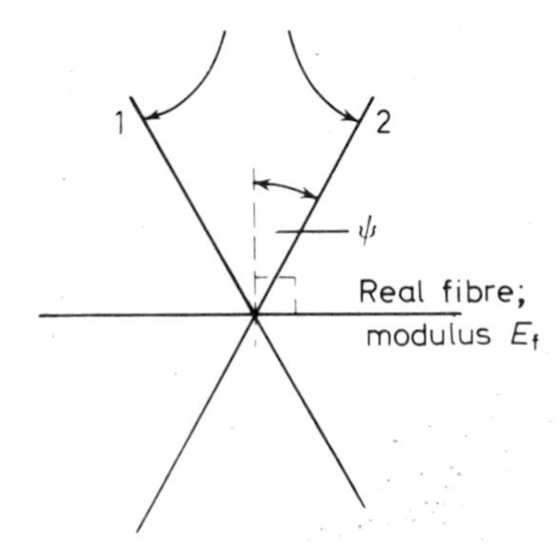

Figure 13.1. Fibre notation

associated with it, and lying at angles $(90 \pm \psi)^\circ$ to it in the plane of the laminate (*Figure 13.1*), a revised netting analysis which takes account of the lateral fibres is easier to adapt to observations.

A further refinement is necessary because the transverse contribution of a real fibre is non-linearly related to volume fraction v_f. This is because, as packing improves, more and more fibres touch each other, i.e. are no longer separated by a film of resin.

This effect can be accounted for empirically if we assign to each lateral fibre the modulus $E_l(1 + \mu v_f)$ where μ is an empirical constant. The contribution of one lateral fibre is thus given by $v_f E_l(1 + \mu v_f)$.

With the three extra parameters ψ, E_l and μ, together with the parameters E_f, E_m and ν of the ordinary netting analysis, it is possible to develop a mathematical model which can be adapted to describe the elastic behaviour of a wide range of practical laminates.

For a given type of fibre, resin and filler, only one set of numerical values for the constants E_f, E_m, ν, ψ, E_l and μ needs to be calculated, even though the observations relate to laminates of widely different fibre orientation, volume fractions and loading conditions. This model is developed in the following section; for practical purposes one can avoid the detailed argument and go straight to the final model in Section 5.

4. MATHEMATICAL DERIVATION OF MODEL TO CORRELATE STIFFNESS AND STRENGTH WITH ORIENTATION

4.1. THE PLANAR MAT OF FIBRES

The theory of this section is based on the analysis by H. L. Cox[3].

Consider a mat of straight fibres consolidated to a volume fraction v_f, of unit thickness and area, and arbitrary orientation. Assume that the orientation can be defined by a fibre distribution function $f(\theta)$ such that the volume of fibre that lies between angles θ and $(\theta + d\theta)$ to the OX axis is proportional to $f(\theta)\, d\theta$. $f(\theta)$ will obviously be a periodic function with period π. Let the factor of proportionality be K; since the total volume of fibre is v_f then

$$K \int_0^\pi f(\theta)\, d\theta = v_f \tag{29}$$

If the fibres lying between θ and $(\theta + d\theta)$ are intercepted by a line at right angles to them, the cross-sectional area of the fibres cut by unit length of this line will be $Kf(\theta)\, d\theta$. (In physical terms, consider a piece of mat in the form of a square of side 1 in, and depth 1 in. The volume of fibre is v_f in^3. If all the fibre lying between θ and $\theta + \Delta\theta$ were removed, its volume to first order would be $Kf(\theta)\, \Delta\theta$ in^3. Because the removed fibres were lying at an angle θ to the lower edge of the square, they are of different length depending on their position in the square. If we required fibres of equal length, we could rotate the square through an angle θ, and remove the fibre lying between the directions at angles θ and $\Delta\theta$ with respect to the lower edge. The measured volume would be unaltered and equal to $Kf(\theta)\,\Delta\theta$ in^3, because the density of the mat is invariant with respect to rotation of the square. Since the fibres are now all of unit length, their cross-sectional area is again given by $Kf(\theta)\,\Delta\theta$ in^2.) The cross-sectional area of fibre, viewed lengthwise, lying between θ and $(\theta + d\theta)$ and intercepted by a line of unit length perpendicular to OX, is $Kf(\theta)\cos A\,d\theta$. The cross-sectional area for fibres intercepted by a line of unit length parallel to OX is $Kf(\theta)\sin\theta\,.d\theta$.

4.2. STRESS–STRAIN RELATIONS (*Figures 13.2 and 13.3*)

If the mat of fibre is subjected to tensile strain ε_x parallel to OX, ε_y parallel to OY, and shear strain ε_{xy} between OX and OY, the strain ε_f in a fibre inclined at an angle θ to OX can be shown to be[3]

$$\varepsilon_f = \varepsilon_x \cos^2\theta + \varepsilon_y \sin^2\theta + \varepsilon_{xy} \sin\theta\cos\theta \tag{30}$$

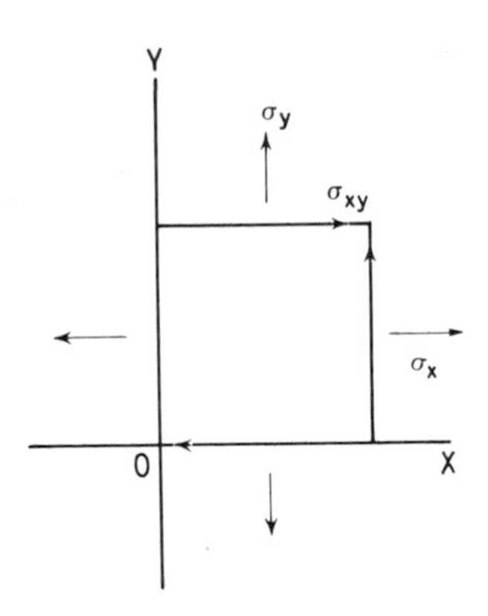

Figure 13.2. Stress notation

If load is proportional to strain, contributions of the fibre in the directions OX and OY respectively are the product of load in the fibre with $\cos\theta$ and $\sin\theta$ respectively. Combining these data and noting that modulus of the fibre is E_f

$$\sigma_x = KE_f \int_0^{\pi} (\varepsilon_x \cos^2\theta + \varepsilon_y \sin^2\theta + \varepsilon_{xy}\sin\theta\cos\theta)\cos^2\theta f(\theta)\,d\theta \tag{31}$$

$$\sigma_y = KE_f \int_0^{\pi} (\varepsilon_x \cos^2\theta + \varepsilon_y \sin^2\theta + \varepsilon_{xy}\sin\theta\cos\theta)\sin^2\theta f(\theta)\,d\theta \tag{32}$$

$$\sigma_{xy} = KE_f \int_0^{\pi} (\varepsilon_x \cos^2\theta + \varepsilon_y \sin^2\theta + \varepsilon_{xy}\sin\theta\cos\theta)\sin\theta\cos\theta f(\theta)\,d\theta \tag{33}$$

where σ_x, σ_y are tensile stresses and σ_{xy} is shear stress. Because $f(\theta)$ is periodic it may be written

$$f(\theta) = a_0 + a_1 \cos 2\theta + a_2 \cos 4\theta + a_3 \cos 6\theta + \ldots\ldots$$
$$b_1 \sin 2\theta + b_2 \sin 4\theta + b_3 \sin 6\theta + \ldots\ldots \quad (34)$$

(The double angles ensure that $f(\theta)$ is periodic over an angular interval of π. Integrals are evaluated from θ to π, so that a fibre contributes only once to an

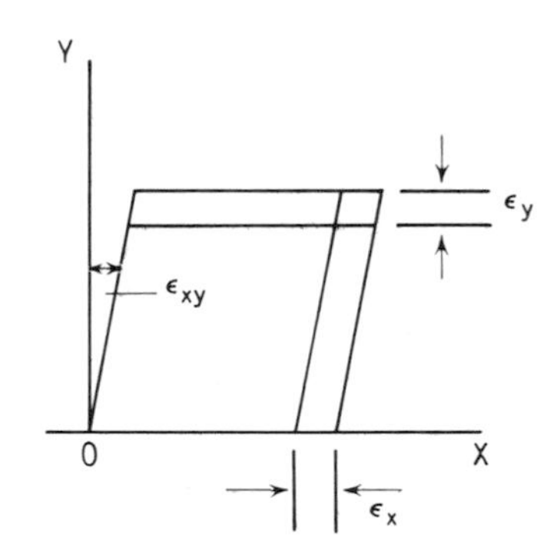

Figure 13.3. Strain notation

integral, and does not make a further contribution when viewed lengthwise in the opposite direction.)

We note

$$\int_0^\pi f(\theta)\, d\theta = \int_0^\pi (a_0 + a_1 \cos 2\theta + \ldots\ b_1 \sin 2\theta \ldots)\, d\theta = a_0\pi \quad (35)$$

whence from equation (29)

$$K = v_f / a_0\pi \quad (36)$$

Equations (31), (32) and (33) may be written

$$\sigma_x = \alpha_1\varepsilon_x + \beta_1\varepsilon_y + \gamma_1\varepsilon_{xy} \quad (37)$$

$$\sigma_y = \alpha_2\varepsilon_x + \beta_2\varepsilon_y + \gamma_2\varepsilon_{xy} \quad (38)$$

$$\sigma_{xy} = \alpha_3\varepsilon_x + \beta_3\varepsilon_y + \gamma_3\varepsilon_{xy} \quad (39)$$

where the coefficients α_1, β_1 etc. are evaluated from two definite integrals, of which the following is typical

$$\alpha_1 = KE_f \int_0^\pi \cos^4 \theta (a_0 + a_1 \cos 2\theta + a_2 \cos 4\theta + a_3 \cos 6\theta + \ldots \quad (40)$$
$$b_1 \sin 2\theta + b_2 \sin 4\theta + b_3 \sin 6\theta + \ldots)\, d\theta$$

$$= a_0 K E_f \int_0^\pi \cos^4 \theta \left(1 + \frac{a_1}{a_0} \cos 2\theta + \ldots\ \text{etc}\right) d\theta \quad (41)$$

Terms higher than suffix 2 vanish during integration, and we obtain

$$\alpha_1 = a_0\pi K E_f(6 + 4a_1/a_0 + a_2/a_0)/16 \tag{42}$$

$$\beta_1 = \alpha_1 = \gamma_3 = a_0\pi K E_f(2 - a_2/a_0)/16 \tag{43}$$

$$\beta_2 = a_0\pi K E_f(6 - 4a_1/a_0 + a_2/a_0)/16 \tag{44}$$

$$\gamma_1 = \alpha_3 = a_0\pi K E_f(2b_1/a_0 + b_2/a_0)/16 \tag{45}$$

$$\gamma_2 = \beta_3 = a_0\pi K E_f(2b_1/a_0 - b_2/a_0)/16 \tag{46}$$

4.3. THE UNIDIRECTIONAL CASE

To represent a system of parallel fibres parallel to the direction ϕ, Cox takes for the distribution function $f(\theta)$ a Dirac delta function[9]:

$$f(\theta) = [1 + 2\cos 2(\theta - \phi_1) + 2\cos 4(\theta - \phi_1) + \ldots.]/\pi \tag{47}$$

which is a step function having the values

$$\theta = \phi_1, \qquad f(\theta) = \infty$$

$$\theta \neq \phi_1, \qquad f(\theta) = 0$$

Although the function of equation (47) contains a discontinuity at $\theta = \phi_1$, it is legitimate to use it as a factor in an integrand[9].

Equation (47) may be written

$$\begin{aligned} f(\theta) = [1 &+ 2\cos 2\phi_1 \cos 2\theta + 2\cos 4\phi_1 \cos 4\theta + \ldots. \\ &2\sin 2\phi_1 \sin 2\theta + 2\sin 4\phi_1 \sin 4\theta + \ldots.]/\pi \end{aligned} \tag{48}$$

By comparison with equation (34)

$$a_0 = 1/\pi \tag{49}$$

$$a_1 = \frac{2}{\pi}\cos 2\phi_1 \tag{50}$$

$$a_2 = \frac{2}{\pi}\cos 4\phi_1 \tag{51}$$

$$b_1 = \frac{2}{\pi}\sin 2\phi_1 \tag{52}$$

$$b_2 = \frac{2}{\pi}\sin 4\phi_1 \tag{53}$$

Substituting these values, and K from equation (36), into equations (42) to (46) the coefficients become

$$\alpha_1 = \frac{v_f E_f}{8}(3 + 4\cos 2\phi_1 + \cos 4\phi_1) \tag{54}$$

$$\beta_1 = \alpha_2 = \gamma_3 = \frac{v_f E_f}{8}(1 - \cos 4\phi_1) \tag{55}$$

$$\beta_2 = \frac{v_f E_f}{8}(3 - 4\cos 2\phi_1 + \cos 4\phi_1) \tag{56}$$

$$\gamma_1 = \alpha_3 = \frac{v_f E_f}{8}(2\sin 2\phi_1 + \sin 4\phi_1) \tag{57}$$

$$\gamma_2 = \beta_3 = \frac{v_f E_f}{8}(2\sin 2\phi_1 - \sin 4\phi_1) \tag{58}$$

If there are n similar bands of parallel fibre lying at n different angles to OX,

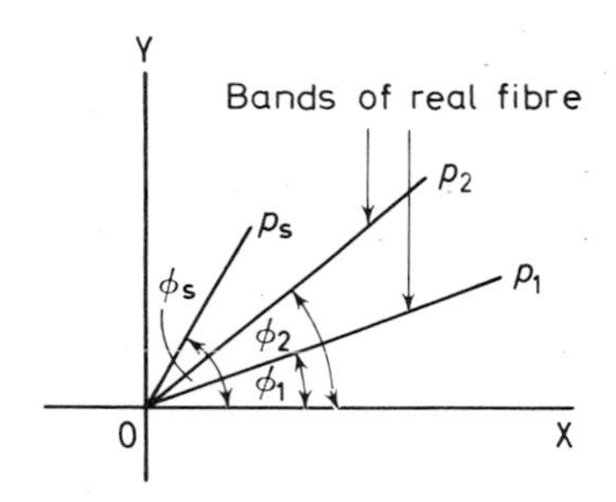

Figure 13.4. Orientation notation

the contribution to stiffness of the sth band of fibre, modulus E_{fs}, fraction p_s, angle ϕ_s, (*Figure 13.4*) is given by coefficients $(\alpha_1)_{fs}$ to $(\gamma_3)_{fs}$ where:

$$(\alpha_1)_{fs} = \tfrac{1}{8}p_s v_f E_{fs} \quad (3 + 4\cos 2\phi_s + \cos 4\phi_s) \tag{59}$$

$$(\beta_1)_{fs} = (\alpha_2)_{fs} = (\gamma_3)_{fs} = \tfrac{1}{8}p_s v_f E_{fs}(1 - \cos 4\phi_s) \tag{60}$$

$$(\beta_2)_{fs} = \tfrac{1}{8}p_s v_f E_{fs}(3 - 4\cos 2\phi_s + \sin 4\phi_s) \tag{61}$$

$$(\gamma_1)_{fs} = (\alpha_3)_{fs} = \tfrac{1}{8}p_s v_f E_{fs}(2\sin 2\phi_s + \sin 4\phi_s) \tag{62}$$

$$(\gamma_2)_{fs} = (\beta_3)_{fs} = \tfrac{1}{8}p_s v_f E_{fs}(2\sin 2\phi_s - \sin 4\phi_s) \tag{63}$$

The contributions of the lateral fibres are given by coefficients $(\alpha_1)_{1s}$ to $(\gamma_3)_{1s}$ where:

$$\begin{aligned}(\alpha_1)_{1s} = \tfrac{1}{8}p_s v_s(1 + \mu v_f)E_{fs} \quad &\{3 + 4\cos 2(\phi_s + \pi/2 - \psi) \\ &+ \cos 4(\phi_s + \pi/2 - \psi) + 3 + 4\cos 2(\phi_s + \pi/2 + \psi) \\ &+ \cos 4(\phi_s + \pi/2 + \psi)\}\end{aligned}$$

$$(\beta_1)_{1s} = (\alpha_2)_{1s} = (\gamma_3)_{1s} = \text{etc.} \qquad (64)\text{–}(68)$$

4.4. CONTRIBUTIONS OF THE MATRIX

Stiffness contributions of the resin matrix are given by classical elasticity theory for isotropic materials. Noting that the matrix volume fraction is $(1 - v_f - v_v)$, the contributions are

$$(\alpha_1)_m = (\beta_2)_m = (1 - v_f - v_v)\, E_m/(1 - \nu^2) \tag{69}$$

$$(\alpha_2)_m = (\beta_1)_m = \nu(1 - v_f - v_v)\, E_m/(1 - \nu^2) \tag{70}$$

$$(\gamma_1)_m = (\alpha_3)_m = (\gamma_2)_m = (\beta_3)_m = 0 \tag{71}$$

$$(\gamma_3)_m = (1 - v_f - v_v)\, E_m/2(1 + \nu) \tag{72}$$

5. THE COMPLETE MODEL (*See Figures 13.2, 13.3* and *13.4*)

The coefficients for the complete laminate, comprising real fibres, lateral fibres and resin matrix, are obtained by summing the individual coefficients, i.e.

$$(\alpha_1)_T = \sum_{s=1}^{n} (\alpha_1)_{f_s} + \sum_{s=1}^{n} (\alpha_1)_{l_s} + (\alpha_1)_m \tag{73}$$

etc.

where subscript T signifies total coefficient. Summing the coefficients and applying trigonometrical simplification we obtain the required model which describes the relation between stress and strain:

$$\sigma_x = (\alpha_1)_T\, \varepsilon_x + (\beta_1)_T\, \varepsilon_y + (\gamma_1)_T\, \varepsilon_{xy} \tag{74}$$

$$\sigma_y = (\alpha_2)_T\, \varepsilon_x + (\beta_2)_T\, \varepsilon_y + (\gamma_2)_T\, \varepsilon_{xy} \tag{75}$$

$$\sigma_{xy} = (\alpha_3)_T\, \varepsilon_x + (\beta_3)_T\, \varepsilon_y + (\gamma_3)_T\, \varepsilon_{xy} \tag{76}$$

where

$$(\alpha_1)_T = \tfrac{1}{8}v_f\{3C_0 + 4C_2 + C_4\} + (1 - v_f - v_v)\, E_m/(1 - \nu^2) \tag{77}$$

$$(\beta_1)_T = (\alpha_2)_T = \tfrac{1}{8}v_f(C_0 - C_4) + \nu(1 - v_f - v_v)\, E_m/(1 - \nu^2) \tag{78}$$

$$(\beta_2)_T = \tfrac{1}{8}v_f\{3C_0 - 4C_2 + C_4\} + (1 - v_f - v_v)\, E_m/(1 - \nu^2) \tag{79}$$

$$(\gamma_1)_T = (\alpha_3)_T = \tfrac{1}{8}v_f\{2S_2 + S_4\} \tag{80}$$

$$(\gamma_2)_T = (\beta_3)_T = \tfrac{1}{8}v_f\{2S_2 - S_4\} \tag{81}$$

$$(\gamma_3)_T = \tfrac{1}{8}v_f\{C_0 - C_4\} + \tfrac{1}{2}(1 - v_f - v_v)\, E_m/(1 + \nu) \tag{82}$$

and

$$C_0 = \sum_{s=1}^{n} p_s\{E_{fs} + 2E_{ls}(1 + \mu v_f)\} \tag{83}$$

$$C_2 = \sum_{s=1}^{n} \{p_s \cos 2\phi_s\}\{E_{f_s} - 2E_{l_s}(1 + \mu v_f)\cos 2\psi\} \tag{84}$$

$$C_4 = \sum_{s=1}^{n} \{p_s \cos 4\phi_s\} \{E_{f_s} + 2E_{1_s}(1 + \mu v_f) \cos 4\psi\} \quad (85)$$

$$S_2 = \sum_{s=1}^{n} \{p_s \sin 2\phi_s\} \{E_{f_s} - 2E_{1_s}(1 + \mu v_f) \sin 2\psi\} \quad (86)$$

$$S_4 = \sum_{s=1}^{n} \{p_s \sin 4\phi_s\} \{E_{f_s} + 2E_{1_s}(1 + \mu v_f) \sin 4\psi\}. \quad (87)$$

6. USE OF THE MODEL

For given ambient conditions and a given type of fibre, resin and filler, the parameters E_f, E_m, ψ, E_1, ν and μ must first be determined by fitting the model to test-data on appropriate laminates. A computer is usually needed to do this. As a typical example, a number of data on miscellaneous glass fibre/epoxy resin laminates were correlated by a least-squares type computer programme based on the model of Section 5. The observations related to fibre volume fractions ranging from 0.25 to 0.71, to unidirectional, cross-ply, and chopped mat laminates, and to direct and shear moduli ranging from 1.0 to 7.5 × 10^6 lbs/in^2. The computed values of parameters giving the best fit between the model and the observations were*:

$$E_f = 1.01289 \times 10^7 \text{ lbf/in}^2$$

$$E_m = 5.73844 \times 10^5 \text{ lbf/in}^2$$

$$\psi = 31.7856^\circ$$

$$E_1 = 7.97516 \times 10^5 \text{ lbf/in}^2$$

$$\nu = 2.96817 \times 10^{-4}$$

$$\mu = 4.85771.$$

The values for the ratio using the fitted model 'predicted stiffness/observed stiffness, for all the laminates lay between 100 ± 17 per cent, a very satisfactory result for data taken from many sources and including different fibre finishes and different types of epoxy resins. One would expect to do better with data from a formal exercise confined to single types of fibre and resin, and would look closely at observations that could not be correlated to better than 10 per cent.

Note that the fitted value of $E_f = 1.01289 \times 10^7$ corresponds closely to the accepted value for E-glass fibre, $E_f = 1.05 \times 10^7$ lbf/in^2. Also $E_m = 5.73844 \times 10^5$ is in good agreement with the value often used for epoxy resin, $E_m = 6 \times 10^5$ lbf/in^2.

The Poisson's ratio $\nu = 2.96817 \times 10^{-4}$ for the resin does not agree with the commonly accepted value $\nu = 0.35$. This is a consequence of the unimportant role it plays in the model; future computations would reasonably fix ν at 0.35. Leaving out the ν term altogether would be equivalent to assuming a resin Poisson ratio of zero, which is impossible.

* Non-significant figures retained throughout calculations to avoid rounding error.

The modulus of the lateral fibres at a volume fraction $v_f = 0.6$ is 7.97516×10^5 $(1 + 0.6 \times 4.85771) = 3.12197 \times 10^6$ lbf/in^2. This is 30 per cent of the modulus of the real fibre, demonstrating that for glass-fibre the lateral contribution is significant.

Strictly speaking one should go through the formal test-programme and fitting operation for each combination of fibre and resin, and environmental condition, but obviously this is not always possible. If one is going to design a Forth Bridge in glass-fibre one would face up to this expense. For lesser problems, one can use a mixture of literature values and earlier fitted values. For example, for a problem in S-glass, one could base preliminary calculations on:

$$E_f = 1.2 \times 10^7 \text{ (nominal value for S-glass fibre)}$$

$$E_m = 0.6 \times 10^5$$

$$\psi = 31.7856$$

$$E_l = 7.97516 \times 10^5 \times 1.20/1.01289 = 9.4484 \times 10^5$$

$$\nu = 0.35$$

$$\mu = 4.85771.$$

Here it has been assumed that lateral modulus is linearly proportional to direct modulus. If design calculations on the basis of the above showed promise, one would then be encouraged to determine the correct constants for the S-glass and the particular resin formulation in hand.

In any event one would expect to make considerably more realistic predictions by this method than by not using any theory at all, or using an over-simplified theory.

7. MAKING A PREDICTION

When the above constants have been evaluated, the stress–strain behaviour for any laminate with specified volume fractions and orientation can be calculated using the same model. This is because the quantities v_f, v_v, p_s and ϕ_s are specified, therefore we have all the information to compute $C_0 \ldots S_4$ (equations 83 to 87) thence $(\alpha_1)_T$ to $(\gamma_3)_T$ (equations 77 to 82). (In general we are concerned with a single type of fibre, so that $E_{f_s} = E_f$ and $E_{l_s} = E_l$, but it is worth bearing in mind that the model will deal with mixed fibres; p_s would be calculated on a volumetric basis, not weight basis.)

Once $(\alpha_1)_T$ to $(\gamma_3)_T$ are calculated, strains can be computed from stresses or vice versa. (Equations 74 to 76). When the strains ε_x, ε_y and ε_{xy} are known, the strains in the fibres can be calculated if necessary from

$$\varepsilon_f = \varepsilon_x \cos^2 \phi + \varepsilon_y \sin^2 \phi + \varepsilon_{xy} \sin \phi \cos \phi \tag{88}$$

Example

What is the modulus along one fibre band of a satin-weave laminate for which $v_f = 0.5$ and $v_v = 0.05$?

Assume $E_f = 1.01289 \times 10^7$; $E_m = 5.73844 \times 10^5$; $\psi = 31.7856$; $E_l = 7.97516 \times 10^5$; $v = 0.35$; $\mu = 4.85771$.

We have

$$p_1 = 0.5 \qquad \phi_1 = 0$$

$$p_2 = 0.5 \qquad \phi_2 = 90^\circ$$

Note $E_{fs} = E_f$ and $E_{ls} = E_l$

Then

$$C_0 = 0.5 \times 1.01289 \times 10^7 + 0.5 \times 2 \times 7.97516 \times 10^5 \times (1 + 4.85771 \times 0.5) + 0.5 \times 1.01289 \times 10^7 + 0.5 \times 2 \times 7.97516 \times 10^5 \times (1 + 4.85771 \times 0.5) = 1.55980 \times 10^7$$

$$C_2 = \{0.5 \cos 2 \times 0^\circ\}\{1.01289 \times 10^7 - 2 \times 7.97516 \times 10^5 \times (1 + 4.85771 \times 0.5) \cos (2 \times 31.7856)\}$$

$$+ \{0.5 \cos 2 \times 90^\circ\}\{1.01289 \times 10^7 - 2 \times 7.97516 \times 10^5 \times (1 + 4.85771 \times 0.5) \cos (2 \times 31.7856)\} = 0$$

$$C_4 = \{0.5 \cos (4 \times 0^\circ)\}\{1.01289 \times 10^7 + 2 \times 7.97516 \times 10^5 \times (1 + 4.85771 \times 0.5) \cos (4 \times 31.7856)\} + \{0.5 \cos (4 \times 90^\circ)\}\{1.01289 \times 10^7 + 2 \times 7.97516 \times 10^5 \times (1 + 4.85771 \times 0.5) \cos (4 \times 31.7856)\}$$

which (taking $\cos (4 \times 31.7856) = -0.60380$) $= 6.8266 \times 10^6$ S_2 and S_4 are zero because $\sin 2 \times 0^\circ = 0$; $\sin 2 \times 90^\circ = 0$; $\sin 4 \times 0^\circ = 0$ and $\sin 4 \times 90^\circ = 0$.

Then:

$$(\alpha_1)_T = \tfrac{1}{8} \times 0.5\{3 \times 1.55980 \times 10^7 + 4 \times 0 + 6.8266 \times 10^6\} + (1 - 0.5 - 0.05)\, 5.73844 \times 10^5/(1 - 0.35^2) = 3.6456 \times 10^6$$

$$(\beta_1)_T = (\alpha_2)_T = \tfrac{1}{8} \times 0.5\{1.55980 \times 10^7 - 6.8266 \times 10^6\} + 0.35(1 - 0.5 - 0.05)\, 5.73844 \times 10^5/(1 - 0.35^2) = 0.65121 \times 10^6$$

$$(\beta_2)_T = \tfrac{1}{8} \times 0.5\{3 \times 1.55980 \times 10^7 - 4 \times 0 + 6.8266 \times 10^6\} + (1 - 0.5 - 0.05)\, 5.73844 \times 10^5/(1 - 0.35^2) = 3.6456 \times 10^6$$

$$(\gamma_1)_T = (\alpha_3)_T = 0$$

$$(\gamma_2)_T = (\beta_3)_T = 0$$

$$(\gamma_3)_T = \tfrac{1}{8} \times 0.5\{1.55980 \times 10^7 - 6.8266 \times 10^6\} + \tfrac{1}{2}(1 - 0.5 - 0.05)\, 5.73844 \times 10^5/(1 + 0.35) = 0.64385 \times 10^6$$

whence

$$\sigma_x = 3.6456 \times 10^6\, \varepsilon_x + 0.65121 \times 10^6 \varepsilon_y$$

$$\sigma_y = 0.65121 \times 10^6 \varepsilon_x + 3.6456 \times 10^6 \varepsilon_y$$

$$\sigma_{xy} = 0.64385 \times 10^6 \varepsilon_{xy}.$$

For a simple tensile test piece $\sigma_y = 0$, i.e. there is no lateral constraint. Thence:

$$\sigma_x = 3.6456 \times 10^6 \varepsilon_x + 0.65121 \times 10^6 \varepsilon_y$$

$$0 = 0.65121 \times 10^6 \varepsilon_x + 3.6456 \times 10^6 \varepsilon_y$$

Then

$$\varepsilon_y = -\frac{0.65121 \times 10^6}{3.6456 \times 10^6} \cdot \varepsilon_x$$

and therefore:

$$\sigma_x = \left\{3.6456 \times 10^6 - \frac{0.65121 \times 10^6 \times 0.65121 \times 10^6}{3.6456 \times 10^6}\right\} \varepsilon_x$$

$$= 3.5293 \times 10^6 \varepsilon_x$$

Hence

$$E = \frac{\text{stress}}{\text{strain}} = \frac{\sigma_x}{\varepsilon_x} = 3.5293 \times 10^6 \text{ lbf/in}^2. \tag{89}$$

An observed value[10], $E = 3.09 \times 10^6$, was available for a laminate with slightly lower fibre volume fraction ($v_f = 0.45$). A similar calculation to the example above, using this value of v_f, gave $E = 3.1914 \times 10^6$ lbf/in^2, which is in good agreement. Note:

$$G = \text{shear stress/shear strain} = \sigma_{xy}/\varepsilon_{xy} = 0.64385 \times 10^6 \text{ lbf/in}^2 \tag{90}$$

An observed value[10] of $G = 1.06 \times 10^6$ for $v_f = 0.652$ was available; a similar calculation to the example above, using this value of v_f, gave $G = 0.953 \times 10^6$, which again is in good agreement.

8. STRENGTH CORRELATION

Up to now, the model has been developed and considered from the point of view of elastic behaviour, but the same model can be used in a similar way to correlate and predict strength data. This application of the model is entirely empirical, but appears to be effective.

One computes a 'best fit' set of empirical constants analogous to $E_f \ldots \mu$, using the observed strength properties of a set of laminates, and basing predictions on these constants. In an exercise using the observed ultimate strengths of various glass-fibre/epoxide laminates, predicted strength/observed strength correlations to better than 100 ± 10 per cent were obtained.

An approach which is physically more meaningful is to calculate the strains in each fibre and to predict failure by comparing these to critical strains. For glass-fibre, critical strains in compression are usually about half critical strains in tension.

To predict failure at the fibre-resin interface, some success has been obtained by imagining the lateral fibres to have a critical strain[8].*

9. OTHER STRENGTH PROPERTIES

There is no reason why the model should not correlate and predict empirically more sophisticated strength properties such as creep and fatigue. As before, one would determine data from a representational range of stress systems and laminates, and use these data to determine a single set of constants which are then available for prediction.

10. MIXED FORMS OF REINFORCEMENT

It is common practice to mix different forms of a reinforcement in the same laminate, for example layers of chopped glass mat with a layer of satin glass fabric to improve finish. Such systems can be analysed by the model as follows:

All reinforcements can be characterised by a weight u g/cm^2 per unit area, and a characteristic thickness t cm depending on laminating pressure. Each reinforcement has its own orientation system, for example satin, $p = \frac{1}{2}$, $\phi = 0°$; $p = \frac{1}{2}$, $\phi = 90°$; 2:1 tape, $p = 0.667$, $\phi = 0°$; $p = 0.333$, $\phi = 90°$. To arrive at the correct orientation terms for the combined laminate, one sets out a table as follows:

Reinforcement type No.	*Number of layers*	*Weight per unit area of 1 ply*	*Thickness of 1 ply*	*Orientation with respect to ref. axis*
1	n_1	u_1	t_1	$(p_1, \phi_1; p_2, \phi_2)_1$
2	n_2	u_2	t_2	$(p_1, \phi_1; p_2, \phi_2)_2$
	etc.			

The total thickness is $n_1t_1 + n_2t_2 + \ldots\ldots$ cm
The total weight of fibre is $n_1u_1 + n_2u_2 + \ldots\ldots$ g/cm^2
Hence the overall volume fraction is:

$$v_f = \frac{1}{\rho_f} \cdot \frac{n_1u_1 + n_2u_2 + \ldots}{n_1t_1 + n_2t_2 + \ldots} \tag{91}$$

Also, weight of fibre at $(\phi_1)_1$ is $(p_1)_1n_1u_1$

$(\phi_2)_1 \quad (p_2)_1n_1u_1$

$(\phi_1)_2 \quad (p_1)_2n_2u_2$

$(\phi_2)_2 \quad (p_2)_2n_2u_2$

and so on.

* In Reference 8 the factor λ has since been found redundant and should be replaced by unity.

The right-hand side sums to $n_1u_1 + n_2u_2 + \ldots$

Angles can be set out on the left-hand side into common groups ϕ_1, ϕ_2 etc. and corresponding fractions p_1, p_2 etc. for the complete laminate can be calculated by adding the appropriate right-hand fractions and dividing by $(n_1u_1 + n_2u_2 \ldots .)$. Having calculated v_f; p_1, ϕ_1; p_2, ϕ_2 .... the model can then be used.

The preferred way to represent random mat is the system

$$\begin{aligned} p_1 &= 0.25 & \phi_1 &= 0 \\ p_2 &= 0.25 & \phi_2 &= 45^\circ \\ p_3 &= 0.25 & \phi_3 &= 90^\circ \\ p_4 &= 0.25 & \phi_4 &= 135^\circ \end{aligned}$$

This automatically confers isotropic planar properties in terms of mathematical representation.

Example

The above method is illustrated in terms of a laminate consisting of four layers of chopped strand mat, faced with three layers of 2:1 fabric and 1 layer of 1:1 fabric on each face. Values for (hypothetical) reinforcements are:

Reinforcement	*Number of layers*	*Wt. per unit area of 1 ply* g/cm²	*Thickness of 1 ply* cm	*Orientation*
(1)2:1 fabric, 9 oz/yd²	6	0.03047	0.0254	$p_1 = 0.67$; $\phi = 0$ $p_2 = 0.33$; $\phi = 90$
(2)1:1 fabric, 2 oz/yd²	2	0.00677	0.0066	$p_1 = 0.5$; $\phi = 0$ $p_2 = 0.5$; $\phi = 90$
(3)chopped mat, 2 oz/yd²	4	0.06095	0.1016	$p_1 = 0.25$; $\phi = 0$ $p_2 = 0.25$; $\phi = 45$ $p_3 = 0.25$; $\phi = 90$ $p_4 = 0.25$; $\phi = 135$

From equation (91):

$$v_f = \frac{1}{2.54}\,\frac{6 \times 0.03047 + 2 \times 0.00677 + 4 \times 0.06095}{6 \times 0.0254 + 2 \times 0.0066 + 4 \times 0.1016} = 0.51162$$

Weights of fibre at the following angles:

0 $\quad 0.67 \times 6 \times 0.03047 + 0.5 \times 2 \times 0.00677 + 0.25 \times 0.06095 \times 4 = 0.190209$

45 $\quad 0.25 \times 0.06095 \times 4 = 0.06095$

90	$0.33 \times 6 \times 0.03047 + 0.5 \times 2 \times 0.00677 + 0.25 \times 0.06095 \times 4 =$	0.12805
135	$0.25 \times 0.06095 \times 4 =$	0.06095
Total		0.44015 g/cm^2

One can then complete the following table:

$p_1 = 0.190209/0.44015$	$= 0.43214$	$\phi_1 = 0$
$p_2 = 0.06095\ /0.44015$	$= 0.13847$	$\phi_2 = 45$
$p_3 = 0.12805\ /0.44015$	$= 0.29092$	$\phi_3 = 90$
$p_4 = 0.06095\ /0.44015$	$= 0.13847$	$\phi_4 = 135$
Total	1.00000	

The computations using the model can now proceed.

Acknowledgement

The author is indebted to Mr. K. F. Rogers for vetting this chapter in detail.

SYMBOLS

A, B	constants in strain condition equation
E	Young's modulus, axes identified by subscript; lbf/in^2
F	denoting function of strain, identified by subscript
G	shear modulus, axes identified by subscript; lbf/in^2
K	proportionality factor relating fibre volume to fibre distribution function
R	weight of resin in 1 g of resin/filler matrix
Q	weight of filler in 1 g of resin/filler matrix
V	volume of matrix in one ply of 1 cm^2 reinforcement; cm^3
$a_0 \ldots\ldots a_n$, $b_0 \ldots\ldots b_n$	coefficients in fibre distribution function
n	integer
p	fraction of total fibre lying in band at angle ϕ to OX
r	residue after heating, identified by subscript; g/g
s	integer
t	thickness of 1 ply of reinforcement, located in laminate; cm
u	weight of 1 ply of reinforcement or one layer of filaments; g/cm^2
v	volume reaction of laminate component, identified by subscript
w	weight fraction of laminate component, identified by subscript
$\alpha_1\alpha_2\alpha_3$, $\beta_1\beta_2\beta_3$, $\gamma_1\gamma_2\gamma_3$	coefficients in stress/strain equations
ε	strain

$\varepsilon_x, \varepsilon_y$	direct strains in OX, OY directions
ε_{xy}	shear strain between YX axes
ε_f	strain in fibre
θ	variable angle between radius vector and X axis
μ	empirical constant allowing for contiguity effects
ρ_L	bulk density of laminate; g/cm^3
ρ_m	density of resin/filler matrix, excluding voids, i.e. intrinsic (not bulk) density; g/cm^3
ρ_q	intrinsic (not bulk) density of filler; g/cm^3
ρ_f	apparent density of fibre; g/cm^3
ρ_{re}	density of cured resin including catalyst, excluding voids; g/cm^3
σ_x, σ_y	direct stresses on OX and OY directions; lbf/in^2
σ_{xy}	shear stress; lbf/in^2
ν	Poissons ratio of resin/filler matrix
ϕ_s	angle between *s*th band of fibre and OX
ψ	half included angle between lateral fibres
ω	angle between major principal axis and OX

Subscripts

f	fibre
m	matrix
l	lateral fibre
L	laminate
q	filler
re	resin
s	ordinal integer
T	total
v	voids
x	in OX direction
y	in OY direction
xy	between OX and OY
1, 2 . . . *n*	ordinal integers

REFERENCES

1. Bishop, P. H. H., Rogers, K. F., Schulz, J. C. *RAE Tech. Rep. 69053* March 1969.
2. Schulz, J. C. '*Netting' Analysis of Filament Wound Pressure Vessels*. A.S.M.E. Paper No-63-WA-223, Winter Annual Meeting, Philadelphia. 17th November 1963.
3. Cox, H. L. 'The Elasticity and Strength of Paper and other Fibrous Materials'. *Br. J. appl. Phys.* **3**, 72–79, March 1952.
4. Tsai, Stephen W. 'Structural Behaviour of Composite Materials'. *NASA Contractor Rep., NASA* CR-71, July 1964
5. Rosen, S. Walter, *et al.* 'Mechanical Properties of Fibrous Composites'. *NASA* 65 16806, July 1963
6. Rosen, S. Walter, *et al.* 'Study of the Relationship of Properties of Composite Materials to Properties of their Constituents'. *Prof. Report No. 1* Contract NAS w-470, R.A.E. unpublished test note.
8. Bishop, P. H. H. 'An Improved Method for Predicting Mechanical Properties of Fibre Composite Materials.' R.A.E. *Tech. Rep.* 66245, August 1966, H.M.S.O.

9. Lighthill, M. J. *An Introduction to Fourier Analysis and Generalised Functions*. Cambridge University Press 1959
10. Bishop, P. H. H. Personal communication.

NOTE

English units have been used in this mathematical treatment because of the preponderance of British and American published works on this subject. The following conversion factors to SI may be found helpful.

CONVERSION FACTORS TO SI

$1\ \text{lbf/in}^2 = 6.894\ 76 \times 10^3\ \text{N/m}^2$
$1\ \text{lb/in}^3 = 2.767\ 99 \times 10^4\ \text{kg/m}^3$
$1\ \text{in} = 2.540\ 00 \times 10^{-2}\ \text{m}$
$1\ \text{lb} = 4.535\ 92 \times 10^{-1}\ \text{kg}$

14

Mechanical Behaviour of Fibre Composites

R. M. OGORKIEWICZ

1. CATEGORIES OF COMPOSITES

Of the two principal components of GRP composites, glass fibres are by far the stiffer and the stronger, as indicated in Table 14.1. The mechanical properties of the composites depend, therefore, to a large extent on the proportion of glass which they contain. The proportion of glass is, in turn, a matter of the arrangement of the fibres—the more orderly the arrangement, the more closely can the fibres be packed and the higher, consequently, the maximum possible proportion of glass. It is appropriate, therefore, as well as convenient, to categorise glass fibre/plastics composites according to the principal modes of the arrangement of the fibres, of which there are three.

The first of the three categories into which glass fibre/plastics composites can be divided consists of that in which the fibres are randomly distributed within the plastics matrix. The fibres are generally short and because they are randomly arranged the maximum proportion of glass is limited to about 50 per cent by weight. In practice, there may only be 10 per cent of glass, or even less, but in general the percentage of glass lies between 30 and 40. The principal examples of this category of composites include laminates made with chopped strand mats, dough moulding compounds and glass fibre reinforced thermoplastics.

The second category consists of composites in which the fibres are orthogonally arranged, generally in the plane of the laminate, by being woven into glass cloth or woven rovings. Because of the more orderly arrangement of the fibres, the maximum percentage of glass can be higher in this category than the first, ranging from 40 up to about 65 per cent by weight.

The third category consists of composites in which the fibres are all laid in one direction. This is the case with rods produced by pultrusion and certain laminates based on filament winding as well as layers of cylindrical vessels made by this process. Since the fibres are laid in one direction they can be packed more closely than before and the percentage of the glass is, therefore, highest in this category of composites. They are also the strongest and stiffest. In fact, so far as their uniaxial tensile strength is concerned, unidirectional glass fibre/epoxy composites are, on a strength-to-weight basis, the strongest industrially used materials. In theory, with the hexagonal packing illustrated in *Figure 14.1*, the maximum possible percentage of glass is 90.67 per cent, by volume. In practice, it ranges from 60 up to about 90 per cent, by weight.

When the glass fibres form such a large proportion of the composite, the term 'reinforced plastics' loses much of its meaning, from the point of view of the mechanical properties at any rate. Clearly, when most of the composite consists of glass fibres its mechanical properties are no longer those of the plastics component 'reinforced' by the fibres, as they are for instance with glass-filled thermoplastics. Instead, they are primarily those of the glass fibres and it is more appropriate, therefore, to describe such composites as 'resin-bonded glass fibres' rather than 'reinforced plastics'.

Table 14.1

MECHANICAL CHARACTERISTICS OF GLASS FIBRES AND PLASTICS

Type of fibre or plastic	*Specific gravity*	*Tensile modulus* $kN/m^2 \times 10^7$	*Tensile strength* $kN/m^2 \times 10^4$
E —glass	2.54	7	350*
A —glass	2.45	7	310*
D —glass	2.16	5	240*
S —glass	2.49	9	450*
M—glass	2.89	11	350*
Epoxide	1.6–1.9	0.07–0.18	4–7
Polyester	1.66–2.25	0.04–0.18	2–4
Nylon	1.09–1.14	0.1–0.2	5–8
Polycarbonate	1.2	0.2	6–7

* Of virgin fibres; tensile strength of E-glass fibres in industrial components is between 140 and 200 $kN/m^2 \times 10^4$

Nevertheless, although the nature of the third category of composites differs considerably from that of the first, there is a thread of continuity running through all three of them. In particular, if their stiffness or strength parameters are plotted against the percentage of glass, they are seen to follow the same general pattern. The pattern is one of the values of such parameters as the tensile modulus or the tensile strength rising continuously with glass content. Thus, to a first approximation, the three categories merely occupy successive and overlapping ranges of glass content and the differences between them appear to be questions of the glass content.

Illustrations of the pattern are provided by *Figures 14.2* and *14.3* and the values of glass content, tensile modulus and tensile strength for the different types of composite are listed in Table 14.2. The curves and the tabulated figures all refer to polyester/glass composites which are made in a greater variety of forms than others and which cover, therefore, a wider range of ratios of glass to plastics.

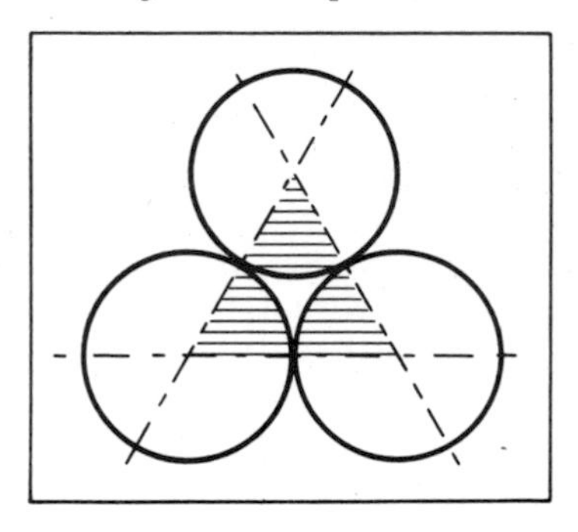

Figure 14.1. Hexagonal packing of fibres: the shaded area represents 90·67 *per cent of the triangle, which corresponds to the maximum volume percentage of the fibres*

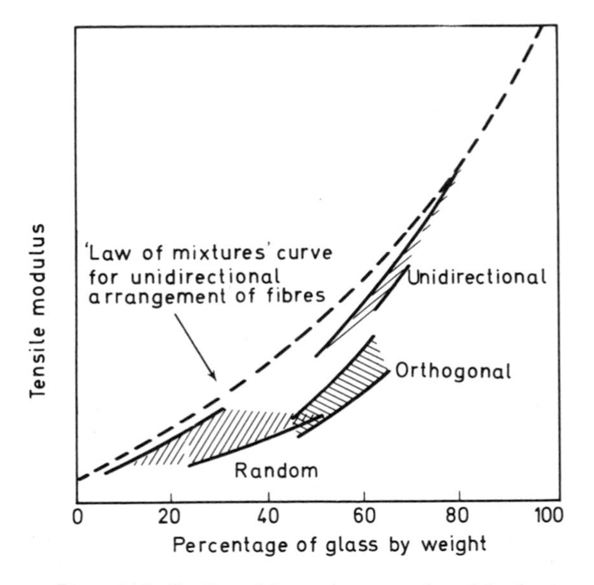

Figure 14.2. Tensile modulus v. glass content by weight, for the three principal categories of glass-fibre/plastics composites

2. ANISOTROPY

While the plots in *Figures 14.2* and *14.3* show a general trend running through the three categories they also indicate that there are differences between them. A major difference which emerges as the three categories of composites are examined more closely is that the stiffness and strength of most of them vary with direction. An exception to this are the composites where the fibres are randomly distributed and where the mechanical properties are, therefore,

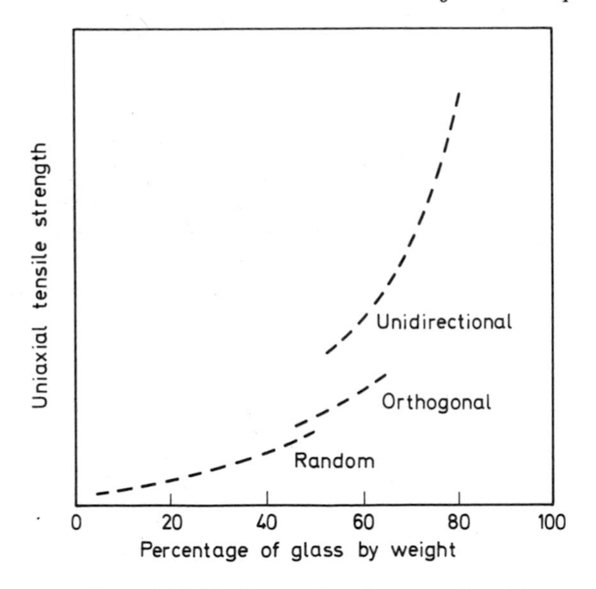

Figure 14.3. Tensile strength v. glass content by weight

the same in all directions. In other words, composites with a perfectly random arrangement of fibres are isotropic.

Other composites are anisotropic or aeolotropic. The anisotropy may be a consequence of the orientation of glass fibres due to flow in the injection moulding of a glass filled thermoplastic article which will result in greater stiffness and strength in the direction of fibre alignment. Alternatively, it may be a consequence of the orthogonal arrangement of the fibres in the glass cloth or woven rovings used in laminates. In this case, the strength of stiffness of the composite is greater in the two orthogonal directions in which the fibres lay than in any other direction. The characteristic which results from the orthogonal arrangement of the fibres constitutes a particular form of anisotropy and materials which exhibit them are called orthotropic.

Composites in which all the fibres are aligned in one direction are also orthotropic. However, their properties in planes perpendicular to the direction of the fibres do not vary with direction, which means that they are also transversely isotropic. In the plane of the fibres, on the other hand, the degree of anisotropy is greater than in any other case. For instance, the tensile modulus in the direction of the fibres can be almost three times that in a direction normal to the fibres. The difference between the tensile strengths in the two directions can be much greater and it is common to treat thin composites with a unidirectional arrangement of fibres as if they were only able to support tensile loads in the direction of the fibres or, in other words, as monotropic membranes[1].

The anisotropic nature of most fibre/plastics composites needs to be considered whenever data on them are presented since, in general, they represent the maximum values of stiffness or strength, as they do in Table 14.2. By the same token they are not necessarily indicative of the stiffness and strength in all directions.

The maximum values which are quoted may also be higher than the maxima often achieved in practice. This is particularly so with reinforced thermoplastics. In their case the tensile stiffness and strength tend to be established with end-gated injection moulded specimens in which the fibres are highly aligned in the direction of the application of loads and of the extension measurements. In the case of other mouldings the alignment is less pronounced and the maximum values of the tensile parameters are less, even though the general nature of the reinforced thermoplastic and its glass content are the same as in the specially moulded tensile specimens.

Table 14.2

TENSILE CHARACTERISTICS OF DIFFERENT TYPES OF GLASS FIBRE/POLYESTER COMPOSITES

Material	*Glass percentage by weight*	*Specific gravity*	*Tensile modulus* $kN/m^2 \times 10^7$	*Tensile strength* $kN/m^2 \times 10^4$
Unfilled resin, as cast	0	1.1–1.3	0.2–0.4	2–7
Dough moulding with filled resin	6–30	1.8–2.1	0.5–1.4	4–12
Hand lay-up with chopped strand mat	25–45	1.4–1.6	0.6–1.1	6–18
Matched die moulding with preform	25–50	1.4–1.6	0.6–1.2	6–20
Hand lay-up with woven rovings	45–62	1.5–1.8	1.2–2.4	20–35
Hand lay-up with plain weave cloth	45–65	1.7–1.8	1–2	25–35
Hand lay-up with unidirectional cloth	62–70	1.7–1.9	3–4	55–60
Unidirectional rovings (filament winding or pultrusion)	50–80	1.6–2.0	2–5	40–125
E-glass strands	100	2.55	7.2	125–185

3. STRESS–STRAIN CHARACTERISTICS

While GRP might differ from other materials, such as ferrous metals, in being highly anisotropic they appear to resemble them in behaving within the range of small strains as if they were linearly elastic. This is particularly so where the matrix is a thermoset and the glass content is high. Where this is the case the tensile stress–strain curves certainly closely approximate to straight lines, up a certain value of strain at any rate. At higher values of strain the slope of the stress–strain curve is generally less than that of its initial portion or, more exactly, there is a significant increase in the stress rate of strain[2]. The point at which this change occurs, or what is often referred to as the 'knee' in the stress–strain curve, is generally associated with crazing or cracking of the matrix material. Once this point is passed the composite is, therefore, permanently damaged and, although it may be capable of considerable further extension, if it is unloaded it does not return to its original dimensions but exhibits a permanent strain. In other words, the behaviour of a composite ceases to be even approximately elastic beyond the knee of the stress–strain curve.

Where the knee occurs on the stress–strain curve depends on the stiffness of the matrix material. In general, the more flexible the resin the higher the strain, and the stress, in the composite at which it begins to crack and the higher, therefore, the strain (and stress) at which the knee in the stress–strain curves occurs. Ultimately, with very flexible matrix materials, there may be no cracking at all[3].

The absence of any pronounced change in the slope of the stress–strain curve or its apparently linear elastic character before the onset of permanent damage does not, however, mean that glass fibre/plastics can be regarded without reservation as linear elastic materials, even in the realm of small strains. In fact, their apparent linear elastic behaviour is confined to the conditions of the conventional tensile tests in which load-extension—and hence stress–strain—curves are established in a relatively short period of time, usually of the order of a few minutes. Under other conditions the behaviour appears different. In particular, when the specimens are loaded for days or, even more, for months, as they are in creep tests, the strain is seen to be a function not only of stress but also of time. This means that glass fibre/plastics are visco-elastic rather than elastic.

The visco-elastic nature of glass fibre/plastics is due to the characteristics of their plastics matrices which are visco-elastic. Glass, on the other hand, is linearly elastic and the higher, therefore, its proportion in any composite the less pronounced its visco-elasticity might be expected to be.

In principle, however, all glass fibre/plastics are visco-elastic. Their mechanical behaviour might be better established, therefore, by means of tests in which constant loads are applied to specimens for long periods of time, that is by means of creep tests, than by means of the conventional tests in which a range of loads is covered in a short period of time. Creep tests do not, of course, directly produce load-elongation curves which might be simply converted into stress–strain curves. Their first result are curves of elongation against time at different values of stress. However, constant time sections through families of such creep curves readily produce isochronous stress–strain curves[4].

Isochronous, or constant time, stress–strain curves derived from creep tests are different in principle from the stress–strain curves derived from the more conventional tests in which specimens are pulled at a constant rate, because they represent the strain response to a different stress input. However, even with unreinforced thermoplastics, the differences between stress–strain curves obtained in different ways over comparable periods of time are not very great and, to a first approximation at any rate, curves of different origins can be used interchangeably. The major advantage of isochronous stress–strain curves derived from creep tests is that the latter can and do cover considerable durations of loading whereas other tests are essentially of a short-term nature.

It is possible, therefore, to obtain from creep tests a series of isochronous stress–strain curves, each for a different time from the commencement of loading, and this provides information directly related to the large number of structural applications of GRP[5].

The slope of the isochronous stress–strain curves decreases with time, since strain increases with time for a given stress. This is tantamount to saying that the tensile modulus, that is the ratio of stress to strain, decreases with time. By the same token it is not possible to characterise the extensional characteristics of a glass fibre/plastics by means of a single-valued tensile modulus. What is required, instead, is a series of values or a curve of modulus versus time; it is possible to consider single values only if the duration of loading is short.

4. STRENGTH AND FACTORS INFLUENCING IT

Just as the tensile modulus varies with time so does the tensile strength, that is the maximum load which a specimen can support before it breaks, divided by its original cross-sectional area. Thus, tensile strength needs to be defined in terms of series of values or curves of the maximum nominal tensile stress versus time, unless, once again, only short duration loading is considered. To establish such curves it is necessary to perform creep tests to the point of specimen rupture and they show that, in general, the tensile strength of GRP decreases with time[2, 6].

What is also important from the point of view of the ability of GRP components to withstand loads is the amount of energy which they can absorb before they break. This is a function not only of the maximum stress but also of the strain to failure. In fact, it is proportional to the product of the stress and strain or, more exactly, to the area under the stress–strain curve. When this is examined it is apparent that GRP fails without much elongation. A concomitant of this is that GRP is relatively brittle. In consequence, in practice its strength is sensitive to the presence of notches and other local stress raisers, particularly under conditions of suddenly applied loads.

Environmental conditions also have an important bearing on the performance of GRP under load. This is especially true of temperature, which has been assumed normal and invariant in the whole of the foregoing discussion. In fact, increases in temperature cause progressive decreases in the stiffness and strength of GRP composites until a limit of mechanical performance is reached in the range of 200–300°C with commonly used thermosetting resins[2, 7]. On the other hand, operating temperatures below normal tend to be advantageous to GRP composites which do not become as brittle as other materials at low temperatures. This makes them attractive for cryogenic applications[8].

Wetting is another factor which has an important effect on the performance of GRP. Thus, immersion in water leads to a much more rapid fall in strength with time than that experienced in air. This is due to the hydrolytic degradation of the resin-to-fibre bond as well as that of the fibres and resin itself in aqueous environments and in particular in boiling water. The degradation involves swelling of the resin and hydrolysis of the resin, fibres and coupling agent, all of which contributes to a gradual debonding between the fibres and the resin and a progressive weakening of the composite[9].

The above and all the earlier points about the mechanical properties of GRP relate to their behaviour under tensile loads. However, many of them are equally applicable to the behaviour under compressive or shear loads. This is especially true of the general effects of glass-to-resin ratio and of time and temperature.

There are, however, some important differences between the behaviour under the different types of loading. In particular, under compressive loads applied parallel to the direction of the fibres, the latter tend to become unstable and to buckle, like long slender struts. This makes high glass content composites with a unidirectional arrangement of fibres relatively less effective under compression than under tension. The high glass content composites are also relatively less effective under shear because the resistance to shear between fibre layers depends predominantly on the plastics matrix.

To define qualitatively all these mechanical characteristics of glass fibre/plastics is difficult and on the basis of experimental observations alone impossible, because of the wide variety of different percentage combination of glass fibres and plastic, the variation of properties with direction and the different load systems to which materials are subjected. However, what cannot be covered by direct experimental observation can be filled in, in a number of instances, by theoretical considerations.

5. THEORY OF LINEAR ELASTIC ORTHOTROPIC MATERIALS

The outstanding example of this is the variation in the tensile stiffness of orthotropic laminates. In this case the stiffness in any direction in the plane of laminate can be calculated with a high degree of accuracy given only the values of four deformational constants.

The derivation of the equation which makes this possible has been given by several authors and it is sufficient here to consider only the underlying assumptions and the final results. The relevant analytical work antedates, in fact, the development of GRP and was prompted originally by the anisotropy of crystals and then of plywood laminates[10–14]. At the outset the material is assumed to have orthogonal planes of elastic symmetry and to be subjected to plane stress in one of these planes: if the plane of symmetry in which the plane stress acts is the X–Y plane, as shown in *Figure 14.4*, then the OX and OY are the orthotropic axes. Then, assuming that the parameters of deformation referred to the orthotropic axes are known, the corresponding parameters referred to any other set of orthogonal axes, such as OA and OB, with OA making an angle θ with OX, can be found by writing the usual three sets of equations:

(i) equations of equilibrium, which relate to each other the stress components referred to the two sets of axes,

(ii) compatibility equations, which similarly relate the strain components,
(iii) constitutive equations, which relate the stress to the strains.

In this case, the material is assumed to be linearly elastic and the constitutive equation becomes a generalised form of Hooke's law, which means that each component of strain is a linear function of all the components of stress. Hence, for the plane stress problem considered, there are three linear equations with nine coefficients but by symmetry the latter reduce to six.

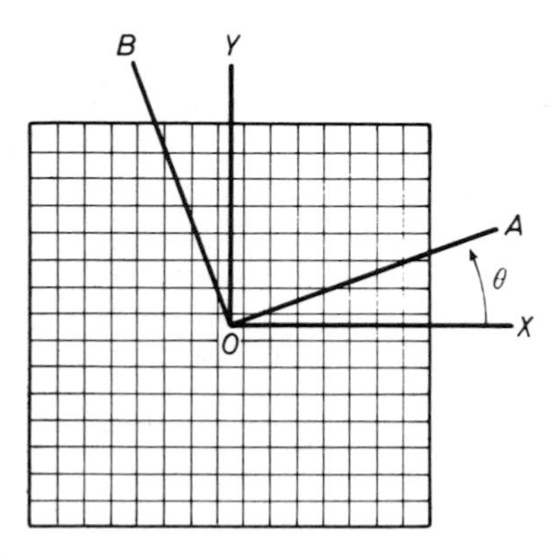

Figure 14.4. Orthotropic axes OXY and arbitrary axes OAB

Then, making use of the fact that, when the stress directions are parallel or perpendicular to the orthotropic axes, the extensional and shear strain components are associated respectively with the extensional and shear stress components only and putting all but one stress component equal to zero, it can be shown that the ratio strain to stress in the OA direction is,

$$\frac{1}{E_A} = \frac{\cos^4\theta}{E_X} + \frac{\sin^4\theta}{E_Y} + \left[\frac{1}{G_{XY}} - \frac{2\nu_{yx}}{E_X}\right]\cos^2\theta.\sin^2\theta \qquad (1)$$

where E_A = tensile modulus in the OA direction
E_X = tensile modulus in the OX direction
E_Y = tensile modulus in the OY direction
G_{XY} = shear modulus in the XY plane
ν_{YX} = Poisson's ratio, the ratio of extensional strain in the OY direction to the strain in the OX direction.

Of the four elastic constants which appear on the right-hand side of equation (1), E_X and E_Y can be readily determined under uniaxial tension and ν_{YX} by taking simultaneous readings of the longitudinal extension and lateral contraction. The shear modulus, or modulus rigidity, G_{XY} can be determined by loading square plate specimens across their diagonals and observing the deflection at the centre[12, 15]. If G_{XY} and ν_{YX} cannot be determined then, instead of an equation for E_A in terms of E_X, E_Y, G_{XY} and ν_{YX}, it is possible to obtain an equation in terms of E_X, E_Y, and E_{45}, where E_{45} = tensile modulus of 45 degrees to the OX (and OY) direction.

$$E_A = \frac{E_X E_Y}{E_Y \cos^4\theta + \left[\frac{4E_Y E_X}{E_{45}} - E_X - E_Y\right] \sin^2\theta \cos^2\theta + E_X \sin^4\theta} \tag{2}$$

Results obtained with equation (1) are illustrated in *Figure 14.5*, in which the curve corresponds to the calculated values of E_A and the circles to the values of E_A determined experimentally with the tensile specimen cut at different angles from the same unidirectional laminate. As can be seen, the agreement between the calculated and experimental values is close. *Figure 14.5* also shows the variation in stiffness in a highly anisotropic glass fibre/plastic and the increase in stiffness over that of the plastic, even at right angles to the lay of the fibres.

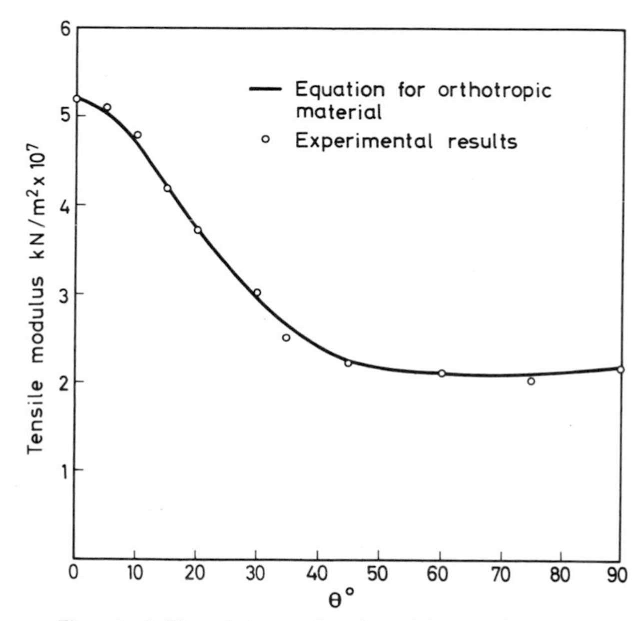

Figure 14.5. Theoretical curve of tensile modulus v. angle and experimental results

Of the other five constants, one is S_{BB}, the modulus of compliance in the OB direction, which is equal to $1/E_B$ and which is given by an equation similar to (1), i.e.,

$$S_{BB} = \frac{1}{E_B} = \frac{\sin^4\theta}{E_X} + \frac{\cos^4\theta}{E_Y} + \left[\frac{1}{G_{XY}} - \frac{2\nu_{YX}}{E_X}\right] \sin^2\theta \cos^2\theta \tag{3}$$

Two others are S_{AB}, the quotient of Poisson's ratio and the modulus of elasticity, and $S_{AB,\,AB}$, the ratio of the shear strain to the shear stress, which are given by the following equations:

$$S_{AB} = -\frac{\nu_{AB}}{E_B} = \left[\frac{1}{E_X} + \frac{1}{E_Y} - \frac{1}{G_{XY}}\right]\sin^2\theta\cos^2\theta - \frac{\nu_{YX}}{E_X}\left[\cos^4\theta + \sin^4\theta\right] \quad (4)$$

$$S_{AB,AB} = \frac{1}{G_{AB}} = 4\left[\frac{1}{E_X} + \frac{1}{E_Y} + \frac{2\nu_{YX}}{E_X}\right]\sin^2\theta\cos^2\theta + \frac{1}{G_{XY}}\left[\cos^2\theta - \sin^2\theta\right] \quad (5)$$

The remaining two compliances are $S_{AB.A}$ and $S_{AB.B}$, which are the constants of proportionality between the shear strain in the AB plane and the tensile stress in the OA and OB directions, respectively. As already implied, shear strains of this kind do not arise when the tensile stresses are parallel to the orthotropic axes and the two constants have no equivalent in the elastic theory of isotropic materials. In consequence, they cannot be expressed in terms of the familiar elastic constants. The equations to them are:

$$S_{AB.A} = \left[-2\left(\frac{\cos^2\theta}{E_X} - \frac{\sin^2\theta}{E_Y}\right) + \left(\cos^2\theta - \sin^2\theta\right)\left(\frac{1}{G_{YX}} - \frac{2\nu_{YX}}{E_X}\right)\right] x \sin\theta\cos\theta \quad (6)$$

$$S_{AB.B} = \left[-2\left(\frac{\sin^2\theta}{E_X} - \frac{\cos^2\theta}{E_Y}\right) - \left(\cos^2\theta - \sin^2\theta\right)\left(\frac{1}{G_{XY}} - \frac{2\nu_{YX}}{E_X}\right)\right] x \sin\theta\cos\theta \quad (7)$$

6. FAILURE UNDER COMPLEX STRESS AND CREEP

In addition to the above equations which make it possible to calculate the moduli of compliance and hence the deformation in any direction in the plane of an orthotropic laminate from the basis of a few experimentally determined constants, it is also possible to predict failure under a complex stress system from knowledge of the stress under which glass/plastics fail when subjected to uniaxial tension, or compression, or simple shear.

Predictions of failure depend on the criteria adopted for it. These include the maximum stress theory which postulates that the material will fail if a stress component in the direction of one of the material axes of symmetry reaches the ultimate strength in that direction. There is also a maximum strain theory which is similar to the maximum stress theory. At present, however, it is the maximum work theory that is favoured. This theory is based on the assumption that the failure condition is a quadratic function of the stress components and leads to a relationship which is a generalisation of the von Mises' criterion for the yielding of isotropic materials under complex states of stress[16, 17]. In particular, for plates in plane stress this theory takes the following form:

$$\left(\frac{\sigma_x}{\sigma(x_u)}\right)^2 - \frac{\sigma_x\sigma_y}{\sigma^2(x_u)} + \left(\frac{\sigma_y}{\sigma(y_u)}\right)^2 + \left(\frac{\tau_{xy}}{\tau(xy_u)}\right)^2 = 1 \tag{8}$$

where σ_x, σ_y = direct stress components in the OX and OY directions, respectively
$\sigma(x_u), \sigma(y_u)$ = ultimate strength in the OX and OY directions, respectively
τ_{xy} = shear stress in XY plane
$\tau(xy_u)$ = ultimate shear stress in XY plane.

The most widely quoted relationship has been the so-called 'interaction formula' originally established at the U.S. Forest Products Laboratories, in work on plywood[18]. In terms of the above symbols, this formula is:

$$\left(\frac{\sigma_x}{\sigma(x_u)}\right)^2 + \left(\frac{\sigma_y}{\sigma(y_u)}\right)^2 + \left(\frac{\tau_{xy}}{\tau(xy_u)}\right)^2 = 1 \tag{9}$$

Equation (9) differs from the more recently derived equation (8) only in not including the second term of the latter. In practice, this term is often small and the two equations are, therefore, approximately the same.

Attempts have also been made to set up mathematical models for the creep of glass fibre/plastics. The resulting equations range from simple empirical expressions, which give tensile creep strain as a power function of time, to more elaborate equations which incorporate deductions from the activation energy theory that the creep rate is proportional to the hyperbolic sine of stress. The last is exemplified by the following equation, which has been found to describe creep of several plastics composites with reasonable accuracy[6]

$$\varepsilon = \varepsilon_0 \sinh\frac{\sigma}{\sigma_\varepsilon} + m\left(\frac{t}{t_0}\right)^n \sinh\frac{\sigma}{\sigma_m} \tag{10}$$

where ε = tensile creep strain
σ = constant tensile stress
t = temperature
and ε_0, σ_ε, m, n, σ_m and t_0 are constants, which may depend on temperature.

Another equation, which has been put forward much more recently and which is claimed to cover longer periods of time than equation (10), is [19]

$$\log_{10}(A - \log_{10} E) = B + C\log t \tag{11}$$

where E = tensile creep modulus, and A, B and C are constants.

7. MICROMECHANICS APPROACH TO STIFFNESS AND STRENGTH

A different and more ambitious objective of analytical studies are mathematical models which, instead of serving to extrapolate or interpolate the behaviour of GRP from experimental observations, would make it possible to predict the behaviour from knowledge of the mechanical characteristics and morphology of their constituents as discussed in Chapter 13.

To establish the required models it is necessary to consider the micro-mechanics instead of the macromechanics, of composite. In the case of composites with a unidirectional arrangement of fibres and under uniaxial tensile loads applied in the direction of the fibres this is easy and a model can be established readily, if two assumptions are made. The first of these is that both constituent materials are linearly elastic, which is true of the glass fibres but only approximately true of the plastics matrices. The second assumption is that the strains in the two materials are equal, which implies perfect bonding between them and is sometimes dignified by the title 'Theory of Combined Action'[20].

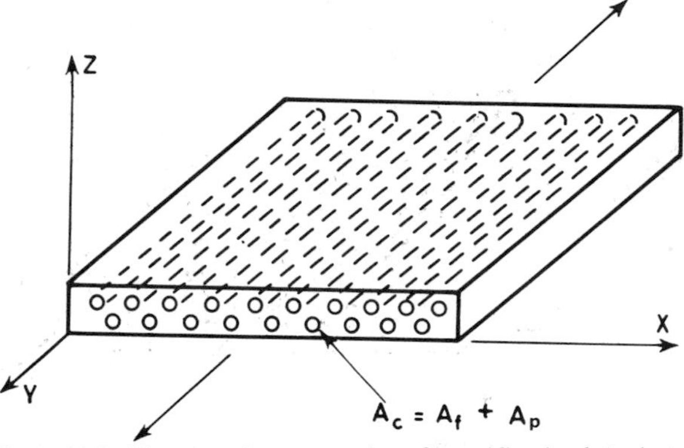

Figure 14.6. Diagrammatic representation of a unidirectional laminate

The first assumption is tantamount to saying that both materials obey Hooke's law, i.e.

$$\varepsilon = \frac{\sigma}{E} \tag{12}$$

where ε = strain, σ = stress, E = tensile modulus.

The second assumption implies that

$$\varepsilon_c = \varepsilon_f = \varepsilon_p \tag{13}$$

where the subscripts c, f, p apply to the composite, fibres and plastic respectively. Combining equations (12) and (13) with the equilibrium equation

$$A_c\sigma_c = A_f\sigma_f + A_p\sigma_p \tag{14}$$

where A = cross-sectional area, leads to the following equation for the modulus of the composite,

$$E_c = E_f\frac{A_f}{A_c} + E_p\frac{A_p}{A_c} \tag{15}$$

This means that the modulus of the composite is the volume weighted average of the moduli of the fibres and the plastic, which is often referred to as the 'Law of mixtures'. Equation (15) is, strictly, only correct if the Poisson's ratios of the constituent materials are the same but although this is not generally the case the differences are small and the 'Law of mixtures' gives a good indication of the longitudinal stiffness of unidirectional composites.

Following a very similar argument, it can be shown that if the plastics matrix is highly compliant, i.e. that its strain to fracture is greater than that of the glass fibres, the ultimate strength of the composite is given by:

$$\sigma_{cu} = \sigma_{fu}\left(\frac{A_f}{A_c} + \frac{E_p A_p}{E_f A_c}\right) \tag{16}$$

where suffix u signifies ultimate. If the plastics matrix has a lower strain to failure than the glass fibres, equation (16) reduces to:

$$\sigma_{cu} = \sigma_{fu}\frac{A_f}{A_c} \tag{17}$$

Transverse and shear properties of composites are more difficult to predict analytically. Several models of different orders of sophistication have been postulated with varying degrees of success[13, 14, 17, 21–23]. In general, predictions of stiffness tend to be more accurate than the predictions of strength. The simplest solution is to assume that the fibres have no influence on properties other than those in the direction of their alignment which was done in some of the early analyses, but this is inappropriate whenever there is a significant percentage of fibres. An example of this is provided by *Figure 14.5* which shows that the modulus in the transverse direction is several times higher than that of the plastics matrix.

Apart from making it possible to predict the gross mechanical properties of composites from knowledge of the properties of their constituents, a study of micro-mechanics of composites has also provided detailed information about stresses in the composite. In particular, they have brought out the existence of stress concentrations in the matrix near the ends of discontinuous fibres, which can lead to debonding at the fibre tips. The stress concentrations at fibre tips and the local failures which they might induce do not, however, govern the strength of composites which depends on the transfer of stresses by shear between the matrix and fibre over the whole length of the fibre. Thus, given sufficiently long fibres, the fibres are loaded up to their breaking point through interface shear stresses which are low except at the fibre tips, because they are distributed over a large area. The ratio of fibre length to diameter at which the loading of the fibres becomes high enough to lead to their fracture is called the critical aspect ratio. Fibres with smaller aspect ratios do not effectively influence the properties of the composite. In fact, subcritical length fibres fail to reinforce the matrix effectively and single fibres, or a few widely dispersed fibres, even decrease the strength of brittle matrices. This happens because the stiffer fibre breaks well before the matrix does and the break initiates a crack in the matrix which propagates resulting in a lower composite strength.

The spacing of fibres also has a significant effect on the local stresses. When the spacing fibres is greater than 5 or 6 times the fibre diameter, there is no significant interaction. When the spacing is closer and the fibre overlap is 100 or 0 per cent, however, which corresponds respectively to the fibres lying completely adjacent to each other and barely overlapping, the stress concentrations at the fibre tips of high stress associated with the tips of individual fibres are brought together and the stress concentration factors can be of the order of ten. However, when the overlap is 50 per cent the stress concentrations are less than with single fibres but, in contrast to the 0 and 100 per cent overlap cases, they increase as the fibre spacing increases[17].

Close spacing corresponding to the maximum packing of fibres, can also reverse the nature of the stresses in the matrix. Thus, instead of the fibres being under pressure from the resin, the fibres are under tension because of the tensile stresses due to shrinking, or cooling, of the 'tricorn' of resin between any trio of glass fibres. The tensile stresses tend to destroy the bond between the glass and the plastics matrix[24].

Further work is, however, required to elucidate this and other aspects of the action occurring at the fibre-matrix interface. This includes the mechanism by which loads are transferred between the fibres and the matrix. At present there are two principal theories. According to one the transfer of loads is due to friction at the interface; according to the other it is due to chemical coupling. In essence, the first theory postulates that during the polymerisation or curing of the plastic, the matrix shrinks around the fibres with sufficient force to provide frictional resistance to the movement of the fibres relative to the matrix. According to the second theory, certain functional groups in the polymeric matrix react with the surface of the fibre to form a chemical bond or, where a chemical coupling agent is used, one part of it reacts with the fibre and another part reacts, or is compatible with the polymeric matrix.

REFERENCES

1. Schuerch, H. U., Burggraf, O. R. and Kyser, A. C. 'A Theory and Application of Filamentary Structures, *NASA tech. Note D-1692*, December, 1962
2. 'Plastics for flight vehicles', Part I, *Reinforced Plastics*, MIL-HDBK-17, Armed Forces Supply Support Centre, Washington, 1959
3. Beek, M. H. B. van den, and Hamm, G. 'The Influence of the Ultimate Strain of a Polyester Resin on Mechanical Properties of Laminates, *5th Int. Reinforced Plastics Conf.*, British Plastics Federation, London, 1966
4. Ogorkiewicz, R. M. (ed.) *Engineering Properties of Thermoplastics*, Wiley, London, 1970
5. Ogorkiewicz, R. M., Culver, L. E. and Bowyer, M. P. 'Stress–strain Characteristics of an Acetal Copolymer under Different Types of Tensile Loading, *Plastics and Polymers*. **37**, June 1969.
6. Findley, W. N. 'The Effect of Temperature and Combined Stresses on Creep of Plastics, *2nd Int. Reinforced Plastics Conf.*, British Plastics Federation, London, 1960
7. Denney, M. A. and Martindale, J. C. 'An Approach to the Tensile Testing of Reinforced Plastics at Elevated Temperatures, *3rd Int. Reinforced Plastics Conf*, British Plastics Federation, London, 1962
8. Hertz, J. 'The Effect of Cryogenic Temperature on the Mechanical Properties of Reinforced Plastics Laminates, *S.P.E.Jl*, **21**, Feb. 1965
9. James, D. I., Norman, R. H. and Stone, M. H., 'Water Attack on the Glass-resin Bond in GRP', *Plastics and Polymers*, **36**, Feb. 1968
10. Love, A. E. H. *A Treatise on the Mathematical Theory of Elasticity*, Cambridge University Press, 1927

11. March, H. W. 'Stress–strain Relations in Wood and Plywood Considered as Orthotropic Materials', *Forest Prod. Lab. Rep.* No. 1503, Madison, Feb. 1944
12. Hearman, R. F. S. *An Introduction to Applied Anisotropic Elasticity*, Oxford University Press, 1961
13. Shaffer, B. W. 'Material Properties of Reinforced Plastics', *S.P.E. Trans.*, October, 1964
14. Baer, E. (ed.) *Engineering Design for Plastics*, Reinhold, New York, 1964
15. Ogorkiewicz, R. M. and Sayigh, A. A. M. 'Mechanical Behaviour of Rigid PVC', *British Plastics*, **38**, October, 1965
16. Tsai, S. W. 'Strength Characteristics of Composite Materials, *NASA CR-224*, Washington, 1965
17. Schwartz, R. T. and Schwartz, N. S. *Fundamental Aspects of Fibre Reinforced Plastic Composites*, Wiley, New York, 1968
18. Norris, C. B. and McKinnon, P. F. 'Compression, Tension and Shear Tests on Yellow Poplar Plywood Panels of sizes that do not Buckle with Tests made at Various Angles to the Face Grain, *Forest Lab. Rep.* No. 1328, 1946
19. McLoughlin, 'A New Creep Law for Plastics', *Modern Plastics,* Vol. 45, February, 1968.
20. Sonneborn, R. H. *Fibreglass Reinforced Plastics,* Reinhold, New York, 1954.
21. Ekvall, J. C. 'Elastic Properties of Orthotropic Monofilament Laminates', *ASME Paper No. 61-AV-56*, 1961.
22. Allison, I. M., and Holloway, L. C. 'The Stresses around Fibres', *Research projects in Reinforced Plastics Conference*, Plastics Institute, London, 1968
23. Holloway, L. (ed.) '*Composite Materials*', Elsevier, Amsterdam, 1966
24. Outwater, J. O. and West, D. C. 'Stress Distribution in the Resin of Reinforced Plastics, *Modern Plastics*, **39**, September, 1961

15

Resin-Glass Interface

R. H. NORMAN, M. H. STONE and W. C. WAKE

1. INTRODUCTION

A bundle of parallel glass fibres is by itself useless as a load-bearing structure, but embedding the glass in a resin matrix gives the necessary stiffness in shear and compression. The glass and resin are mutually reinforcing: the strong, stiff glass carries most of the stress and the polymer matrix distributes the external load to all the fibres, while at the same time protecting them. This load sharing requires that stress be transferred across the interface between glass and resin, and we are thus concerned to know the physical and chemical properties of the interfacial region. There is no sharp, well-defined interface between glass and resin, for the glass is coated with a heterogeneous mixture itself resinous in nature and this is coated with the resin forming the matrix of the composite structure. The constitution and properties of the resin next to the coating on the glass may differ from those of the bulk material. The whole interfacial region is about 10^{-2} μm thick or greater (*Figure 15.1*).

Our interest therefore covers three main components and the two boundaries between them. The main concerns will be the chemical reactions at these boundaries and the degree of molecular contact that precedes or follows the reactions. Our account will take the form of a life history of the interfacial region, starting with the deposition of surface treatment on the glass and continuing with impregnation of the strands by resin. When the resin is cured, the interfacial region is fully formed and is then subject to ageing; this commonly leads to rupture of the glass-to-resin bond.

2. ADHESION, WETTING AND ADSORPTION

Stress can only be transferred across an interface between two materials if they are in intimate molecular contact, separated only by about the same distance as the molecules inside bulk materials. The materials are then

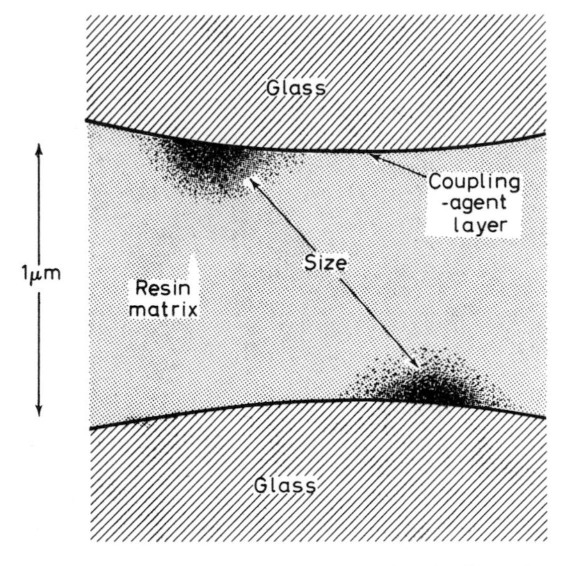

Figure 15.1. Section of a composite normal to the fibre axis: an area of the interfacial region about 2 × 2 μm *between two* 10 μm *fibres* 1 μm *apart* (*to scale*). *Diffuse boundaries of size particles indicate partial mixing with resin*

adhering to each other. There need not be any chemical linking of the materials; they merely have to be so close that the normal inter-molecular forces are operative. This simple picture underlies the Adsorption Theory of Adhesion, the term *adsorption* denoting the sticking of molecules of a fluid to a solid surface. This adsorption theory is appropriate for glass-to-resin adhesion in GRP although another theory of adhesion based on molecular diffusion may be relevant to the interaction of the surface-treatment with the resin. In this theory the usual inter-molecular forces are equally important; but instead of a sharp interface there is a region, thick relative to molecular dimensions, over which long molecules have intermingled to form a solid solution.

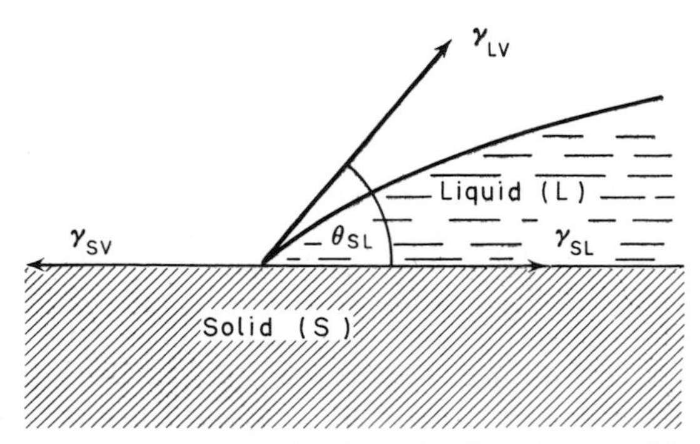

Figure 15.2. Balancing of surface tensions in wetting of a solid by a liquid

In making GRP we apply two liquids in turn to the glass and so we must briefly consider the process of wetting, which determines how well they are adsorbed. The completeness and ease of wetting depends on the geometry of the system, the amount of liquid available, and the contact angle of the liquid on the solid. The balance between the surface tension forces, which determines the contact angle (θ) (*Figure 15.2*), is given by Young's equation,

$$\gamma_{SV} = \gamma_{SL} + \gamma_{LV} \cos \theta_{SL}$$

If $\theta > 0$, there are two practical consequences that can be important. Firstly, a thin liquid film will tend to form separate droplets and this causes uneven deposition of any solute or dispersant the liquid contains. Secondly, the liquid is less likely to wet all the surfaces of a rough or porous structure such as a bundle of glass fibres, particularly if θ is about 90 degrees or more. This important idea is given empirical expression in Zisman's Critical Surface Tension (γ_c): i.e. the value of γ_{LV} for which θ is just zero on a particular surface[1]. This is largely a property of the solid surface although it can depend on the solid-liquid combination. Thus for good adhesion we hope for a system where γ_c is greater than γ_{LV}. For GRP materials this cannot always be satisfied although fortunately the contact angle is not usually very high.

However, in practice it is often the advancing dynamic contact angle that determines wetting behaviour in GRP forming processes. This is invariably higher than the equilibrium angle and can be very large for viscous liquids since it is a combined rheological and surface property[2]. There is little information on this important aspect.

It has been stated in this section that physical intermolecular forces can explain adhesion but, as will be shown later, chemical bonds are needed between resin and glass for the best results. This arises from the ease with which physical bonds can be displaced by water so that the strengths ordinarily measured are determined by partial degradation with this almost universal contaminant. Chemical bonds are much more slowly attacked and bond strength is the better retained.

3. SURFACE TREATMENT OF THE GLASS

We shall emphasise treatment by aqueous mixtures containing silane coupling-agents because these give a good bond, have been most studied, and the principles involved are relevant to other systems. We prefer the descriptive term *coupling-agent* to the vague *finish* or the misleading *keying-agent*, although these are equivalent terms denoting the material used to improve the bond of resin to glass. Treatment mixtures also contain a binder/lubricant but very little has been published on this subject. For convenience we shall call this constituent the *size* although in common usage *size* is often a complete treatment containing both coupling-agent and binder/lubricant.

3.1. THE GLASS SURFACE

The bulk composition of glasses suggests that they should have high surface energy and so be wetted by most liquids. Real surfaces, however, are always contaminated, initially with water vapour and later with traces of organic materials if the surface remains exposed. The contamination with water occurs between bushing and sizing point. During this short period the glass surface is struck by water molecules 10^5 times the monolayer capacity although, of course, the surface does not reach equilibrium in this time. Fortunately this adsorbed water is unlikely to impede wetting by aqueous treatments because γ_c for the surface presented to these treatments is about 70 dyne cm^{-1}; i.e. greater than their surface tension[3]. Coupling-agents are also applied to glass that retains a carbonaceous and hydrophobic residue from starch/oil size incompletely removed by heating in air. Unfortunately, there are no data on the wettability of such surfaces.

The chemical nature of the glass surface is not known in detail. The surfaces of the network-forming oxides are normally composed mainly of hydroxyl groups and the glasses are probably similar. Ions such as Na^+, K^+, Ca^{2+}, Mg^{2+}, Fe^{2+}, Fe^{3+} may also be present in hydrated form at the surface, and there is evidence that some constituents tend to concentrate there[4–6]. Some glasses tend to separate into phases differing in SiO_2 content and so giving a patchy surface[7,8]. Finally, we should recall that the surface hydroxyl groups will be less reactive than those of small molecules in solution because of immobility and steric hindrance.

3.2. NATURE OF THE TREATMENT MIXTURE AND DEPOSITION ON THE GLASS

Some kind of size is always needed to aid processing through all stages up to and including impregnation by resin, and it does two main jobs. Firstly, it sticks the fibres together weakly and maintains them in the form of a strand; without this action the fibres fray out and the strand becomes unmanageable. Secondly, the size prevents fibres coming into intimate contact and abrading one another, which would weaken them by creating surface defects. Size is therefore always applied to the newly-formed fibres as they come together just below the bushing.

If the size is needed mainly as a weaving aid it is usually a starch/oil aqueous emulsion without a coupling-agent. Before the cloth is treated with the coupling-agent the size is removed by heating, but any residue may impede wetting by the coupling-agent solution and may also decrease chemical linkage between glass and coupling-agent.

The other common type of size, the aqueous dispersion of polymers such as poly(vinyl acetate) (PVAc) is usually mixed with the coupling-agent in a single treatment liquid. The presence of size probably reduces chemical linking of glass to coupling-agent but there is little published work on this and its importance is difficult to gauge. The amount of size deposited is commonly around 1–2 per cent by weight of the glass or about 2–4 per cent by volume, and if present as a uniform layer would increase the filament radius by 1–2 per cent or 0.05–0.1 μm. However, the size particles in the dispersion are spheres about 1 μm in diameter and if they do not spread cannot cover the surface. Moreover, when the treatment liquid first contacts the glass only a minor fraction of the size particles will be close enough to the surface to be adsorbed before there is appreciable adsorption of the much smaller and more mobile coupling-agent molecules. Thus, probably only a minor fraction of the glass would be prevented from initially adsorbing the coupling-agent.

The other main constituent of the treatment liquid, and sometimes the only one, is the coupling-agent. We shall concentrate on the alkoxy silanes of general form $R'Si(OR)_3$, where R′ is a group that can react with the polymer matrix, and R is usually a short alkyl chain ($—CH_3$, $—C_2H_5$). These are commonly applied at about 0.5 per cent concentration in aqueous solution or dispersion, in which they hydrolyse and condense to give polysiloxanes of a low degree of polymerisation which retain many active silanol groups.

The thickness of the initial layer[9–12] is 0.003–0.03 μm and adsorption is near complete in a few minutes[9, 10, 12]. Coupling-agent solutions have surface tensions in the range 40–70 dyne cm^{-1} and they should wet bare glass[9, 13, 14], although residues from starch/oil size could give a surface having γ_c less than 40 dyne cm^{-1}, with consequent incomplete wetting.

3.3. THE SURFACE LAYER AFTER DRYING

Consider firstly the layer given by coupling-agent alone. The total amount is usually in the range 0.1–1.0 per cent by weight of the glass and would be 0.005–0.05 μm thick on 10 μm diameter filaments if uniformly deposited. The deposit tends to collect along lines where adjacent filaments have been closest[15], and so the thickness elsewhere will be rather less. Early electron-micrographs showed apparent agglomerates of coupling-agent large enough to account for the whole deposit, with large areas of apparently bare glass between them[15, 16]. The authors of this work showed due caution in their conclusions, but have since been uncritically quoted. Most later evidence suggests that coverage is essentially complete; or at least there are no large areas of bare glass. For example, the contact angle with water is high, particularly where the coupling-agent has been heated to increase polymerisation[13, 17–22], and the character of dye adsorption also suggests the glass surface is screened[18, 23]. The only later work suggesting incomplete coverage is one example of a low contact angle with water, which is inconsistent with the same authors' electron-micrographs[9], and titration curves of surface hydroxyl groups[6, 24] which are difficult to interpret unequivocally. We conclude that hydrolysed alkoxy silanes cover the glass completely although there may sometimes also be agglomerates superimposed on a more uniform layer. The surface of the coupling-agent layer is hydrophobic, with γ_c 25–40 dyne cm^{-1} appropriate for a mainly hydrocarbon surface[13, 17, 18–22].

There is no direct evidence that hydrolysed alkoxy silanes react with the glass surface under typical treatment conditions. However, much indirect evidence suggests this: for example, the difficulty of desorbing monofunctional silanols; and the lower composite strengths if the coupling-agent cannot react with the glass. Thus we see the coupling-agent as a polymeric layer of variable thickness, several tens of molecules deep, more or less cross-linked and reacted to form Si—O—M bonds with the glass, where M is Si, Al, or B of the glass surface.

When size is also present, it will be clear from earlier comments that we know much less about the surface layer. Total solids on the glass are now 0.5–2.5 per cent by weight, of which most is size. Somtimes this deposit is so heavy and uneven as to be visible under the microscope[23], and its outer surface is mainly size[18]. Thus we can picture a relatively thin coupling-agent layer containing partially embedded size particles that have softened and flowed out over the coupling-agent. Both size and coupling-agent may be soft at the same time during drying, permitting their interdiffusion.

Finally, we must note that minor constituents in the size may also conflict with the aim of having a good glass-to-resin bond.

4. IMPREGNATION WITH RESIN: WETTING, INTERACTIONS AND STRUCTURE

4.1. WETTING OF THE SURFACE TREATMENT BY RESIN

The critical surface tension of surface treatments is 25–40 dyne cm^{-1}, while the surface tension of resins is 35–50 dyne cm^{-1}, with polyester towards the lower end of this range and epoxy towards the upper[13, 17, 25–27]. Thus resins should have low or even zero contact angles on the surface treatments, but in fact diverse and sometimes high values are found[2, 9, 13, 17, 27, 28]. Possibly preferential adsorption of one resin component creates in some systems an unexpectedly low energy surface (*see* Section 4.2).

However, in practice it is the advancing dynamic contact angle that is important and this can be greater than 90 degrees even for low interface velocities[2]. This leads to entrapment of air in the strand, and the resulting voids have a marked effect on composite strength[17, 29–34]. The dynamic angle is largely a rheological rather than a surface property[2, 17, 34], suggesting improvement by reduced resin viscosity.

4.2. PHYSICAL INTERACTION OF THE LIQUID RESIN WITH THE SURFACE TREATMENT

The resins contain at least three components, resin, cross-linking agent and catalyst. Preferential adsorption from among these components has been demonstrated in two cases[35, 36] but the effect on the properties of the composite is unknown. Concentration of one component at the interface may affect chemical reaction between coupling-agent and resin, or the resin curing reaction.

The other possible physical interaction of importance is interdiffusion of resin with the materials of the surface treatment, giving a region of varying composition rather than a sharp interface. Interdiffusion is unlikely if only fully cross-linked coupling-agent is on the surface, but if size such as PVAc is present then interdiffusion of this with the resin undoubtedly does occur. At the extreme the size might be completely dissolved by the resin. Parts of the interfacial region may therefore consist of a solution of size in resin, and other parts of size swollen by the resin or one of its components. The size is a significant proportion of the resin, about 5–10 per cent by volume, and locally near the interface will form a much higher proportion.

4.3. CHEMICAL INTERACTIONS AT THE INTERFACE

The curing reactions of the resins are well described elsewhere and it remains to consider the reaction of coupling-agent with resin.

There is now no doubt that the highest composite strengths and water-resistances are given by those coupling-agents that should react most readily with the resins[11, 16, 37–47]. In particular, styrene copolymerises less readily with vinyl silanes than with methacryloxy silanes; and this is matched by lower strength and water-resistance of polyester resin composites incorporating vinyl silanes[16, 41–44, 46–53]. The more direct evidence on chemical bond formation is of uncertain interpretation, but is consistent with there being reaction between resins and some coupling-agents[11, 12, 42, 49, 54, 55]. Nothing is known, however, about the extent of chemical bond formation in actual composites; in particular, size may remain as an undissolved layer partly covering the coupling-agent and thus substantially reducing reaction.

5. FAILURE AT THE INTERFACE

A mass of indirect evidence already cited suggests that properties of composites, particularly mechanical properties, are often limited by failure at the interface, and this is supported by direct evidence of separation at the interface by visual observation, photo- and electron-micrography (*Figure*

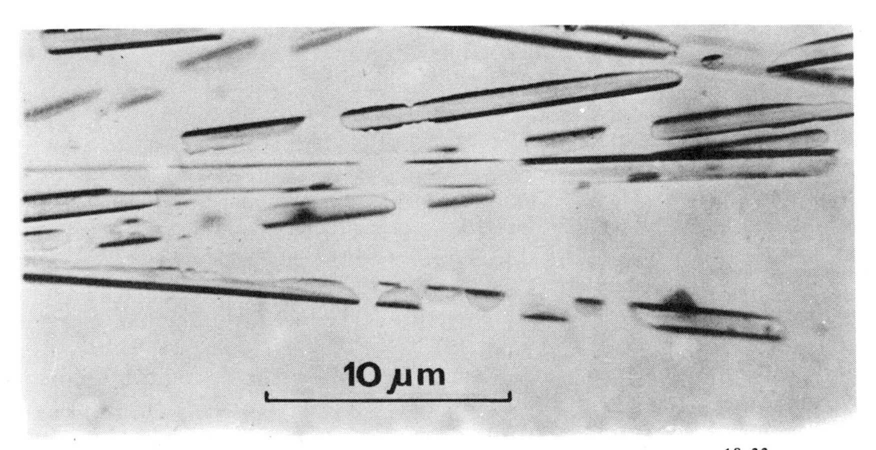

Figure 15.3. Bond failure on 10 μm *fibres; sample exposed to boiling water* [18,22]

15.3)[18, 20, 22, 23, 39, 40, 56–61]. We end this chapter by considering aspects of interfacial failure, with particular emphasis on the important case of attack by water.

5.1. TRANSPORT OF WATER TO INTERFACE

Water can only reach a bonded interface by diffusion both through the matrix and along the interfacial layer. The path length for diffusion may, however, be shortened when ingress into a composite is assisted by capillary flow along channels, for example voids or stress cracks[62–66]. The word *wicking* has been correctly used to describe this process; but it has also been used rather loosely to denote the accumulation of water in the channels formed around filaments that are debonding. If the resin was initially in intimate contact with the filaments, then this latter process can hardly be termed wicking, which suggests capillary flow in pre-existing channels whose width is much larger than molecular size. The *appearance* of wicking is given by the spread of debonding from exposed fibre ends in some systems. This occurs either because there is only a physical bond that can be rapidly displaced by water; or because diffusion along the interfacial layer is much faster than diffusion through the matrix[13]. Wicking in these senses, however, is by no means general and diffusion of water through the matrix can equally well cause debonding[18, 22, 61].

The attainment of equilibrium water content in the resin is fast compared to the progress of debonding, provided an effective coupling-agent is used[18, 22]. The water sorption is essentially an osmotic effect, and is increased by soluble impurities in the resin and decreased by solutes in the external aqueous medium[67].

5.2. CHEMICAL ACTION OF WATER AT THE INTERFACE, AND THE SITE OF DEBONDING

In Sections 3 and 4 we emphasised the importance of chemical linkage between glass and resin via the coupling-agent. It follows that debonding necessitates chemical bond cleavage, and in the presence of water this is a hydrolytic reaction. Indeed, water will be the active agent for slow breakdown under moderate stress, although the overall process is accelerated by the stress[68].

In the usual GRP materials two types of chemical bond are susceptible to hydroylsis, siloxane and ester linkages. The former occur in the glass, between glass and coupling-agent, and within the coupling-agent. The ester linkages occur in polyester resins, in anhydride-hardened epoxies, and in some coupling-agents.

Hydrolysis reactions are, in principle, reversible and equilibrium does not necessarily favour scission. In GRP the hydrolysis fragments are attached to different networks and would tend to be separated by any stress at the interface. Thus the hydrolytic cleavage becomes irreversible. The actual site of final complete failure is not in general known, and probably varies according to the materials present and the conditions of attack. Susceptibility to hydrolysis has been qualitatively demonstrated for all three main components, glass, coupling-agent, and resin. This is well known for glass but hydrolysis of coupling-agent adsorbed on glass is difficult to demonstrate because the glass substrate is simultaneously attacked, with consequent detachment of the coupling-agent. Degradation of the coupling-agent is no doubt a partial explanation for some of the observations[9, 12, 15, 18, 22, 54, 69–73]. The hydrolytic degradation of polyester and epoxy resins is well established. It is accompanied by the extraction of low molecular weight materials and the formation of characteristic circular cracks[18, 22, 67, 74–76]. The rates of hydrolysis can be appreciable in the context of the life desired for a GRP structure, although they are difficult to measure for the separate materials when combined in a composite. Degradation products of glass and resin found in the water in which a polyester composite had been immersed for one month at 50°C suggested breakdown of a layer hundreds of Ångstroms thick at the interface[77]. In boiling water glass loses a layer 0.01 μm thick in 28 h extraction, and in similar conditions a coupling-agent layer on glass is partly degraded in a few hours[9, 12, 18, 22]. Simple esters in aqueous solution reach hydrolytic equilibrium in 8–10 days at 100°C and 30–40 days at 78°C[78, 79].

The activation energy for ester and siloxane hydrolysis is 11–24 kcal mole^{-1} and the times for cracks to appear in polyester resin at different temperatures imply an activation energy of 24 kcal mole^{-1} for the overall process[76, 80–89].

Bond lives in epoxy and polyester composites at 60°C and 100°C give an overally activation energy[90]* of 15–33 kcal mole^{-1}.

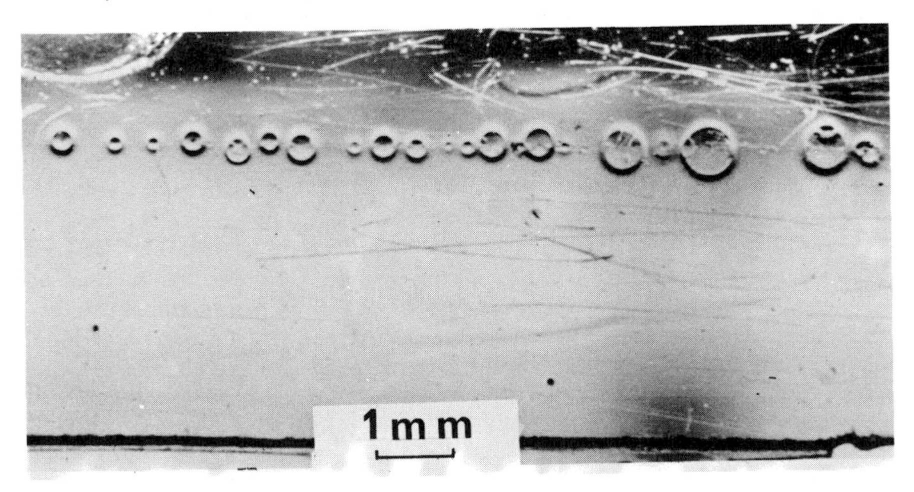

Figure 15.4. Resin cracks associated with a well-bonded fibre; sample exposed to boiling water[18, 22]

This information does not indicate the site of failure, however, because we do not know the relative rates of hydrolysis or the initial crosslink concentration in the various parts of the interfacial region.

Direct evidence on the failure locus is sparse and contradictory. Visual

* An activation energy of 27 kcal mole^{-1} implies halving the reaction rate for every 5 degC fall in temperature.

observations, photo- and electron-micrographs sometimes show cracks in the glass, but this is mainly from dynamic tests on dry specimens; in any case, the cracks may not have started in the glass[20, 39, 57]. Changes in the composition of the glass affect both wet and dry strengths of composites, and also the ease of removal of coupling-agent. This suggests that failure starts at the interface between glass and coupling-agent although the effect may be an indirect result of altered alkalinity over the whole interfacial region[9, 12, 54, 70, 91, 92].

Failure within the coupling-agent layer is a possible interpretation of high contact angles for water on the cleavage surfaces[13]; and similar evidence could equally well indicate failure at the interface between coupling-agent and resin, or within the resin[13, 28]. Other evidence suggests failure in the resin. Increased bond life in boiling water obtained by a better coupling-agent caused resin cracks to concentrate at a filament (*Figure 15.4*)[18, 22] and bond or composite strengths depend on resin type and cure[14, 18, 32, 42, 54, 93–99].

Thus the interfacial region harbours no obvious and universal 'weak-link'.

5.3. ENERGY CONSIDERATIONS IN BOND FAILURE

Evidently debonding is often a spontaneous process, and so must be favoured by a consequent decrease in the energy of the system. Unfortunately, this is impossible to demonstrate in detail because real composites are too complex and the data are too sparse. So in this final section we shall simply note the main energy changes and some of the implications.

In the absence of an external load there are three main sources of energy change: strain energy in resin and glass due to their differential volume changes; the chemical energy change of network scission (e.g. hydrolysis); and surface energy changes as debonded interfaces separate.

The resin undergoes volume changes relative to the glass because of polymerisation shrinkage, thermal shrinkage on cooling from cure, and swelling by absorbed liquids (e.g. water). Polymerisation shrinkage usually gives only small stresses because it occurs mainly at temperatures above T_g. The net effect of thermal shrinkage and swelling by absorbed liquids may however, give a relatively large stress (at temperatures below T_g), which can be a substantial fraction of the ultimate stress of the resin. A crude estimate of the corresponding strain energy gives 5 ergs cm^{-2} of interface, and much of this energy could be released by debonding.

A comparable value is obtained for the decrease in chemical energy caused by hydrolytic scission of ester groups. The equilibrium constant for this reaction indicates that a substantial fraction of the attachments between resin and glass would eventually be cleaved in composites exposed to water.

Surface energy changes will occur if the physically adhering interfaces that remain after hydrolysis are separated and a layer of water enters the gap. Even where the two surfaces are similar the sign of the energy change is uncertain, but we can be reasonably certain that ± 50 erg cm^{-2} is a maximum. It can thus be comparable to the strain and chemical energy changes in size although it may be opposed to them in sign.

Several practical implications are apparent for composites working in reactive environments. For example, the attachments between resin and glass should preferably not contain susceptible linkages; if this is unavoidable, then the concentration of the attachments should be as high as possible. The rate of attack on susceptible groups may be substantially reduced by modifying adjacent parts of the molecule. The resin strain energy may be substantially reduced by curing in such a way that swelling caused by reactant uptake balances the original thermal shrinkage of the resin. In practice, such general desiderata may turn out to be mutually contradictory or economically unattainable. Nevertheless, we feel that in materials development and selection it may be helpful to keep in mind the basic energy considerations that govern interfacial breakdown.

6. CONCLUSIONS

Understanding of the GRP interface lags far behind its technology, as is so often the case. Not surprisingly most of the evidence is a by-product of practical improvement, with very few thorough studies of well-characterised systems. This has given a mass of data difficult to correlate, but two facts do stand out. Firstly, breakdown of the interface is often critical for composite properties; and secondly, a chemical bond between resin and glass is needed. The importance of the interfacial region, really a third phase in the composite, is therefore evident. A prerequisite for chemical bonding across this region is that glass, coupling-agent and resin must be in intimate molecular contact whereas, in fact, air entrapment and the presence of size combine to prevent this over part of the interface.

Composites possess stored mechanical energy and are not in chemical equilibrium with their service environment so that degradation is occurring throughout their working life, in particular degradation at the interface. It is therefore equally important to consider this stage, because improvements in performance may stem as much from reducing degradation rates as from better initial bonding.

REFERENCES

Notes:

1. The annual reinforced plastics conferences of the Society of the Plastics Industry, Inc., New York, are abbreviated 'S.P.I., R.P. Conf.'.

2. The biennial reinforced plastics conferences of the British Plastics Federation are abbreviated 'B.P.F., R.P. Conf.'.

3. A six-figure serial number after authors' names, prefixed by 'AD', indicates a U.S. Government report. These are published by the Clearinghouse for Federal Scientific and Technical Information of the U.S. Department of Commerce. In Great Britain microfiche copies are available from the National Lending Library, Boston Spa.

1. Gould, R. F. (Ed.) 'Contact Angle, Wettability and Adhesion', *Adv. Chem. Ser.* No. 43 (1964)
2. Peters, R. H., White, E. F. T. and Inverarity, G. *B.P.F., R.P. Conf.*, No. 6, Paper 23 (1968)
3. Olsen, D. A. and Osteraas, J. A. *J. phys. Chem.* **68**, 2730 (1964)

4. Eakins, W. J. *S.P.I., R.P. Conf.*, No. 18, Sect. 9C (1963)
5. Yates, P. C. and Trebilcock, J. W. *S.P.I., R.P. Conf.*, No. 16, Sect. 8B (1961)
6. Eakins, W. J. *S.P.I., R.P. Conf.*, No. 17, Sect. 10C (1962)
7. Levine, S. N. and Lacourse, W. C. *AD 641500, St. Univ. N.Y., Dep. Mater. Sci.*, Contract NONR 480 3 (00), (1966)
8. Prebus, A. F. and Michener, J. W. *Ind. Engng Chem. ind. Edn*, **46**, 147 (1954)
9. Johannson, O. K., Stark, F. O. and Baney, R. *AD 629777, US Air Force Mater. Lab. tech. Rep.* 65–303, Part 1 (1965)
10. Tutas, D. J., Stromberg, R. R. and Passaglia, E. *S.P.I., R.P. Conf.*, No. 19, Sect. 21C (1964); *S.P.E. Trans.* **4**, 256 (1964)
11. Young, F. M., Roskos, T. G. and Radloff, H. R. *S.P.I., R.P. Conf.*, No. 17, Sect. 14C (1962); *S.P.E. Trans.* **2**, 326 (1962)
12. Vogel, G. E., Johannson, O.K., Stark, F. O. and Fleischmann, R. M. *S.P.I., R.P. Conf.*, No. 22, Sect. 13B (1967)
13. Laird, J. A. and Nelson, F. W. *S.P.E. Trans.* **4**, 120 (1964)
14. Matting, A. and Ehrenstein, G. W. *Kunststoff-Rdsch.* **14**, 413 (1967)
15. Sterman, S. and Bradley, H. B. *S.P.I., R.P. Conf.*, No. 16, Sect. 8D (1961); *S.P.E. Trans.* **1**, 224 (1961)
16. Sterman, S. and Marsden, J. G. *S.P.I., R.P. Conf.*, No. 18, Sect. 1D (1963); *Mod. Plast.* **40**, No. 11, 125 (1963); *Mod Plast.* **41**, No. 2, 254 (1963)
17. Bascom, W. D. *S.P.I., R.P. Conf.*, No. 20, Sect. 15B (1965)
18. James, D. I., Norman, R. H. and Stone, M. H. *Plastics Polym.* **36**, 21 (1968)
19. Lee, L. *S.P.I., R.P. Conf.*, No. 23, Sect. 9D (1968); *J. Colloid Interface Sci.* **27**, 751 (1968)
20. Laird, J. A. and Nelson, F. W. *S.P.I., R.P. Conf.*, No. 19, Sect. 11C (1964)
21. Trostyanskaya, E. B., Poimanov, A. M. and Kazanskii, Yu. N. *Plast. Massy*, No. 7, 52 (1964); *Soviet Plast.*, No. 7, 55 (1965)
22. Hands, D., James, D. I., Norman, R. H. and Stone, M. H. *B.P.F., R.P. Conf.*, No. 5, Paper 16 (1966)
23. Wake, W. C. and Norman, R. H. *B.P.F., R.P. Conf.*, No. 4, Paper 1 (1964); James, D. I., Stone, M. H. and Wake, W. C. *Appl. Plast.* **8**, No. 1, 49 (1965)
24. Levine, S. N. and Lacourse, W. C. *AD 652408, St. Univ. N.Y., Coll. Engng.*, Contract NONR 4803 (00) (1967)
25. Matting, A. and Ehrenstein, G. W. *Kunststoff-Rdsch.* **14**, 361 (1967)
26. Schlabach, T. D. and Klosowski, S. *S.P.I., R.P. Conf.*, No. 20, Sect. 19C (1965)
27. Throckmorton, P. E. and Browne, M. F. *S.P.I., R.P. Conf.*, No. 20, Sect. 15A (1965)
28. Trostyanskaya, E. B. and Gunyaev, G. M. *Plast. Massȳ*, No. 5, 30 (1967); Soviet Plast., No. 5, 28 (1967)
29. Fried, N. *S.P.I., R.P. Conf.*, No. 20, Sect. 1C (1965)
30. Kohn, E. J., Sands, A. G. and Clark, R. C. *Ind. Eng. Chem., Prod. Res. Develop.* **7**, 179 (1968)
31. Paul, J. T. and Thompson, J. B. *S.P.I., R.P. Conf.*, No. 20, Sect. 12C (1965)
32. Petker, I. *S.P.E. Trans.* **5**, 49 (1965)
33. Hand, W. *S.P.I., R.P. Conf.*, No. 20, Sect. 1E (1965)
34. Bascom, W. D. and Romans, J. B. *Ind. Eng. Chem., Prod. Res. Develop.* **7**, 172 (1968)
35. Bucki, L. *Trans. J. Plast. Inst.* **34**, 311 (1966)
36. Erickson, P. W., Volpe, A. and Cooper, E. R. *S.P.I., R.P. Conf.*, No. 19, Sect. 21A (1964); *Mod. Plast.* **41**, No. 12, 141 (1964)
37. Trivisonno, N. M., Lee, L. and Skinner, S. M. *Ind. Engng Chem. ind. Edn.* **50**, 912 (1958)
38. Baigozhin, A. and Sergeyev, L. V. *Vȳsokomolek. Soedin.* **4**, 972 (1962); *Polym. Sci. U.S.S.R.* **4**, 291 (1963)
39. Kenyon, A. S. *J. Colloid Interface Sci.* **27**, 761 (1968)
40. Kenyon, A. S. and Duffey, H. J. *Polymer Eng. Sci.* **7**, 189 (1967)
41. Plueddemann, E. P., Clark, H. A., Nelson, L. E. and Hoffman, K. R. *S.P.I., R.P. Conf.*, No. 17, Sect. 14A (1962); *Mod. Plast.* **39**, No. 12, 135 (1962)
42. Wende, A. and Gähde, J. *B.P.F., R.P. Conf.*, No. 5, Paper 15 (1966)
43. Holt, J. S. *B.P.F., R.P. Conf.*, No. 4, Paper 3 (1964)
44. Plueddemann, E. P. *S.P.I., R.P. Conf.*, No. 20, Sect. 19A (1965)
45. Vanderbilt, B. M. and Simko, J. P. *Mod. Plast.* **38**, No. 4, 135 (1960)
46. Vanderbilt, B. M. *S.P.E. Tech. Pap., A. Tech. Conf.*, No. 22, **12**, Session 23, Paper 1 (1966)

47. Ziemianski, L. P. *B.P.F., R.P. Conf.*, No. 6, Paper 10 (1968)
48. Moebes, W. and Wende, A. *Plaste Kautsch.* **9**, 232 (1962)
49. Clark, H. A. and Plueddemann, E. P. *S.P.I., R.P. Conf.*, No. 18, Sect. 20C, (1963)
50. Alekseeva, I. A., Semerneva, G. A., Samarina, L. A., Bulatov, M. A. and Spasskii, S. S. *Vysokomolek. Soedin.* **6**, 265 (1964); *Polym. Sci. U.S.S.R.* **6**, 306 (1964)
51. Greber, G. and Metzinger, L. *Makromolek. Chem.* **39**, 217 (1960)
52. Scott, C. E. and Price, C. C. *J. Am. chem. Soc.* **81**, 2670 (1959)
53. Pike, R. M. and Bailey, D. L. *J. Polym. Sci.* **22**, 55 (1956)
54. Golubenkova, L. I., Nikonova, S. N., Shabadash, A. N. and Akutin, M. S. *B.P.F., R.P. Conf.*, No. 5, Paper 24 (1966)
55. Gutfreund, K. and Weber, H. S. *S.P.I., R.P. Conf.*, No. 16, Sect. 8C (1961); *S.P.E. Trans.* **1**, 191 (1961)
56. Throckmorton, P. E., Hickman, H. M. and Browne, M. F. *S.P.I., R.P. Conf.*, No. 18, Sect. 14A (1963); *Mod. Plast.* **41**, No. 3, 140 (1963)
57. Broutman, L. J. *S.P.I., R.P. Conf.*, No. 19, Sect. 9C (1964); *Mod. Plast.* **42**, No. 8, 143 (1965)
58. McGarry, F. J. and Willner, A. M. *S.P.I., R.P. Conf.*, No. 23, Sect. 14B (1968)
59. Owen, M. J., Dukes, R. and Smith, T. R. *S.P.I., R.P. Conf.*, No. 23, Sect. 14A (1968); *Plast. Inst. Conf.*, 'Research Projects in Reinforced Plastics', London, Paper 1 (1968)
60. Smith, T. R. and Owen, M. J. *B.P.F., R.P. Conf.*, No. 6, Paper 27 (1968)
61. Norman, R. H., James, D. I. and Gale, G. M. *Chem. Engr. Lond.*, No. 182, 243 (1964)
62. Gähde, J. and Pohl, G. *Plaste Kautsch.* **13**, 207 (1966)
63. Fried, N., Kaminetsky, J. and Silvergleit, M. *S.P.I., R.P. Conf.*, No. 21, Sect. 14A (1966)
64. Desai, M. B. and McGarry, F. J. *S.P.I., R.P. Conf.*, No. 14, Sect. 16E (1959)
65. Krolikowski, W. *S.P.E. Jl* **20**, 1031 (1964)
66. Brelant, S., Petker, I. and Smith, K. W. *S.P.E. Jl* **20**, 1019 (1964)
67. Ashbee, K. H. G., Frank, F. C. and Wyatt, R. C. *Proc. R. Soc.* **300A**, 415 (1967)
68. Bartenev, G. M. and Zuyev, Yu. S. *Strength and Failure of Viscoelastic Materials*, p. 30 et seq., Pergamon Press, London (1968)
69. Bascom, W. D. *Am. chem. Soc.*, Division of Organic Coatings and Plast. Chem. **27**, No. 2, 27 (1967)
70. Nikonova, S. N., Golubenkova, L. I., Shabadash, A. N. and Akutin, M. S. *Plast. Massy̆*, No. 2, 27 (1966); *Soviet Plast.*, No. 2, 32 (1967)
71. Kolb, K. E. and Koelling, J. G. *S.P.I., R.P. Conf.*, No. 21, Sect. 13D (1966)
72. Schrader, M. E., Lerner, I., D'Oria, F. J. and Deutsch, L. *S.P.I., R.P. Conf.*, No. 22, Sect. 13A (1967); *Mod. Plast.* **45**, No. 1, 195 (1967)
73. Stober, W. *Kolloidzeitschrift* **149**, 39 (1956)
74. Holtmann, R. *Kunststoffe*, **53**, 22 (1963)
75. Holtmann, R. *S.P.I., R.P. Conf.*, No. 21, Sect. 13C (1966); *Kunststoffe* **57**, 110 (1967)
76. Steel, D. J. *Trans. J. Plast. Inst.* **35**, 429 (1967)
77. Stone, M. H. Rubber and Plastics Research Association, unpublished work (1965)
78. Cantelo, R. C. and Billinger, R. D. *J. Am. chem. Soc.* **50**, 3212 (1928)
79. Williams, R. J., Gabriel, A. and Andrews, R. C. *J. Am. chem. Soc.* **50**, 1267 (1928)
80. Grubb, W. T. and Osthoff, R. C. *J. Am. chem. Soc.* **77**, 1405 (1955)
81. Gimblett, F. G. R. *Inorganic Polymer Chemistry*, p. 386 et seq., Butterworths, London (1963)
82. Ingold, C. K. and Nathan, W. S. *J. chem. Soc.*, 222 (1936)
83. Newling, W. B. S. and Hinshelwood, C. N. *J. chem. Soc.*, 1357 (1936)
84. Evans, D. P., Gordon, J. J. and Watson, H. B. *J. chem. Soc.*, 1430 (1937)
85. Timm, E. W. and Hinshelwood, C. N. *J. chem. Soc.* 862 (1938)
86. Tommila, E. and Hinshelwood, C. N. *J. chem. Soc.* 1801 (1938)
87. Harned, H. S. and Ross, A. M. *J. Am. chem. Soc.* **63**, 1993 (1941)
88. Smith, H. A. and Steele, J. H. *J. Am. chem. Soc.* **63**, 3466 (1941)
89. Smith, H. A. and Myers, R. R. *J. Am. chem. Soc.* **64**, 2362 (1942)
90. Stone, M. H. Rubber and Plastics Research Association, unpublished work (1968)
91. Vanderbilt, B. M. and Jaruzelski, J. J. *S.P.I., R.P. Conf.*, No. 17, Sect. 10D (1962); *Ind. Eng. Chem., Prod. Res. Develop.* **1**, 188 (1962)
92. Andreevskaya, G. D. and Shiryaeva, G. V. *Vy̆sokomolek. Soedin.* **5**, 1733 (1963); *Polym. Sci. U.S.S.R.* **5**, 854 (1964)
93. Eakins, W. J. *S.P.E. Jl* **19**, 379 (1963)

94. Sheppard, H. R. and Sampson, R. N. *S.P.E. Jl* **20**, 347 (1964)
95. Nikonova, S. N., Golubenkova, L. I., Shabadash, A. N. and Akutin, M. S. *Plast. Massy*, No. 3, 45 (1966); *Soviet Plast.*, No. 3, 51 (1967)
96. Cole, L. F. and Mulvaney, W. P. *S.P.I., R.P. Conf.*, No. 22, Sect. 4C (1967); *Mod. Plast.*, **44**, No. 10, 151 (1967)
97. Trent, P. J., Edwards, D. L. and Von Bramer, P. *S.P.I., R.P. Conf.*, No. 21, Sect. 1B (1966)
98. Van Der Beek, M. H. B. and Hamm, G. *B.P.F., R.P. Conf.*, No. 5, Paper 12 (1966)
99. Andreevskaya, G. D., Gorbatkina, Yu. A., Zamotova, A. V., Kiseleva, R. L., Odnoletkova, T. V. and Khvilivitskii, R. Ya. *Mekhan. Polimerov,* Akad. Nauk Latv. S.S.R., **1**, No. 1, 93 (1965); *Polym. Mech.* **1**, No. 1, 68 (1965)

BIBLIOGRAPHY

Patrick, R. L. (Ed.) *Treatise on Adhesion and Adhesives*, Edward Arnold, London (1967)
Houwink, R. and Salomon, G. (Ed.) *Adhesion and Adhesives*, Vol. 1, Elsevier, London (1965)
S.C.I. Monogr. No. 25, *Wetting,* Soc. Chem. Industry, London (1967)
Baer, E. (Ed.) *Engineering Design for Plastics*, Chapter 14, Reinhold, New York (1964)
Broutman, L. J. and Krock, R. H. (Ed.) *Modern Composite Materials,* Addison-Wesley, London (1967)
Holister, G. S. and Thomas, C. *Fibre Reinforced Materials,* Elsevier, London (1966)
Kelly, A. *Strong Solids,* Clarendon Press, Oxford (1966)
Schwartz, R. T. and Schwartz, H. S. (Ed.) *Fundamental Aspects of Fibre Reinforced Plastic Composites,* Interscience, London (1968)

16

Weathering

K. A. SCOTT and JACOB MATTHAN

1. INTRODUCTION

Glassfibre reinforced plastics are employed in a variety of outdoor applications. Possibly one of the most interesting is one listed as 'Playground Equipment'. A recent report[1] on the use of GRP in playgrounds stated that the material offered new design freedom, improved performance flexibility, reduced manufacturing and maintenance costs, increased resistance to vandalism and greater safety. Stating the requirements for playground equipment, the report said: 'Playground equipment must be rugged, attractive and completely safe . . . no lacerations, splinters and abrasions. It must stand up not only to normal wear and tear, but to additional punishment dished out by kids. With a few exceptions, it is outside most of the year, so it must withstand exposure to sun, wind and rain—possibly even snow and ice. Moreover, such equipment, particularly if colourful, will draw vandals like a magnet. What's called for is a material that is childproof, weatherproof, vandalproof and foolproof. In a nutshell—reinforced plastics'.

GRP can indeed claim some degree of weatherability in drastic environments. However, some applications require a guarantee of 60 years of life, which is quite impossible to substantiate at these early days of development of these materials.

The subject of weathering has been widely reported[2–7]. In the case of polyester resins, good weathering properties are often offered as a selling point for the material. Epoxy resins, however, do suffer from yellowing due to ultraviolet light. In the case of glass reinforced resins, however, besides the problem of weathering of the resins themselves, one is faced with the problem of determining the effects of weather on the mechanical, physical and chemical properties which are dependent on the efficiency with which the resin/glass composite is prepared. One of the major difficulties encountered in assessing published weathering data is that the resin and glass types used and the structure of the laminate are not always specified.

2. POLYESTER RESINS

Considerable changes in the weather resistance of polyester resins can be made by the choice of the raw materials and the method of production. Requirements of laminates to be produced from the resins and the cost of the raw materials very often determines the choice. For instance, curtain walling panels are generally opaque and pigmented, and retention of gloss is an important property. A good general purpose laminating resin would probably satisfy these requirements. On the other hand, roofing panels could require good initial clarity which general purpose resins do not completely satisfy, since good wetting-out characteristics and a controlled refractive index are required of the resin.

In many areas heat resistant or self-extinguishing resins are required, and it is generally recognised that some of the special resins have relatively poor weathering characteristics.

2.1. GENERAL PURPOSE LAMINATING AND ROOFING RESINS

'Good quality' resins have a sufficiently high molecular weight and employ raw materials which minimise water absorption. Good weathering properties can then be expected. Flexible resins are less resistant to weathering than rigid resins. Plasticisers and many fillers also detract from good weathering properties. Excessive quantities of monomers have a similar effect.

High water absorption and rapid deterioration in electrical properties after 2 h immersion in boiling water, suggests that the resin would have poor stability on exposure[8–9]. For a given formulation, the molecular weight of the resin must be sufficiently high to ensure a minimum number of water susceptible end groups consistent with the other requirements of the resin such as viscosity. Modification of end groups by the use of mono basic acids and/or alcohols, could also influence and improve durability. Excess glycol used during manufacture generally compensates for losses during processing. However, too great an excess of glycol could lead to a low molecular weight resin or the presence of unwanted residual glycol in the product.

Acid and glycol types and quantities employed play important roles as they influence cross-linking density and water-absorption of the resin[10–12]. Isophthalic modified resins are used to give improved weathering over the normal phthalic/maleic acid resins, particularly with respect to gloss retention. Less yellowing is claimed to occur because of the improved light stability[13]. These resins have been shown to be an improvement over the orthophthalic type[14].

Adipic, sebacic and HET acids tend to yield resins with poorer weathering characteristics. However, in the case of gel coats some degree of flexibility may be needed to minimise cracking during service. In this instance, a flexible resin would have better durability than a rigid one which, in theory, has better weathering characteristics.

Neopentyl glycol has been shown to have considerable advantage over propylene glycol[15]. Glycols such as diethylene and dipropylene glycol tend to yield resins that are more flexible and absorb more water than those containing propylene glycol, and hence possess poorer weathering characteristics.

Weathering applications generally require retention of colour and it is believed that resins which have minimum colour initially perform better than those with a pronounced yellow/orange colour. Obviously, avoidance of contamination will yield resins of low initial colour and therefore better weatherability.

2.2. SELF-EXTINGUISHING RESINS

None of the systems used to obtain self-extinguishing resins provide good weathering characteristics to the resins. Not only does the appearance deteriorate, but also in the case of the trichloroethylphosphate containing resins, self-extinguishing properties have been found to diminish[16], although this observation is not supported by Crowder[17].

A comparison of fire-retardant resin with a general purpose resin[18] showed that the former, even after 2 years, showed much greater surface deterioration than the latter. Flexural strength of the fire-retardant resin decreased while no significant change for this time period occurred in the general purpose resin. A weight decrease of the fire-retardant resin was noted in contrast to a weight increase of the general purpose resin, and this has been interpreted as meaning that the surface erosion more than compensated for the water absorption. It was also noted that the addition of antimony oxide to the fire-retardant resin resulted in better retention of flexural strength till the end of the first year of exposure, but by the end of 3 years, values were approximately equal to values of the resin not containing antimony oxide.

Two years exposure of glass mat reinforced polyester laminates to which 5 per cent oxide was added showed little change in mechanical properties, but during the third year, tensile strength, elongation and work value showed a significant decrease[19]. Flame resistant polyester has been shown to exhibit a weight loss of about 10 per cent in 5 years compared to an average weight loss of slightly less than 4 per cent in a light-stabilised resin and more than 1 per cent in an acrylic-modified resin[20]. Rugger[21], however, points out that caution should be exercised in using these figures since the data represents findings on one flame-resistant resin, while the value for the light-stabilised resin was an average for thirty-one materials.

Although weight loss differed significantly, the white light transmission for the flame-resistant resin was found to be nearly as good as the acrylic-modified resin[20]. Rugger[21] states that this could well have been due to the reduced thickness of the flame-resistant sections due to erosion.

Whitehouse[22] has studied the light transmission characteristics of various laminates containing 70 per cent resin, after exposure, and has shown that self-extinguishing resins are inferior. The effect of natural exposure on the discoloration of sheeting with differing chlorendic (i.e. HET) acid contents was investigated, and it was shown that the weathering character was dependent on the amount of HET acid present, i.e. the greater the HET acid in relation to the phthalic acid, the more the discoloration. Allowance, however, does not appear to have been made for the differing densities of the resin systems used. Although it is valid to use the resin/glass ratio as a guide when the same resin is used, it is probably better to make comparisons on a resin volume basis when differing resins are employed.

The durability of non-chlorine containing fire-resistant resins is still to be assessed. As a general rule, it can be said that fire-resistance can only be obtained at the expense of durability. The same can probably be said of some heat-resistant resins. Rugger[21] states that an educated guess would be that loss of mechanical properties of fire-retardant resins would be about twice that suffered by general purpose resins.

2.3. MONOMERS

Resistance of styreneated polyester to weather depends on chemical composition and molecular weight of the resin, the degree of cure, the types of initiator and accelerator used, and the presence or absence of glass-fibres[23–26].

In addition to the effects of moisture and fluctuating temperature, polyester resins are susceptible to photo-oxidation[27, 28], and maximum yellowing in natural sunlight is produced by radiation of the wavelength 325–330 mμ, virtually identical results being obtained from resins prepared from adipic, phthalic and succinic acid[28].

The mechanism of photo-oxidation is little understood, but it is suggested that generation of hydroxyl radicals or other reactive species probably gives rise to a host of free radical reactions[29]. Polyesters, like polystyrene, give a yellow-green fluorescence after a short exposure to radiation[27], and it is suggested that styrene groups are the main points for attack[23, 24]. Degradation appears to be due mainly to light absorbed by ketone groups, for the wavelength producing maximum yellowing corresponds not to maximum absorption by the resin but to maximum absorption by acetone[27, 30]. A post-irradiation effect of the absorbance at 435·8 nm, decreasing during storage in the dark has also been reported[28].

Polyester resins are claimed to be more resistant to yellowing than polystyrenes because they contain less than 50 per cent styrene, and because each cross-link is comprised of only one or two styrene units and not long polystyrene chains[25]. Even so, styrene containing polyesters need to be stabilised against yellowing by the addition of about 0.25 per cent of a benzotriazole or a hydroxy-benzophenone ultraviolet absorber which is dissolved in the styrene monomer before it is added to the resin[31]. The absorber imparts a slight yellow tinge to the polyester but helps to preserve the light transmission characteristics of the resin over long periods[32–34].

Probably the most spectacular improvement in general weathering properties arises from the use of mixed monomer systems. Replacement of part of the styrene monomer with methyl methacrylate leads to better clarity and better retention of clarity, improved gloss retention, reduced surface failures such as crazing, less yellowing, better resistance to erosion, and generally referred to as 'fibre pop-out'[23–35], over a period of 5 years of weathering.

The optimum ratio of styrene to methyl methacrylate appears to be 1:1. Greater quantities of methyl methacrylate adversely affects cure of the resin and gives inferior weather resistance to the product.

2.4. CATALYST-ACCELERATOR SYSTEM

The use of the benzoylperoxide/amine system of cure is objectionable from the point of view of weathering since the yellow colour produced in the resin deepens with ageing. Hot cured benzoyl peroxide systems give no marked change in colour during cure and the weatherability is good. Less yellowing is experienced with cobalt-based accelerator systems, but they are themselves highly coloured and can change the colour of some resins to an orange-yellow shade. Inadequate mixing of the components could impart a greenish shade to the resin due to oxidation of the cobalt salt by the peroxide catalyst. However, the colour stability is superior to that obtained with the benzoyl peroxide/tertiary amine system.

Benzoyl peroxide and 1-hydroxy cyclohexyl hydroperoxide-1 were investigated at different concentrations, but the durability characteristics of the resins cannot be compared since the latter was cured at room temperature with an accelerator[36].

It has been stated that acyl peroxides are best from the point of view of durability[24] if the peroxides are well dispersed. However, for easy dispersion a paste is used, and the dispersion agent can have an adverse effect on the weatherability of the resin.

In general, decreasing the amount of catalyst would increase the average molecular weight, since there would be fewer competing initiating points. This in turn would influence the number of water susceptible end-groups as the molecular weight increased.

Amine type accelerators, particularly dimethyl aniline, are claimed to be undesirable in polyester resins[24]. Thiols and soluble metal soaps have some yellowing tendency, while quaternary ammonium salts are among the better accelerators from the point of view of weatherability.

3. EPOXY RESIN

While the majority of epoxy resins in use are the bisphenol 'A' type, newer types based on novolak resins or on aliphatic or cycloaromatic structures are also in use, but weathering data on these is limited. Cycloaromatic resins have been prepared which are claimed to show superior resistance to degradation by ultraviolet light than normal epoxy resins[37].

Although epoxy resins are used fairly extensively, yellowing due to ultraviolet light has always been a problem. Little information is available on light degradation. Information on yellowing under the action of natural and artificial light sources has been reviewed and summarised[38], and in almost all cases of epoxy resins, yellowing on exposure to ultraviolet light is reported.

The effects of weather on mechanical and electrical properties has been studied[7], and properties such as tensile strength and flexural strength were found to deteriorate in the majority of resins. In most cases surface fissures were observed and the degree of degradation was interpreted by the relative size of the fissures, and results were compared with the measured change in mechanical properties.

The dielectric properties of epoxy resins change on ageing[39]; ageing at low temperature causing an increase in cross-linking and reduction of ionic conduction, dielectric constant, and at most temperatures, dielectric loss.

While ultraviolet absorbers are widely used in plastics, in the case of epoxy resins none have been found to be completely satisfactory in reducing degradation and yellowing. This can be partly attributed to epoxy resins, particularly those containing aromatic rings, which being very absorbent in the ultraviolet region, cause yellowing of the resin both during and after cure. Substituted hydroxy benzophenone and similar compounds have, however, been used to stabilise epoxy resins.

3.1. CURING AGENTS

Not much difference was noted between triethylenetetramine and an amine type safety hardener system[7], and it was also noted that aromatic diamine and anhydride systems showed least change in mechanical properties with exposure.

A study of seventeen curing agents[38] in a commercial diglycidyl ether of bisphenol A (DGEBA) showed that amine hardeners which gave the lightest initial colour were most resistant to discoloration on exposure to ultraviolet light. Amines in which nitrogen atoms were not situated on an aromatic ring and were separated from each other by methylene or propylene groups rather than ethylene groups resulted in colour stable resins. Similarly, resins cured with amines of lower functionality showed low colour formation. In general, discoloration ranged from dark brown to a faint yellowing. It has also been shown that trimethoxyboroxine cured DGEBA is relatively more resistant to degradation under black light and a sun-lamp than resin cured with hexahydrophthalic anhydride[40].

Curing at high temperatures has been shown to yield products which show less change in mechanical properties after exposure than resin products cured at low temperatures.

4. RESIN/GLASS COMPOSITE

If, in a glass fibre reinforced product, the resin/glass interface adhesion is inadequate, a variety of effects can be noted. Loss of mechanical strength and properties, loss in light transmission, accentuation of the glass fibre pattern in and at the surface of the laminate are a few of the predominant effects. The major problem is the wicking of water between the glass fibre and the resin.

In polyester resins, if the gel coat is absent, the first effect of weather on translucent panels is the accentuation of the fibre pattern. This is due to the resin layer at the top being eroded away so that the fibre stands proud of the surface. If a gel coat is used, this effect is avoided or delayed for a considerable period of time.

More advanced weathering results in the accentuation of glass fibres by a whitening effect inside the laminate. This effect is primarily due to the ingress of moisture and the breakdown of the resin-glass interface, causing internally reflecting surfaces. This phenomenon was thought to be not very widely found in epoxy resins, but Rugger [21] has shown the profound effect on tensile values when epoxies are immersed in water.

Without doubt the glass-resin interface plays the most significant role in the weathering of laminates. If the breakdown of the interface adhesion is avoided or delayed, the strength characteristics of the laminate can be maintained. However, to maintain this strength characteristic the use of gel coats or a surface seal is required, besides the use of coupling agents which promote the bond strength. So long as the surface resists the effects of weather, the problem of weathering remains an aesthetic one. The correct maintenance procedures can therefore promote the life of a reinforced plastic product.

The type of glass generally employed when a product is to be exposed to the weather is E glass. However, many manufacturers and fabricators feel that this is not a critical factor for British climates. The reinforcement is often in the form of chopped strand mat with a binder present. It has been found that unbonded mat increases the tendency of the product to show more fibres on the surface[41], and it is suggested that the binder on the mat helps to keep the glass strands away from the surface of the product and help preserve the integrity of the strands during fabrication.

The above conclusion is supported by the initial gloss obtained when bound mat is employed since any tendency for the strands to break down into filaments would reduce gloss. This is not necessarily true with respect to retention of properties, since, if the surface of the mat broke down to single filaments, the effect would be similar to that obtained if a surface tissue had been used. A resin rich surface and a reduced risk of strands standing proud from the surface would result.

The chopped strand mat now available is in finer and more compacted form than early ones, and the tendency of strands to lay proud of the surface of laminate is reduced. Even so, for weathering applications, a gel coat or surfacing tissue is desirable.

Many products have cut edges or are drilled, and most workers recommend that such exposed fibres should be avoided as they could be weak points at which attack by the elements could occur. However, it has been shown[42] that these exposed edges should not have any effect on moisture absorption or related problems.

Any approach that reduces the ingress of moisture should be helpful in aiding durability. Rugger[21] states, as an example, that woven rovings are manufactured so that the filaments are tightly compacted and this makes penetration of the resin less effective and has a tendency to leave minute paths of water. Fluffing of the strands would permit better resin penetration. The coarse weave would foster a tendency to entrap air, leaving voids. Also coarse weaves could cause resin rich areas among bundles and layers. These would be more susceptible to crazing, again favouring water penetration. Proper fabrication would minimise all these problems and give a durable product.

Other examples of the above are related to spun rovings and satin weave. With the former, the more open structure should permit better wet-out, but wet-out is a function of resin viscosity and speed of impregnation, and hence use of spun roving alone does not ensure better weatherability. Similarly, satin weave would have fewer crimps in the strands and should be less prone to initiate crazing when a load is applied.

Normally reinforcement choice is not made on the basis of desired weatherability, hence crazing tendencies, and void formation need to be controlled by adjusting variables other than the reinforcement. Fabrication control is probably of prime importance.

4.1. RESIN/GLASS RATIO

The effect of resin content on weatherability of a laminate has been clearly shown by Whitehouse[22], when a marked improvement in light transmission properties was noted as the resin content was increased. A resin content of not less than 75 per cent is usually quoted by resin suppliers, the actual value being determined by the type of reinforcement employed. Coarse mats are likely to require more resin than the finer compressed mats.

4.2. GEL COATS

It is preferable to apply gel coats to both surfaces of a sheet product in order to obtain a balanced laminate with minimum warpage due to differential shrinkage, and to ensure that an unprotected surface is not accidentally exposed.

With the use of surfacing tissue, although the initial light transmission value of the laminate is reduced, the loss in transmission after exposure is lower than laminates with no surfacing tissue. Care should be exercised in the choice of the reinforcement for the gel coat, as Crowder[17] reports disastrous breakdown in a panel with a gel coat reinforced with a thin fibre cloth. A light monofilament tissue is advisable. In the case cited above the resin cloth bond was affected by water and a breakdown of the surface with dust pickup occurred in a few months.

Although E glass reinforcement is considered to give superior weathering resistance to that of A glass reinforcement, surfacing tissues are frequently made from A glass and yet effect a marked improvement in weathering[43].

Unreinforced gel coats enhance the weathering of polyester laminates. They are generally applied on smooth surfaces at a consistent thickness. Such a gel coat may be prone to cracking, not necessarily due to deterioration of the resin because of exposure to ultraviolet light, rain or heat, but because of flexing in the wind or other mechanical means. This could well be overcome by the use of a crinkle or pinhead surface finish.

5. EFFECT OF WEATHER ON PROPERTIES

5.1. OPTICAL PROPERTIES

The conclusion that can be drawn from a variety of studies on polyesters[18,20,44–48] is that with developments in resins, resin quality, reinforcements, fabrication techniques, the optical properties of polyesters have improved such that for the price of the product, appropriate optical properties are obtainable. In the case of glass-fibre reinforced epoxy resins, no work which has been primarily devoted to determining the change of optical properties on weathering is available, and undoubtedly work in this field is required. The yellowing of the resin has been described earlier.

5.2. MECHANICAL PROPERTIES

Determination of different properties can lead to different conclusions. Data exists which show decrease of tensile strength while compressive strength increases. This is because in the tensile test the glass fibres are stressed, the resin playing a minor role as a stress transfer medium distributing the load. Change in tensile strength on weathering therefore reflects a change in the stress transferring ability. In the compression test the fibres play a minor role as they serve as columns which would fail by buckling. The flexural test differs from both the above as failure is generally a tensile failure which originates in the outer fibres. However, before the tensile forces develop, a a deflection must occur. Hence the effects of weather on both resin and the reinforcement will have an effect on the values recorded.

A difference in the magnitude of change of flexural strength would be expected, based on the reconditioning procedure employed[21]. As one of the effects of weather is cracking or crazing of the surface, penetration by water vapour is likely. Hence wet reconditioning could reflect more drastic changes than would reconditioning in a standard laboratory atmosphere. Water is considered as the most damaging of the elements and water penetration with a wicking action along the glass/resin interface could disrupt the bond and result in a decrease in flexural strength.

All the data available on glass-fibre reinforced polyesters indicates that significant loss of mechanical properties can occur on weathering. However, most of the data has been obtained on concentrated exposure and on resins and systems which are now out of date. The data can be used to gauge what could happen to panels which were produced several years ago, but they can also indicate where improvements in recent years have been directed.

With glass fibre reinforced epoxy resins, Rugger[21] states that tensile properties can be expected to change by about 10 per cent and compressive strength can increase up to 10 per cent in 3 years. The latter should be regarded as a warning signal that decreased resistance to impact loading is likely. The value of flexural strength of epoxy resins is queried as the test serves to distinguish the various degrees of durability, but the value for designers is doubtful. A loss of 10–20 per cent when laminates are tested dry after exposure is considered normal.

The change of polyester resin during exposure by use of the colour reaction with *N*,*N*-dimethyl-p-phenylene diamine has been suggested[49]. Oxidation products at the surface of the laminate are determined by this method.

6. MAINTENANCE

With early glass-fibre reinforced polyester products used outdoors, within 2–3 years fraying or blooming of the fibre, yellowing of the laminate, dust accumulation, and effects such as loss in light transmission were evident. It was clear that maintenance was required if the properties were to be retained over long periods. The techniques of ensuring minimum maintenance involve correct choice of resin, reinforcement, fabrication techniques and control of the fabrication process.

Colour change can be due to one of two reasons. The first is chemical modification of the molecular structure of the resin. The only way of preventing this is to ensure that the resin used is stable to the effects of weather. The second reason for colour change is by erosion of the resin, exposure of glass fibre and therefore differing patterns of light transmission and reflection.

The method by which a panel is maintained usually involves thorough cleaning of the affected surface with water and steel wool. The steel wool ensures that all the exposed strands are brushed even and the loose fibres are removed. The water helps remove the dirt and fibres. The cleaned surface is then allowed to dry and a refinisher of the acrylic, polyester or polyurethane type is applied.

Three finishing techniques have been compared[50] using a panel that had been exposed for 5.5 years. One third of the panel was not cleaned in any way. Two-thirds of the panel was cleaned with trisodium phosphate solution and a scrub brush. All the remaining loose fibres were removed with coarse steel wool and the panel was thoroughly washed with water. After drying, one half of the cleaned panel was recoated with an acrylated refinisher applied with a brush.

Total heat transfer through the uncleaned section was 26.8 per cent less than the unobstructed solar radiation and 20 per cent less than through a new acrylated glass fibre panel. Washing and cleaning with solution and steel wool increased transmission 14.1 per cent over the uncleaned section. Application of refinisher increased transmission another 2.7 per cent. The refinished section transmitted 98·5 per cent as many Btu h^{-1} ft^{-2} as the new acrylated glass fibre panel.

Regarding spectral transmission, the decrease was evident with the uncleaned section showing definite effects which appeared in the 500 nm and 600 nm areas. The greatest decrease in intensity occurred in the area between 500 and 700 nm. Washing increased transmission evenly throughout the area between 450 and 850 nm, and the decreased intensities around 500 and 600 nm almost disappeared entirely. Application of the refinisher increased transmission in the range from 650 to 1050 nm to that of a new panel. The transmission from 450 to 600 nm was not increased as effectively. Most of the yellowish colour in the weathered panel disappeared when the material was washed and refinished.

Hence, regular maintenance by washing with water and loosening of any adhered dirt will prolong the life of a component till the first major maintenance requiring refinishing. Some manufacturers feel that hosing down of exposed products is more than sufficient for the first 7 years, after which a refinisher followed by the same washing down periodically is adequate for the next 7 years.

7. LIFE OF GRP

No satisfactory answer to the question 'How long will a moulding last?', is as yet available. Opinions vary from 10 to 15 years of satisfactory performance to 30 years or more in a temperate climate. It is probably correct to state that with reasonable care a good product is likely to last more than 30 years, but to obtain a guarantee to this effect will be virtually impossible. The scene in reinforced plastics for external use is changing and with these changes more confidence is emerging and undoubtedly within a few years a 60 year guarantee of life will indeed become commonplace.

REFERENCES

1. Anon, *Modern Plastics,* Feb. 1967, pp. 94–97, 173
2. Yustein, S. E. *ASTM Bull.* No. 173, April 1951, p. 31; *ASTM Bull.* No. 196, Feb. 1954, p. 29
3. Yustein, S. E. *Tech. Rep. P.B.I.* 57236, 19.9.52
4. Rugger, G. R. *S.P.E. Trans, 20,* No. 7, July 1964, pp. 236–40
5. Haferkamp, H. *Beckacite Nachrichten,* 1963, 22, No. 120, pp. 40–54
6. Lane, E. W. *Plastverarbeiter,* 1963, 14, No. 10, pp. 642–9
7. Pschorr, F. E. and Cianciarulo, A. N. *Polym. Engng. Sci.,* 1965, 5, No. 3, pp. 166–70
8. Smith, A. L. *13th Ann. Meeting of Reinf. Plast. Div of S.P.I.* Section 6B
9. *Handbook of Reinforced Plastics of the S.P.I.* Ed. by Oleesky, S. and Mohr, G. Reinhold 1964
10. Gilman, L. *S.P.E. Journal,* 13, No. 11, Nov. 1957, pp. 33–8
11. Boenig, H. B. *Unsaturated Polyester Resins,* Elsevier 1964
12. Anon, Amoco Chemical Co., *Bull IP17,* 1963
13. Patterson, W. R. *18th Ann Meeting of Reinf. Plast. Div. S.P.I.* Paper 5E
14. Euchner, E. B. *et al., S.P.I. Reinforced Plastics Div, 20th Ann. Meeting,* Section 13B
15. Edwards, P. L. *et al. S.P.I. Reinforced Plastics Div. 20th Ann Conf.* Section 13A
16. Parkyn, B. *British Plastics,* 32, No. 1, Jan 1959, p. 29
17. Crowder, J. R. *Int. Conf. Reinf. Plast. B.P.F., Paper 27,* also *Bull RILEM* No. 25, Dec. 1964, pp. 87–91
18. Yustein, S. E. *US Naval Appl. Sci. Lab., Lab. Project 6035,* Progress Rept 4, Aug. 1963
19. McNally, C. Picatinny Arsenal, *Tech. Rep. 2120,* Addendum 4, June 1963
20. Isham, A. B. and Szpak, W. E. *Owens-Corning Fibreglass Corp. Tech. centre Rep.* No. TC-RPI-63-3.
21. Rugger, G. R. *et al. Weathering of Glass Reinforced Plastics,* Jan. 1966, Plastics Tech. Evaluation Centre, Picatinny Arsenal, Dover, N.J.
22. Whitehouse, A. A. K. and Wildman, D. *4th Int. Conf. Reinf. Plast. BPF,* 1964, Paper 28
23. Smith, A. L. and Lowry, J. R. *Plast. Tech.,* 5, No. 6, June 1959, pp. 42–50
24. Smith, A. L. and Lowry, J. R. *Plast. Tech.,* 35, No. 7, March 1958, pp. 134–42, 200

25. Bucknall, C. B. Chapter 6, *Weathering and Degradation of Plastics,* Columbine Press, 1966
25. Boenig, H. V. and Walker, N. *Modern Plastics,* 38 (6), 1961, p. 123
27. Hirt, R. C., Searle, N. Z. and Schmitt, R. G. *S.P.E. Trans.* 1961, 1, 21
28. Hirt, R. C., Schmitt, R. G. and Dutton, W. L. *Solar Energy,* 1959, 3, (2), 19
29. Kamal, M. R. and Saxon, R. *Applied Polymer Symposia* No. 4, Feb. 1967, 'Weatherability of Plastic Materials'
30. Weicksel, J. A. *Modern Plastics Encyclopaedia,* 1961, p. 401
31. Penn, W. S. *Rubb. Plast. Weekly,* 6th Jan. 1962, 142, 10
32. Coleman, R. A. and Weicksel, J. A. *Modern Plastics* 1959, 36, (12), 117
33. Penn, W. S. *Rubb. Plast. Weekly,* 13th Jan 1962, 142, 40
34. *Antioxidants in Food Regulations,* 1958, HMSO, London
35. Smith, A. L. and Lowry, J. R. *Plast. Tech.,* 6, No. 8, Aug. 1960
36. Mohaupt, A. A. and Freas, A. D. *Forest Prod. Lab. Rep.* 1825, March 1952
37. Trigaux, G. R. *Mod. Plast.,* 1960, 38, No. 1, pp. 147–8, 53, 226
38. Lee, E. H. and Watson, F. T. *S.P.E. 20th Annual Tech. Conf.* Jan 1964, Session 20, Paper 1
39. Olyphant, M. *Insulation,* 1966, 12, No. 8, pp. 42–6; No. 9, p. 42
40. Christie, H. and Medved, T. High Temp Resist Trans Plastics, Midwest Res. Inst. Project, Rep. 2491-C, Oct. 1960
41. Crenshaw, D. B. and Smith, D. *17th Ann. Meeting of Reinf. Plast. Div. S.P.I.,* Paper 11E
42. Pusey, B. B. and Carey, R. H. *Modern Plastics,* 32, No. 7, March 1955, p. 139
43. Schlarb, J. A. *19th Ann. Meeting of Reinf. Plast. Div. S.P.I.* Section 4A
44. Warren, F. and Heebink, B. G. *US Dept of Commerce Tech. Rep.* 55–319, PB 121390
45. Ministry of Aviation, UK, *Reports on Plastics in the Tropics*—14. Low Pressure Glass Fibre Laminates Bonded with Polyester Resins'
46. Yustein, S. E. *New York Naval Shipyard, Lab. Project* 4860-G-3, Final Report, 16th June 1961
47. Sonneborne, R. H. *14th Ann. Meeting of Reinf. Plast Div. S.P.I.* Section 12F
48. Sonneborne, R. H. and Bastone, A. L. *15th Ann. Tech. and Management Conf. S.P.I.,* Feb. 1960
49. *Plastics Institute Conf. Plastics in Building Structures,* London 14th–16th June 1965
50. Goldsberry, H. L. *Florists Review,* April 20th, 1967

17

Chemical Resistance

P. C. OLIVER

1. INTRODUCTION

Glass reinforced plastics can be considered as self-supporting, super surface coatings, when it is necessary to have a critical look at their chemical properties. This is a logical line to take, since the resins which are used as the support matrix for the glass reinforcement have been basically derived from developments within the paint industry, suitably adapted to the special requirements of conversion to the solid state from a mass of considerably greater thickness than is normally encountered in surface coatings. Indeed, the paint industry has even returned the compliment to the techniques developed by GRP fabricators by accepting the use of glass tissue reinforcement, with 'high build' systems, as a convenient answer to the age-old problem of renovating cracked walls and ceilings with a minimum of preparation.

However, it is also necessary to consider the influence of the glass reinforcement on the overall performance of the laminate in order to obtain a true understanding of the parameters essential for the full chemical resistance of GRP under aggressive environmental service conditions. Since the vast majority of GRP laminates employ unsaturated polyester resins as the support matrix, it is this class of materials to which greatest reference is made in this chapter; other resins will also be considered in respect of their special attributes.

Glass reinforced plastics comprise layers of fibrous glass, within a matrix of resin polymer. The glass may be in the form of a 'mat' comprised of individual strands of glass of approximate length 2 in., continuous filaments, woven roving or random chopped strands. The resin matrix may, or may not, contain organic or inorganic fillers. During manufacture of the laminate polymerisation of the resin results in a measurable volume shrinkage. It has been well established that in order to achieve an acceptable adhesion between the glass reinforcement and the resin matrix, it is essential to treat the glass fibre with a suitable coupling agent—this is discussed in another chapter—but for the best chemical resistance of a laminate to be achieved, it is of paramount importance that the optimum resin/glass bond exists in the first instance.

It is also well accepted that the chemical resistance of GRP can be adversely affected by the phenomenon of 'wicking', i.e. the ingress of liquids along the glass fibre strands by capillary action, because of the structure of the strands, which generally comprise 204 individual filaments.

Even if the glass/resin interface bond is ideal, the internal capillaries will still be receptive to liquid ingress and it is therefore essential to ensure that every practical step is taken to prevent such ingress occurring. The use of monofilaments will eliminate interfilamentary capillaries but this places a practical limitation on the application of GRP fabrications to many chemical environments. With chopped strands, either deposited from a 'gun' or in the form of a mat, any wicking which does occur will initially be limited to the length of the individual strands involved. In the case of continuous filament roving, or woven roving, however, the capillary action will progress until a discontinuity in the glass is reached. This probably explains the occasional failure which has occurred with filament wound or woven cloth laminates in combination with completely inert resins, when resins of poorer chemical resistance reinforced with chopped strands have given superior performance under the same conditions. It is now common practice, and indeed an excellent one, to ensure that a relatively thick, resin rich gel coat layer is applied to all surfaces of laminate employed under conditions of chemical attack. The layer is generally reinforced with a thin tissue of glass monofilaments, or a scrim of woven or non-woven acrylic or polyester fibre, in order to eliminate the possibility of glass fibres from the main laminate protruding through the surface and hence presenting an inherent point for liquid ingress.

2. EFFECT OF CURE

Whilst unsaturated polyester resins 'cure' by an addition reaction across double bonds of the resin and the monomer solvent, the other resins normally encountered in reinforced plastics become rigid by condensation, i.e. by a reaction in which small molecules such as water, ammonia or carbon dioxide are produced as by-products. Although these reactions can readily take place at, or near, room temperature, the application of heat accelerates the conversion rate.

Condensation reactions are reversible, i.e. they can proceed in either direction. To consider a simple example such as the formation of an ester,

$$R{\cdot}CH_2OH + R'COOH = RCH_2OCOR' + H_2O$$

$$R{\cdot}CH_2OCOR' + H_2O = RCH_2OH + R'COOH$$

the formation of the ester $R{\cdot}CH_2O{\cdot}COR'$ can only proceed to completion if the water (H_2O) is continuously removed as it is formed. In the practice of ester manufacture, this is achieved either by the use of a desiccant, or by distillation. Curing of resins which can evolve small molecules at temperatures above the vaporisation temperature of these small molecules will be effective—but only if the vapours can escape will the conversion be complete.

Thus, venting of moulds for condensation polymer compositions is essential where hot curing is used.

Addition reactions do not result in the formation of small molecule by-products and they are not reversible, but the rate of reaction is dependent upon

(a) concentration of non-reacted polymerisation points,
(b) concentration of 'reactive centres' (influenced by catalyst/accelerator concentration),
(c) temperature. At a given temperature, the reaction rate reduces logarithmically with time.

In order to attain the best possible resistance to a wide range of chemical environments, it is necessary to ensure that the degree of cure achieved is the most complete consistent with other factors such as ease of fabrication and economics. In general, heat cured laminates will be superior to their cold cured homologues, despite the fact that exotherm has occurred with the latter.

The degree of cure is commonly assessed in practice by measurement of surface hardness, or by solvent extraction—the former, since it is a non-destructive test, being useful as a quality control guide.

Heat distortion point determinations on resin castings show quite clearly the importance of 'post cure' on the attainment of the highest degree of molecular complexity and this is supported by some work which has been reported by Alt[1] on the correlation between heat distortion point, unreacted styrene monomer and residual hydroperoxide for a series of unsaturated polyester resin castings.

Table 17.1.

EFFECT OF POST CURE ON HEAT DISTORTION TEMPERATURE OF TYPICAL UNSATURATED POLYESTER-RESINS

Post cure of 15 h at 100 C given to castings after (*a*) 24 h (*b*) 252 days at room temperature

Polyester	*A*			*B*			*C*		
Cure system	1	2	3	1	2	3	1	2	3
After 24 h									
without post cure	45	37	36	61	59	57	40	39	48
with post cure	59	58	58	101	98	104	128	129	126
After 252 days									
without post cure	63	61	52	79	73	74	80	73	72
with post cure	59	59	60	93	94	83	124	122	124

Key: All figures quoted are heat distortion temperature °C.
Polyester A orthophthalic polyester. low reactivity
Polyester B orthophthalic polyester. high reactivity
Polyester C isophthalic polyester. high reactivity.
Cure system 1 4% of 60% methylethyl ketone peroxide + 0.25% of 1% cobalt naphthenate.
Cure system 2 4% of 50% cyclohexanone peroxide + 0.25% of 1% cobalt naphthanate.
Cure system 3 3% of 50% benzoyl peroxide + 0.1% of dimethylaniline.

Table 17.2.

EFFECT OF ROOM TEMPERATURE CURE ON POLYESTER A CURE SYSTEM 1.

Cure time (days)	*Free styrene* %	*Active oxygen* %	*Undecomposed peroxide (% of quantity used)*
1	3.5	0.115	35.0
3	1.3	0.065	19.5
7	1.1	0.065	19.5
14	1.0	0.055	16.5
28	0.95	0.040	12.0
56	0.55	0.035	10.5
Sample post cured 15 h at 100°C after 24 h at 20°C	0.05	less than 0.002	less than 0.6

Table 17.3.

EFFECT OF POST CURE ON POLYESTER/RESINS A AND B CURED WITH 3 PER CENT OF 50 PER CENT BENZOYL PEROXIDE FOR 3 h AT 75°C

Post cure conditions	*Percentage of styrene*		*Weight loss* %		*Heat distortion temperature* °C		*Residual benzoyl peroxide* %	
	A	*B*	*A*	*B*	*A*	*B*	*A*	*B*
None	1.1	2.6	—	—	46	78	0.35	0.35
15 h at 60°C	0.75	2.6	0.25	0.15	52	91	0.20	0.25
24 h at 60°C	0.65	2.4	0.25	0.15	53	95	0.20	0.20
5 hr at 80°C	0.50	1.8	0.20	0.10	53	95	0.20	0.15
15 h at 80°C	0.45	1.4	0.30	0.20	55	95	0.05	0.05
5 h at 100°C	0.35	0.50	0.40	0.20	56	108	0.02	Less than 0.02
15 h at 100°C	0.30	0.40	0.50	0.20	59	105	less than 0.02	less than 0.02

Table 17·4. EFFECT OF POSTCURE ON FLEXIBILITY OF GEL COAT

Post cure	*Resin A*	*Resin B*	*Resin C*	*Resin D*
None	1.1% +	1.1% +	1.1%	1.1% +
3 h at 80°C	0.76%	0.25%	0.68%	0.84%
16 h at 80°C	0.44%	0.39%	0.54%	0.81%
1 h at 100°C	0.86%	0.24%	0.51%	0.63%
6 h at 100°C	0.43%	0.28%	0.46%	0.40%
1 h at 120°C	0.47%	0.20%	0.40%	0.61%
3 h at 120°C	0.44%	0.25%	0.50%	0.54%
1 h at 150°C	0.43%	0.27%	0.33%	0.41%

Resin *A* General purpose orthophthalic/maleic/ethylene glycol.
Resin *B* Higher heat distortion orthophthalic/maleic/propylene glycol.
Resin *C* Medium heat resistant isophthalic/maleic/propylene glycol.
Resin *D* Bisphenol 'A'/fumaric.

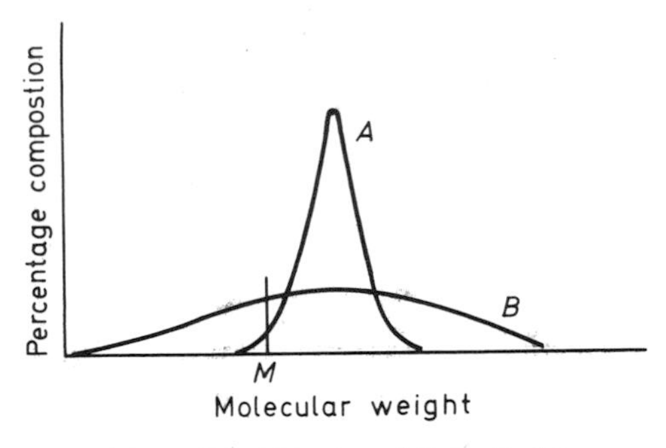

Figure 17.1. Molecular weight distribution

The degree of cure also influences the flexibility of a resin gel coat, this property reducing as the cure advances. It is therefore important that cure has advanced to such an extent before exposure to a chemical environment that little, if any, embrittlement can occur by ageing or chemical action in service.

Some recent, and hitherto unpublished work, has been carried out by the author on the effect of post cure on the flexibility of the resin gel coat of laminates. The laminates were prepared from six layers of 0.45 kg/m^2 chopped strand mat, with a surface tissue reinforced gel coat approximately 0.6 mm thick. Average glass content was 30 to 33 per cent. All samples were allowed to cure for 14 days at ambient temperature before post curing, and subsequently tested 24 h after post cure. The criterion was the gel coat strain in tension before cracking occurred. (*See* Table 17.4)

Molecular weight distribution can also exert an important influence on flexibility. Two samples of a given resin could, under certain circumstances, give rise to polymers whose number average molecular weight were identical, but whose molecular weight distribution was widely different. (*See* Figure 17.1)

The polymer represented by the Gaussian distribution *A*, having a narrower molecular weight distribution, would have superior retention of flexibility than the polymer represented by Gaussian distribution *B*, on the assumption that only a polymer of molecular weight less than *M* would be affected by the solvent action of the environment or would volatalise under conditions of heat. This phenomenon may be compared to the selection of plasticisers of differing molecular weights for polymers such as PVC—low molecular weight plasticisers, although of low cost, tend to be leached out; polymeric plasticisers are relatively expensive but must be employed for applications such as foodstuffs containers where the plasticiser must be completely permanent.

3. MODE OF CHEMICAL ATTACK

The resistance of any composite material to chemical attack is determined by *(a)* the inertness of each component part of the composite *(b)* the mode of protection afforded by the most chemically resistant component to those components less resistant *(c)* the effectiveness of the physical bonding forces between the component parts of a composite under any given set of exposure conditions to an aggressive environment.

As has been stressed earlier, for optimum chemical resistance it is accepted practice to ensure a continuous resin gel coat skin. This implies that, in general, for a given chemical exposure condition the resin system selected will have at least a chemical resistance equivalent to that of the other components in the laminate. Apart from chemical attack, which can be demonstrated on thin resin castings fairly readily, in some environments it is quite common for diffusion into the resin to occur without any chemical attack having occurred. This has been studied by Judd[2]. It is in these cases particularly, that the influence of fillers and fibrous reinforcements becomes important in assessing the overall performance of a given laminate construction. In certain instances a phenomenon known as 'ring cracking' is observed and this has been discussed by Steel[3].

4. INFLUENCE OF FILLERS

The fillers most commonly encountered in GRP fabricating practice are clays, chalks and silicas, in various physical forms but invariably as a powder. Other fillers, both organic and inorganic, may be used where special properties can be imparted to the laminate. The use of fillers tends to be for economic reasons, although a most useful reduction in shrinkage can result, increases in modulus of rigidity are also achieved. Other mechanical properties may or may not be affected, dependent upon the actual filler used, its percentage loading, physical form etc. Often the use of fillers will improve the heat resistance, or impart fire resistant properties. However, it is not infrequently the case that the incorporation of fillers, especially in appreciable quantities, results in a reduction in chemical resistance. This is particularly the case in those environments where diffusion into the resin tends to be favoured—even with those environments to which a given filler has complete chemical inertness it appears that a reduction in adhesion between the filler surface and the resin matrix occurs, resulting in progressive loss of strength and in extreme cases where complete adhesion failure occurs it is even possible to isolate discrete pockets of liquid within the laminate. It should be appreciated that a high filler loading, although reducing the shrinkage of the laminate system during cure, does not in fact reduce the shrinkage of the resin: because there is effectively a lower volume of resin present, the percentage shrinkage of the mass is reduced. The resin will tend to shrink uniformly and thus away from all fibrous and non-fibrous fillers.

5. REINFORCEMENT

Fibrous glass is available to the reinforced plastics industry in a number of physical forms and in various chemical compositions, of which two general types are commonly employed. One is a low alkali, borosilicate glass which was developed originally for the electrical industry and in consequence it is commonly referred to as 'E' glass. The other glass is a typical lime-soda composition with an alkali content of 10–15 per cent and referred to as 'A' glass.

In selecting the type of glass fibre for any application, the conditions of possible chemical environment need to be considered carefully. 'E' glass shows excellent resistance to water, both as a liquid and vapour, and is unaffected by most chemicals except strong mineral acids, which cause breakdown of borosilicate to boric and silica acids. The resistance to hydrofluoric acid is also poor. 'A' glass has relatively poor moisture resistance, but is superior to 'E' glass in the presence of strong mineral acids. It is therefore good practice to select 'E' glass for aqueous and alkaline media contact and 'A' glass where strong mineral acids will be encountered. Where aqueous solutions of mineral acids are involved, it is necessary to consider the concentration of the acid when selecting the reinforcement type—in general, high concentrations favour the use of 'A' glass, with the superior water resistance of 'E' glass becoming more important as the concentration is reduced.

Special high strength glass compositions are available, but are rarely used in conditions where long-term chemical inertness is likely to be an important selection consideration, being mainly used in aerospace and hydrospace applications.

Surface tissue is prepared from 'C' glass, which is similar to 'E' glass in composition and gains its nomenclature as a chemically resistant surface veil. Where strong mineral acids, and in particular hydrofluoric acid and acid fluorides are encountered, the reinforcement of the surface resin, or gel coat, should preferably be composed of a non-glass fibre.

It is very common for a polyester or modacrylic (Acrylan, Dynel) fibre to be used in such cases—where hydrofluoric acid is present the use of such fibres should be considered as essential in order to provide adequate protection to the main glass reinforcement.

The selection of glass reinforcement will also be influenced on economic grounds. 'A' glass is less expensive than 'E' glass, but has marginally lower strength. The correct selection of coupling agent will influence the efficiency of the resin/glass interface bond. With chopped strand mat, it is necessary to use a binder resin to hold the material together sufficiently for ease of handling before and during impregnation with resin, yet which permits rapid wetting of the glass fibres. Commonly such binders are either thermoplastic polyesters, which dissolve in polyester/styrene solutions, or vinyl emulsions which are insoluble. The latter generally are easier to handle than the former, but with certain laminating resins may make the individual fibres visible due to differences in refractive index, even when the refractive index of the cured laminating resin is very close to that of the glass fibre itself. Where opaque laminates are involved, this is obviously of no consequence but where no fillers are used the appearance may be more acceptable if soluble resin binder mat is used.

There is no evidence to indicate that the type of binder used has any effect upon chemical resistance and the choice is one which must be made by the individual fabricator.

6. ASSESSING CHEMICAL RESISTANCE

With the ever expanding use of reinforced plastics in the chemical and allied industries it is becoming continually more important for the fabricator to examine the chemical resistance of the laminating systems chosen for use in these industries. The raw materials suppliers, particularly the resin suppliers, have evaluated a wide range of chemical corrodents in order to promote the market penetration and acceptance of their products, but of necessity these have examined simple chemicals only rather than the multifarious mélanges which are commonly encountered in practical chemical processes.

Thus, whilst the raw materials supplier can give guide lines along which the selection of a suitable laminate composition may be confidently predicted, it is essential that in order to further the acceptance of GRP by the Chemical Industry at large, the fabricator must be able to undertake independent evaluation related to any new set of corrosive conditions which he may encounter. It may well be necessary to evaluate a range of polyester, epoxy, phenolic or furane resins, together with possible thermoplastics 'lining' materials in order to decide upon the best economic composite which will satisfactorily meet the technical requirements for a given application. However, this task need not be as daunting as may appear at first sight and some of the methods commonly used to make such an assessment have been discussed by Oswitch[4–7]. It is appropriate to review these here, together with one or two additional test methods.

It is relatively easy and inexpensive to place a few square inches of a laminate or resin casting in a bottle containing some of the corrodent under review and observe the more obvious visible effects as the sample ages. Perhaps the bottle and contents could be kept at the proposed operating temperature for an extended period of time, or the whole procedure might be speeded up by increasing the temperature on the well-known basis that any chemical reaction will increase its rate by a factor of 2 to 3 for a temperature increase of 10°C. However, such temperature acceleration testing can be misleading, since it is often found that a corrodent which is aggressive above a certain temperature is to a large degree nascent below this temperature. This may well be due to the physical significance of heat distortion temperature in addition to the chemical attack, particularly where diffusion into the resin matrix is necessary to initiate the chemical action on a filler. Even non-accelerated tests sometimes call for a high degree of experience in their interpretation, especially where dynamic loading or intermittent contact is experienced.

It is also often unwise to correlate the results of total immersion testing to service conditions where only one surface is exposed.

Another point to be considered is how long, under non-accelerated conditions, should a test be allowed to run before a decision may be made concerning the chemical resistance of the material under evaluation. Some authorities consider one month ample, whilst others would prefer to extend their tests over a period of 3 or 6 months or even longer. Obviously the fabricator has some excuse for confusion in this field where even at this time there is little established procedure or tradition. Even the various National Standards do little or nothing to resolve this confusion, since the adaption of 1, 2, 24, 72 and 168 h 'boiling water' tests would indicate the acceptability of short-term testing, whereas specific chemical resistance evaluation is generally required over a longer period. A recent paper by Williams[8] discussed the effect of exposure of three polyester resins in water at varying temperatures over a period of time and showed quite clearly that a resin exhibiting low weight increase and strength loss in the short term could quite easily be inferior in weight increase and strength loss in the long term. This would clearly seem to indicate that exposure over several months is essential if pertinent predictions are to be made with any degree of confidence.

7. BASIC TEST METHODS

Of the large number of possible techniques available for the assessment of chemical attack upon laminates or resin castings the following are perhaps the least difficult to apply.

7.1. VISUAL INSPECTION

Here a casting often reveals the commencement of attack more readily than does the normal laminate. For the purpose of this test a casting about 2.5 cm square and approximately 3 mm thick is first checked and if seen to be quite free from any flaws, is totally immersed in the corrodent.

If on inspection, possibly using low magnification, minute cracks or pinholes become visible, this may be taken as an incipient sign of chemical attack. With some resins a micro-blistering effect may become apparent—this is also a sign of corrodent ingress.

In many cases gross flaking or cracking indicates severe attack and hence complete unsuitability; many resins rapidly behave in this manner with highly polar ketone and chlorinated solvents such as acetone, chloroform and methylene chloride.

Discoloration of a casting generally indicates some degree of chemical reaction. However, this visual observation alone is generally a less serious condition than cracking or pinholes. For example, hydrochloric acid will cause a green coloration with any resin which incorporates a cobalt soap as an accelerator for the polymerisation stage due to absorption of hydrochloric acid into the matrix/corrodent surface, which reacts with the cobalt soap giving the green colour characteristic of cobalt chloride. In fact, GRP has been proved in service to be an ideal material for the long-term storage of hydrochloric acid in all concentrations; therefore colour changes need not be significant.

7.2. CHANGES IN SURFACE HARDNESS

Of the many instruments used for detecting changes, the one which is most commonly encountered is the Barcol Impressor, Model GYZJ, 934–1, manufactured by the Barber Colman Company, Rockford, Illinois, U.S.A. This is a small, portable instrument which operates on the principle of a needle point being pressed against the surface of the laminate or resin casting reacting through a spring and lever mechanism to give a pointer readout on a scale of 0–100, on which a tempered steel 'standard' represents a value of 85–87. Generally, castings of cured unsaturated polyester resins show Barcol hardness values of 30–50 depending upon the nature of the resin. Laminates will give higher values according to the glass content. Where a sample loses 10 or more units of Barcol hardness after an immersion test it may be assumed that serious softening of the resin has occurred.

Sometimes an increase in Barcol hardness is observed. This is explained by an increase in the state of cure of the resin in the absence of chemical attack, or the removal from the surface of some low molecular weight 'plasticising' material. It is usually safer to use castings for surface hardness measurements than to employ laminates, since the concentration of reinforcement and its depth below the surface can greatly influence the values obtained.

An average of at least ten readings should be taken on each face of a resin casting—in the case of laminates the number of readings averaged should be at least twenty on each surface. Readings in the immediate vicinity of local obvious chemical attack, for example blisters, are best ignored. In any event, the presence of such areas would indicate chemical incompatibility without the necessity to confirm this by changes in surface hardness.

7.3. CHANGES IN WEIGHT

Weight should be checked after immersion periods of 1, 2, 4, 8, 24, 48, 72 and 168 h ideally, then weekly for a further 3 weeks. Subsequent observations may then be extended to monthly intervals. It is a useful practice to plot the change in weight as a function of time. Over the 168 h period any weight increase over 5 per cent should be regarded with some concern. It is very difficult to suggest precise levels of corrodent absorption which correspond to known degrees of chemical attack; however, as a general guide the 5 per cent figure has been found useful and in most cases has been confirmed by other tests, such as retention of strength. In the majority of instances where completely satisfactory chemical performance will be forthcoming, weight increase will become asymptotic at a value probably below 3 per cent.

Loss in weight generally denotes that attack has occurred through a disintegrative mechanism which may be possibly accompanied by flaking of the test sample. Also, erosion by hydrolysis could be shown by a loss in weight. If a weight loss does occur and continues after the first week or so, it is unlikely that the resin, and hence the derived laminate, will perform satisfactorily in the long term. An initial weight loss may well be allowable, particularly if the test sample happens to be under-cured. If this loss is due only to removal of low molecular weight fractions this will rapidly stabilise.

7.4. CHANGES IN DIMENSION

This is an uncommon manifestation but when dimensional changes do occur it is usually a swelling of the sample which is first noticed. Any dimensional change is indicative of chemical attack or physical absorption. This is more generally encountered in laminates and is caused by absorption leading to swelling in the sample. With unsaturated polyesters some organic solvents give rise to a rapid break up of laminates by this mechanism, for example acetone and trichloroethylene illustrate this mode of attack dramatically.

8. ADDITIONAL TEST METHODS

The following tests require the use of some special, although relatively inexpensive, equipment and should be considered essential, at least in part, if the above simple tests are to be interpreted accurately even by the most experienced observer.

8.1 LOSS IN FLEXURAL STRENGTH

For the purpose of this test, laminate samples rather than resin castings should be used. A convenient instrument with which to carry out the evaluation of strength is a Hounsfield Tensometer. Test specimens 0.5 in (1.27 cm) wide are used, of length about thirty times the thickness. Generally, all cut edges should be sealed with resin prior to immersion in the corrodent—it is, however, convenient to include some specimens with unsealed edges if it is considered useful to compare the effect of cut edges.

The loss of strength with time is determined according to the method described in British Standards[9]. An arbitrary loss of strength figure, say 25 per cent, may conveniently be selected as being the point at which an unacceptable degree of chemical corrosion to the laminate has occurred. Ideally, twelve samples should be used for replicate testing at each withdrawal period—a minimum number of four must be tested on account of the accepted scatter of mechanical property test values, which is rarely better than 10 per cent of the arithmetic mean.

Using three point loading, on a beam supported over a length of sixteen times its own thickness, the ultimate flexural strength is calculated as

$$\frac{1.5 \times W \times L}{b \times t^2}$$

Where W = breaking load
L = distance between supports
b = width of specimen
t = thickness of specimen.

It should be stressed that the criterion of 25 per cent weight loss is purely arbitrary. It has been found that plotting strength retention against immersion time often shows a strength change over an initial period, after which the strength becomes constant. This phenomenon obviously leads to considerable confidence in using the constant strength value as a long-term design parameter.

Obviously, the stress/strain plot which may be obtained during the determination of flexural strength enables the modulus of elasticity to be calculated and this can also be used as a useful parameter in assessing chemical resistance (*see* Subsection 8.3).

8.2 GEL COAT EMBRITTLEMENT

This test is carried out upon 11 in × 2 in (28 cm × 5 cm) samples cut from laminate sheet and is specifically aimed at determining loss in flexibility of gel coat resins.

The laminate is prepared as follows: A film of the gel coat resin under examination, at a spreading rate of approximately 0.3 kg/m^2, is laid on a flat glass surface and reinforced with a single layer of either glass tissue, non-woven polyester such as 'Bondina' or a fine woven scrim cloth of an acrylic fibre such as 'Dynel' or Courtelle'. When this has cured, it is backed up with six layers of 0·45 kg/m^2 chopped strand glass mat impregnated with the same resin, or a medium flexibility resin of known resistance to the environment under examination.

Using a two support system with central loading, i.e. the same basic support/load configuration as for the flexural strength test—the gel coat is stressed in tension until failure by transverse cracking occurs. This is generally an extremely positive failure. Using a support span kept constant at 8 in (20 cm) and thus keeping the thickness/span ratio under 1:24, pure bending tending towards uniform curvature is assumed. The percentage gel coat strain at failure is given by the expression

$$S = \frac{td \times 100}{d^2 + 16}$$

Where S = Percentage gel coat strain
t = thickness of sample at the point of failure
d = deflection at the centre of the laminate at failure
both t and d being measured in inches.

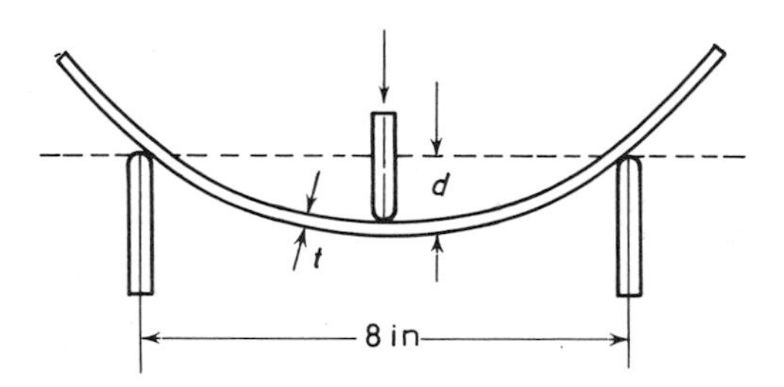

Figure 17.2. A two support system with central loading

A simple piece of equipment suitable for carrying out this determination may be easily made from slotted angle, loading at the centre being by a simple thumbscrew with the deflection measured continuously with a dial micrometer. This test is normally carried out upon a minimum number of four laminate samples and the mean result is taken if the scatter lies within 20 per cent. Gel coat flexibilities can be determined at ambient temperatures or at elevated or reduced temperatures. In a specially designed apparatus tests have been successfully carried out at temperatures as low as −196°C. The actual values of gel coat flexibilities suitable for specific duties depend upon such parameters as operating temperature, nature of corrodent where applicable, thickness of both gel coat and laminate and the stress variables to be encountered in service.

8.3. MEASUREMENT OF MODULUS UNDER LOW STRAIN CONDITIONS

The above tests generally refer to specimens which can be most conveniently 'conditioned' by complete immersion, or exposure, to the environment under examination. Whilst such information is obviously useful, it is however a fact, that the majority of practical applications of GRP involve exposure to the environment on one side only. As mentioned earlier, the elastic modulus of a laminate can also be employed as a measurable property to assess whether physical or chemical attack is taking place, and provided that relatively low stresses are involved, the measurement of strain can be used as a non-destructive test on laminates with only one side exposed. A proprietary instrument[10] is available for this purpose, under the trade name of 'The Deflectron'.

This may be used in the laboratory by measuring the changes in modulus of a laminate clamped to the end of a tube filled with a corrodent. Alternatively, it may be used to determine whether a storage tank or piece of process equipment in service is still satisfactory after a long period—it is thus useful as both an R & D tool and as a piece of 'preventive maintenance' equipment.

8.4. HEAT DISTORTION TEMPERATURE OF RESIN CASTINGS

Perhaps the most confusing and least understood property of binder resins is the Heat Distortion Temperature; the temperature at which a resin casting will deform by a specified amount under standard conditions of stress. Since the presence of a reinforcing fibre and/or granular fillers will invariably raise the observed value of Heat Distortion Temperature, often to well above 200°C irrespective of the H.D.T. of a simple resin casting, one may well query the significance of this property. The H.D.T. may be considered as the temperature at which the molecular attraction of adjacent polymer chains, or the forces within the three dimensional structure, is reduced to a level corresponding to the applied force/deflection conditions. Thus, it enables a comparison to be made of the resistance of the resin system to diffusion, when compared with a similar alternative resin system under identical physical conditions. It can be deduced that, for a family of resins which are resistant to a given environment at ambient temperature, in general for a high H.D.T. resin, the maximum operating temperature will be greater than for a medium H.D.T. resin. As a rule of thumb in aqueous media, an orthophthalic modified polyester having H.D.T. of 55/60°C will be acceptable for use up to 30°C; a similar resin having H.D.T. of 90/100°C can be used up to 50°C. If the resin is isophthalic modified, then at an H.D.T. of 75°C the useful temperature ceiling is 50°C; where the H.D.T. is 90/100°C, operation at 70/75°C may be considered as the ceiling. For aqueous operation above 75°C, the hydrophobic nature of bisphenol modified polyesters, heat cured epoxies, phenolics and furanes are essential at present due to hydrolysis of 'conventional' polyester resins.

The test method for Heat Distortion Point is fully described in B.S. 2782:1965[11] but basically comprises loading a resin casting as a beam with an applied stress of 18.5 kg/m^2 in a liquid bath whose temperature may be raised at a specified rate.

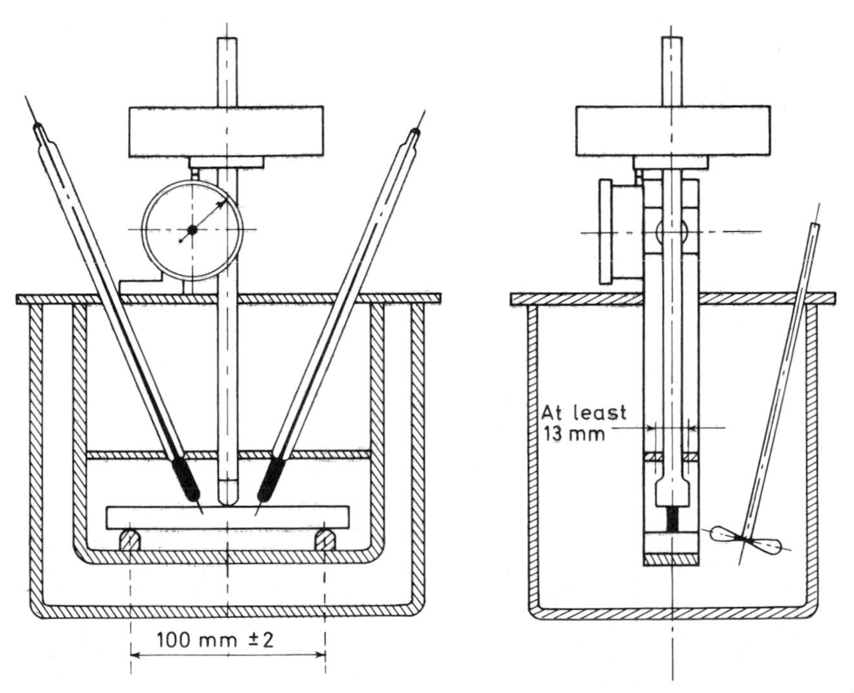

Figure 17.3. Diagrammatic representation of heat distortion temperature apparatus

8.5. A SIMPLE STRESS-ACCELERATED CORROSION TEST

Since acceleration of chemical attack by the use of higher temperatures or concentrations than will be encountered in service may well produce conditions which are beyond the threshold of chemical attack, it is useful to have available some form of acceleration which does not involve these parameters. It has been shown by Oswitch[5] that the application of a stress approximately 50 per cent of the ultimate in flexure enables a rapid 'screening' to be made of the chemical resistance of gel coats in tension. The result of corrodent attack generally appears within 7 days under these conditions, the form of failure depending upon the type of corrodent.

In general, in the presence of acids failure is first shown by surface cracking, followed by the rapid splitting up of the laminate. Bases soften the laminate appreciably but do not cause cracking—thus a gross permanent deformation

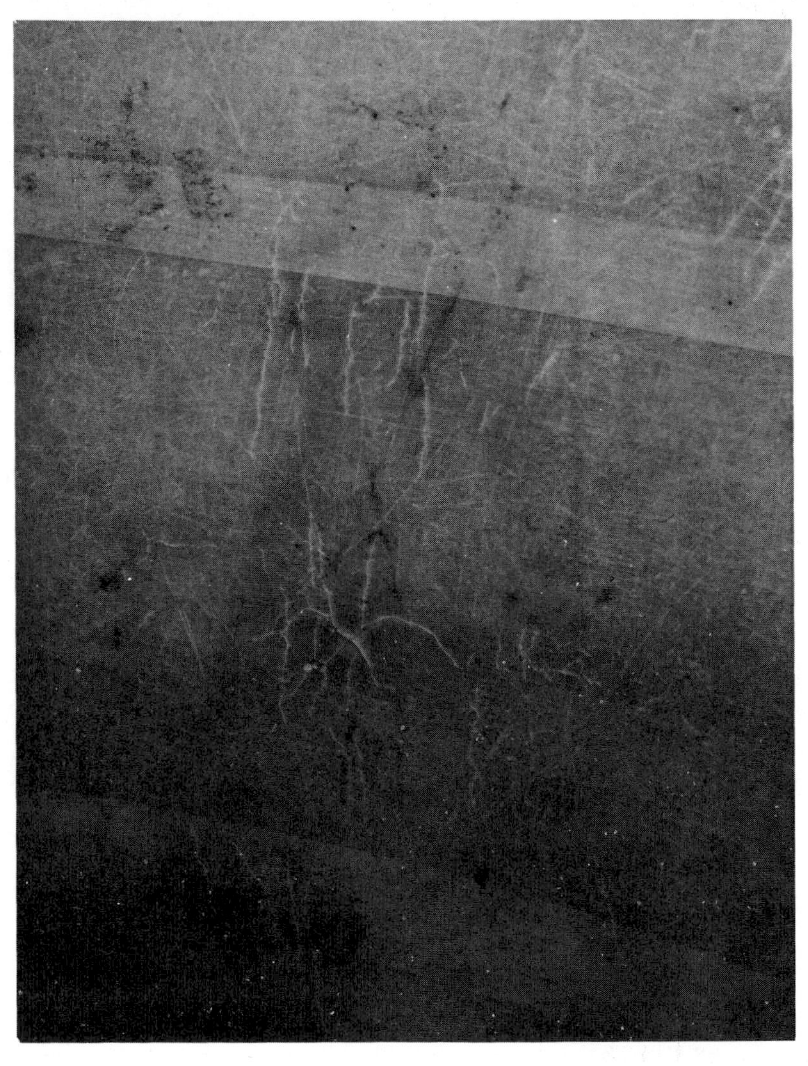

Figure 17.4. Damage caused to interior of storage tank under influence of $3{\cdot}5\ \mathrm{kN/m^2} \times 10^4$ *stress in water at* 70°C *for* 6 *months*

Figure 17.5. 4800 Imperial gal. (21·5 m^3) *road tanker used for transporting hydrochloric acid at all concentrations, and sulphuric acid at concentrations up to* 77 *per cent* w/w. (By courtesy of Whessoe Ltd, Plastics Division)

is seen on releasing the stress. Polar organic materials tend to bring out a pronounced fibre pattern accompanied by softening; non-polar organic solvents also attack by a softening process but little fibre pattern is seen. It is a reasonably safe assumption that laminate samples which can survive a test of this type for 14 days under the proposed conditions of environmental service with no defect are probably possessed of a very long life under these conditions.

9. TYPICAL CHEMICAL RESISTANT PROPERTIES

In general, it is found that heat cured laminates have superior chemical resistant properties to cold cured laminates made from the same resin system which have been post cured. Such post-cured samples will in many instances have superior resistance than other identical samples which have not been post cured. This generalisation would appear to hold true for furanes, epoxies, phenolics and polyesters.

The furanes probably show the best all-round chemical resistance but are normally available only as solutions in inert solvents—thus only thin films can be cast and these resins are generally encountered as surface gel coats. They must be heat cured in order to develop resistance to organic solvents and strong acids and alkalis.

Epoxy resins, when heat cured, have a very good range of chemical resistance over a wide temperature range, with amines and very strong oxidising acids being just about the only common chemical agents which have appreciable effect. These are formulations cured with anhydrides or amides. Formulations using amines, or BF_3 complexes, are cold curing but even after extended post

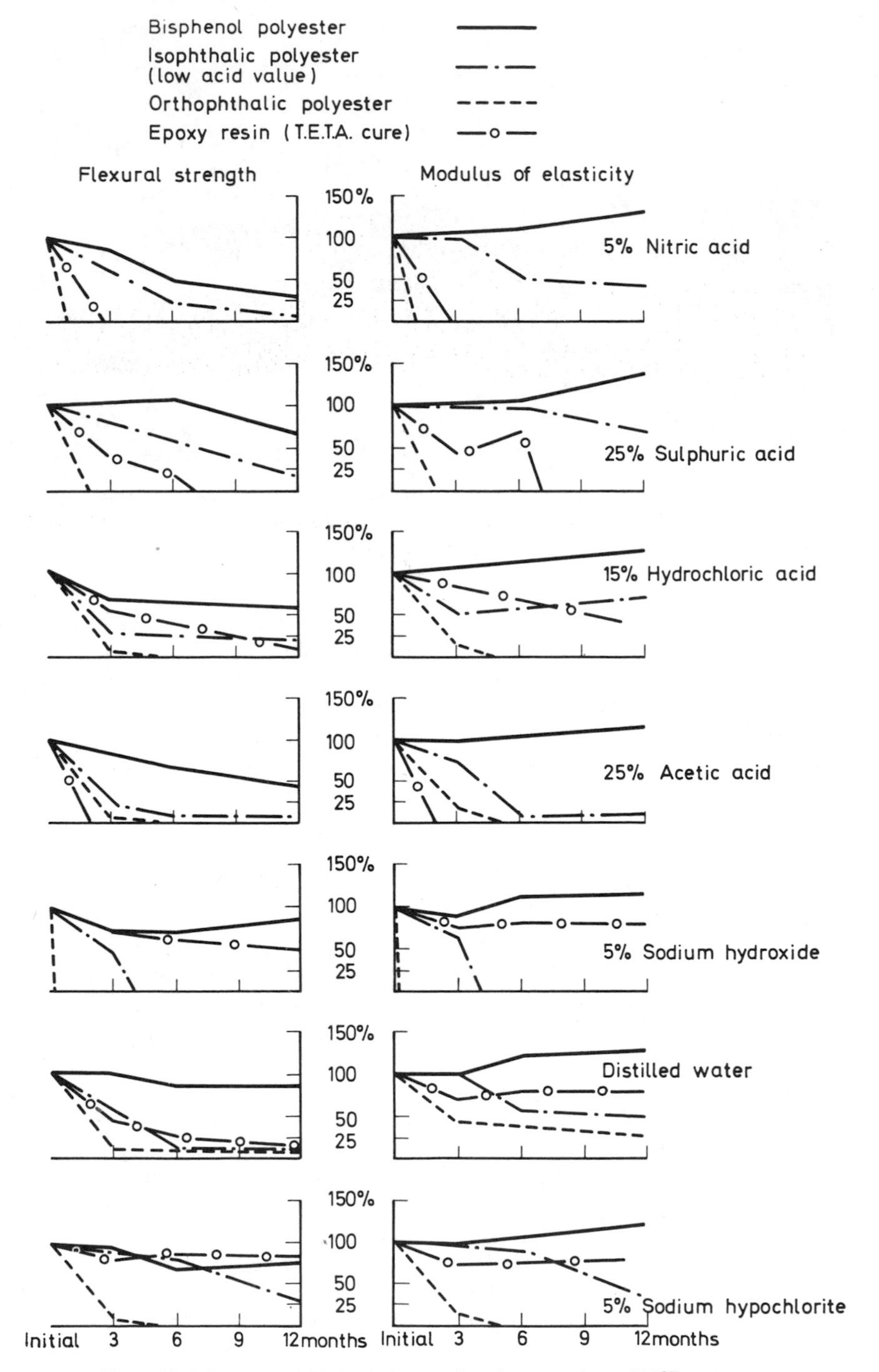

Figure 17.6. Retention of Mechanical properties after immersion at 210°F (99°C)

curing the chemical resistance is much inferior to the hot-cured formulations. Epoxy resins are also of high viscosity compared with polyesters and it is perhaps significant that, even in chemical plant applications, less than 5 per cent of laminates are based on epoxy resins at the present time.

Phenolic resins have been employed in chemical resistant applications for over 50 years, but since they are converted by a condensation reaction it is preferable to employ a hot press moulding technique for their fabrication into laminates. Cold curing by means of an acid catalyst is sometimes employed, but the presence of non-reactive solvents to keep the viscosity acceptably low presents certain fabricating problems. Post curing is essential for developing a wide range of chemical resistance.

Chemically resistant flooring compositions are available based on cold curing furane, epoxy, phenolic and polyester resin compositions which give adequate resistance to the spillage of chemicals. Thus, it is not essential in all applications to attain the highest possible degree of conversion in order to give satisfactory service.

Because of their general versatility, the unsaturated polyester resins find greatest use in the manufacture of GRP laminates. It is therefore important that the general features of these resins are understood. The chemistry of the resins is dealt with in detail elsewhere, but, in general, an unsaturated polyester resin comprises: (*a*) an unsaturated acid (*b*) a glycol (*c*) a vinyl-containing monomer as a solvent *(d)* modifying acid(s)/glycol(s).

The typical 'general purpose' polyester comprises maleic or fumaric acid, orthophthalic acid and an ethylene or propylene glycol, in styrene. This will have good resistance to non-polar solvents, water and aqueous solutions, and weak mineral acids.

Using isophthalic acid in place of orthophthalic acid, the structure of the molecule is altered to give a certain degree of steric protection to the ester groups, thus improving the resistance to hydrolytic attack and also increasing the heat distortion temperature.

Certain glycols also bring about a steric effect, for example, neopentyl glycol and 'Bisphenol A' (2:2 di-4′-hydroxycyclohexyl propane) with the latter being the source of the most highly chemical resistant of the commercial resins. Unfortunately, the latter lose some of the resistance to hydrocarbon solvents which is the feature of a general purpose polyester.

The ratio of styrene to unsaturated double bonds in the polyester resin structure itself also has some influence on chemical resistance, as does the acid value of the resin, particularly in respect of aqueous solutions. Low acid value and relatively large styrene ratios are both advantageous.

For aqueous solutions over about 70°C, there is little doubt that 'Bisphenol A' polyesters give superior performance compared to isophthalic or orthophthalic polyesters. Orthophthalic polyesters in general perform well at ambient temperatures, or if they have heat distortion temperatures above 90°C may well be satisfactory up to 50°C. In the 50–70°C range, isophthalics are generally adequate and show a worth-while economic advantage over the 'Bisphenols'. These 'Temperature Brackets' can be used as a rule of thumb where short-term tests show negligible effect both at room temperature and at the proposed operating temperature.

There is a paucity of information on the behaviour of GRP laminates at temperatures approaching the boiling point of water, but the data in *Figure 17.6* reproduced by permission of Atlas Chemical Industries Inc.[12] shows the general excellent performance, at 210°F (99°C) of bisphenol resin, with isophthalic resins showing up considerably better than general purpose resins, where evaluation was carried out on hot cured resin castings.

REFERENCES

1. Alt, B. *Kunststoffe,* March 1962.
2. Judd, N. C. W. *R.A.E. Tech. Rep.* 67042 February 1967.
3. Steel, D. J. Transactions, Plastics Institute, April 1967.
4. Oswitch, S. *Reinf. Plast.*, December 1963.
5. Oswitch, S. *Reinf. Plast.,* February 1964.
6. Oswitch, S. *Reinf. Plast.,* May 1964.
7. Oswitch, S. *Reinf. Plast.,* February 1965.
8. Williams, H. 6*th Int. Reinf. Plast. Conf.,* London, 1968: Paper No. 8.
9. *B.S.* 2782: 1965 Method 304B.
10. The Deflectron Company, P.O. Box 3751, St. Louis 22, Missouri, U.S.A.
11. *B.S.* 2782: 1965 Method 102G.
12. Atlas Chemical Industries Inc., Wilmington, Delaware 19809, U.S.A. *Leaflet No.* LP-29.

18

Fatigue

M. J. OWEN

1. INTRODUCTION

When engineering materials are subjected to repeated applications of load they tend to fail even when the individual loads are substantially below the short-term ultimate strength. This phenomenon has been a problem to engineers since the early part of the nineteenth century and is called 'fatigue'. Originally attention was focused on metals, but the phenomenon is now recognised in virtually all engineering materials including concrete, asphalt mixes and plastics. Glass reinforced plastics of all kinds have been found to be susceptible to failure under repeated loading.

When considering the behaviour of GRP under repeated loading, there is a slight risk of confusion in terminology between engineers and glass technologists. Engineers apply the term 'fatigue' to the process of permanent structural damage which occurs in materials under repeated or fluctuating loads. Glass technologists have applied the term 'fatigue' to the phenomenon of time dependent failure of glass under sustained loading, i.e. the phenomenon which is called 'stress-rupture' by engineers. Throughout this chapter the engineering terminology will be employed.

2. FAILURE MECHANISMS IN GRP

It will be shown later that the mode of failure of glass reinforced plastics has an important bearing on both the presentation and utilisation of fatigue data. There have been a number of optical microscope studies of GRP which have been subjected to both single and repeated loads[38]. In most applications GRP laminates contain multidirectional reinforcements, for example random mat, woven roving or cloth, or cross-plied non-woven materials in sheet or filament wound form. When mat or fabric reinforced laminates are subjected to a conventional tensile test, the first signs of damage occur at the knee of the stress strain curve. Optical micrographs show that at this stage damage takes the form of separation between glass filaments and the resin matrix within strand groups which lie perpendicular to the line of load. At this stage, no resin cracks are visible. If the load is increased to about 70 per cent of the ultimate strength, depending on the flexibility of the resin matrix, then the debonds spread into the matrix to form resin cracks. At this stage the fibres aligned with load become affected. Where the resin cracks cross the aligned fibres, debonding around the circumference of the fibres occurs on either side of the crack. The ability to debond in this way in preference to crack propagation through the filaments appears to be vital to the non-brittle behaviour of the composite as a whole. Further loading produces a general intensification of damage until eventually filaments fail. Under repeated loading the same type of damage occurs but is cycle dependent as well as load dependent, i.e. if a composite is stressed repeatedly to a stress level which does not produce debonding at the first application of load then in due course, debonding will appear followed much later by resin cracking and ultimately separation of the specimen. If the stress level is sufficiently high to produce debonding, damage at the first application of load then damage is present from the first cycle of the fatigue test and is followed by resin cracking and failure. Laminates with continuous non-woven reinforcements (which generally have high glass contents) show a slightly different behaviour. Again the first signs of damage are in the form of debonding between filaments lying perpendicular to the line of load. Initially these debonds extend from one ply interface to the next. The next stage of damage, however, is for the debonds to turn and propagate along the ply interface. The vulnerability of the ply interface appears to be due to several factors; a slight tendency to resin richness; the tendency for small voids to accumulate at the interface; and because the strain compatibility requirement between adjacent plies causes shear stresses at the interface. At a later stage damage appears among the aligned fibres.

The macroscopic appearance of a fatigue test specimen may be described as follows. An initially translucent specimen gradually becomes opaque. At first this opacity may only be present whilst the specimen is under load and may disappear again if the load falls to zero. Progressively the opacity becomes permanent and more intense. When there is permanent opacity, resin cracks are visible at low magnifications. Subsequently, there is little change in appearance until towards the end of the test; localised intense damage leads to final separation of the specimen. Just before separation the matrix often appears to have been reduced to a white powder. The appearance of the final fracture surface depends on the nature of the stress cycle and the type of reinforcement. With mat and fabric reinforced laminates subjected to wholly tensile stress cycles, the final fracture surface is usually of a fibrous appearance approximately normal to the line of load and generally very similar to the appearance of a failed conventional tensile test piece. If the stress cycle includes a compressive component, the final fracture is usually fairly smooth and inclined at an angle to the central plane of the specimen. Sometimes multiple fracture surfaces form 'V' or 'W' shapes. Final fractures in non-woven continuously reinforced materials tend to have a delaminated appearance, with the filament failures widely separated giving a brush-like effect.

An essential feature of fatigue damage in GRP is that it is progressive and extends throughout the stressed region of the specimen. If the stress level is high enough it may be present from the first cycle of load. After visible damage occurs the specimen may survive tens or even hundreds of thousands of cycles before final separation. For many applications, however, it is the onset of damage which is relevant rather than final separation. Once there is significant damage in the material, a structure or component may, for all practical purposes, have failed through loss of integrity or through deterioration of the other properties. In this respect, GRP differ greatly from the common structural metals where fatigue failure is initiated by microscopic slip processes which do not affect the structural integrity of components until such time as a propagating crack occurs, and this is usually after about 90 per cent of the life. In metal components failure is virtually always initiated at the surface and takes the form of a single crack, or at most a small number of cracks, propagating into the body of the material.

3. TEST METHODS AND EQUIPMENT

In the majority of cases test methods and procedures for GRP have been carried over from the metal fatigue field. The basic choice is between flexural and axial loading. Axial load fatigue testing requires more complicated equipment and more elaborate specimen shapes and is thus more expensive to carry out than flexural testing. However, the results obtained under axial load conditions are readily interpreted and used for design purposes, whereas flexural testing usually provides only comparative data. Whilst there have been a number of individual papers based on flexural results contributed to the literature, by far the majority of data have been obtained under axial loading conditions. Because of the occurrence of progressive damage, flexural specimens do not always fail completely, and it is necessary to adopt some criterion such as the observation of resin cracks, or loss of load-carrying capacity to define failure. However, in the case of axial loading, complete separation of the specimens usually occurs, and this has almost always been adopted as the criterion of failure. Progressive damage has usually been ignored in presenting data.

The design of fatigue specimens involves a number of compromises. One of the theoretically attractive features of axial load testing is the possibility of having a uniform stress distribution across the specimen. However, if a parallel strip specimen is used, it will virtually always fail at the grips. It is thus necessary to form a reduced working cross-section in the specimen. For laminar materials this reduction must be edgewise in order that the material at the neck may be representative. If the reduced portion itself has a parallel working section blending into some curved profile, failure often occurs at the blend between the parallel and curved sections because of the slight stress concentration. If it is intended to use compressive loads in the cycle, it is necessary to restrict the length of the specimen in order to avoid buckling effects. The majority of workers now use a specimen shape consisting of a rectangular blank, edge notched by large radius circular arcs giving a nominal stress concentration at the neck not greater than 1.1 (based on isotropic material). Gripping arrangements vary considerably but an essential feature is for load transfer to take place on the faces of the shanks. If there is a hole in the shank it can only be for location purposes. Any attempt to load substantially through the periphery of the hole leads to failure at the hole rather than failure at the test section. Pin and button grips can successfully be used for mat and fabric based materials but for stronger materials it is usually necessary to clamp over the whole face of the shank. For predominantly unidirectionally reinforced materials, it is often necessary to bond aluminium tabs to faces of the shanks. Completely compressive load cycles usually require special alignment fixtures.

It is well known that there can be significant heating effects in GRP under repeated loading conditions. The heating effect depends on a number of factors including frequency, stress level and specimen thickness. Dally and Broutman[20] have recently conducted a systematic survey of heating effects in two materials. American workers have generally favoured relatively high testing speeds. The extensive work carried out by Boller and his associates[2–13] has been mainly carried out at 900 c/min with forced cooling by means of a fan. The majority of British workers have tried to avoid heating effects, which means testing at speeds below 100 c/min in many cases. If internal heating of specimens occurs, there is undoubtedly an effect of speed on the fatigue properties[20, 31]. There is also a strain rate effect on short-term tensile properties affecting both modulus and the debonding point[16]. However, if the stress cycle has an amplitude less than that necessary to produce debonding, hysteresis effects are relatively small and it is possible to test at cyclic speeds as high as 6000 c/min without noticeable heating. From experiments with carbon fibre reinforced plastics, which have better heat transfer properties, the author believes that frequency effects are small in the absence of internal heating.

A fatigue testing machine for reinforced plastics should preferably be an axial load machine with adjustable mean load and load amplitude, capable of loading accurately in tension and compression. Overall alignment of specimens, grips and machine is very important with flat short specimens, especially in compression. The machine should have sufficient operating stroke to cope with materials whose moduli range down virtually to resin modulus ($3.5\ \mathrm{kN/m^2} \times 10^6$). The machine should be of the variable speed type, so that economical testing speeds can be selected according to specimen shape, material characteristics and stress level.

Usually constant load characteristics are required and if the machine is not inherently of the constant load type, then automatic load adjustment should be capable of coping with substantial changes of specimen stiffness and permanent changes of length during the progressive damage phase. The loading frame should be sufficiently spacious to accommodate various designs of specimen grips including flexural and interlaminar shear attachments, environmental chambers, microscopes, etc. At the lower cyclic speeds several loading frames are desirable thus placing added limitation on cost. Instrumentation should include cycle counting and rapid load measurement facilities. For some purposes a stress–strain display is desirable to determine the progress of damage through hysteresis or modulus determinations, etc.

The best available type of machine now appears to be a closed loop hydraulic servo system which can provide all the necessary speed range, load control, etc. Programmed cyclic loading and random loading are possible together with conventional constant amplitude loading to failure. Measurement and display facilities appear to be available according to demand. However, such machines are prohibitively expensive for most purposes and very few of them are yet available. Clearly most test programmes will adopt a number of compromises in the selection of equipment, test procedures and data obtained. A number of special purpose fatigue machines for GRP have been designed and are available[28, 37]. Most of the machines are mechanical or hydraulic and do not have automatic load adjustment facilities. The quality of the results obtained depends on regular adjustment of load and an understanding of the material characteristics by the test laboratory staff. Most large organisations with mechanical test laboratory facilities possess fatigue machines intended for metals testing. With care and understanding it will often be possible to use these machines for GRP. Rotating bending machines are of little use. Plane bending machines can be used for comparative tests. With constant load machines a deflection criterion can be used to define failure and with constant deflection machines either a loss of load or the appearance of the first crack. Flexural specimens are usually cheap to prepare and higher speeds are practicable because the greatest stresses occur at the surface. Most axial load machines of the resonance or rotating weight type are inherently high-speed machines. In many cases they can only cope with small displacements and have very stiff mean load springs. They are thus very sensitive to changes of modulus and permanent extension in the specimens. The difficulties of using these machines tend to increase with decreasing modulus in the material under test.

4. PRESENTATION OF DATA

Generally the classical method of fatigue testing and data presentation has been adopted. If the stress cycle is represented in terms of mean stress and stress amplitude (half the stress range), usually a set of 10 or 12 specimens have been tested at the same mean stress but each at a different stress amplitude. The stress amplitude is then plotted against the logarithm of the life to obtain the well-known fatigue curve or 'S–N' curve. *Figure 18.1* is based on data by Boller[6] and shows the typical features of the S–N curve. Testing is usually restricted to 10^6 or 10^7 cycles for economic reasons. On the semi-logarithmic plot the fatigue curve is usually concave upwards between 10^3 and 10^7 cycles. Curves are usually drawn in by eye. In some cases straight lines are fitted by least square techniques, especially when testing is discontinued at 10^6 cycles. In most test programmes the average ultimate strength is determined, but few results are obtained between one quarter and 10^3 cycles. The upper part of the curve is thus not well defined. The fatigue strength at 10^6 or 10^7 cycles (the stress corresponding to a stated life read off from the curve) is usually between 0.2 and 0.3 times the u.t.s. There has been no convincing evidence of the existence of a genuine endurance limit which appears to be confined to the

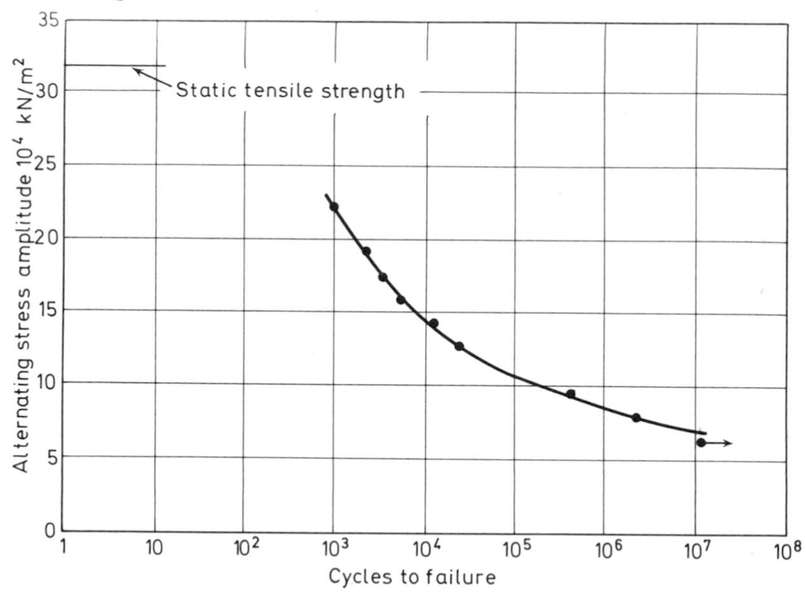

Figure 18.1. S-N *curve of unnotched specimens of polyester resin reinforced with 181 glass fabric, Volan A finish, and tested at* 0° *to warp,* 73°F (23°C), *and* 50 *per cent relative humidity, and zero mean stress* (From Boller[6] by courtesy of Modern Plastics Inc.)

ferrous metals. With only one test point for each stress level the true extent of scatter is suppressed.

Figure 18.2 is based on data obtained by the author and his co-workers[40, 41] for chopped strand mat/polyester resin laminates. Three stages of failure are defined: debonding, resin cracking, and total separation of the specimens. The vertical line plotted at $\frac{1}{4}$ cycle shows the scatter about the mean u.t.s. (two standard deviations on either side of the mean). Six specimens were tested at each of four stress levels and together with single test at several other stress levels define the fatigue properties at total separation. A straight line has been fitted to the medians of the replicate tests. An envelope containing all the u.t.s. and total separation results has been drawn in. Whilst the scatter about the average u.t.s. is far greater than would be expected for metals, the scatter of fatigue lines about the median is only about the same. The band representing the onset of debonding was determined by detecting a loss of modulus and the resin cracking zone was detected by microscopic observation of resin cracks.

Figures 18.1 and *18.2* represent data obtained under alternating stress (zero mean stress). In practice, the mean stress is frequently non-zero. Families of S–N curves can be obtained to indicate the effect of mean stress and stress amplitude. For several mean stresses conventional S–N diagrams are obtained. For design purposes it is convenient to replot these to show the relationship between mean stress and stress amplitude for stated lives. *Figures 18.3a* and *18.3b* show a family of S–N curves and the corresponding master diagram or Goodman Diagram. The results are taken from Smith and Owen[41] and are for a cross-plied non-woven 'E' glass epoxy resin system, Permaglass XE6.

5. AVAILABLE PUBLISHED DATA

There are by now well over 100 individual references relevant to fatigue of GRP in the English language literature alone. Forty-seven selected references are listed at the end of this chapter and their general content summarised in Table 18.1 for the convenience of readers who may be looking for specific information. The selected references contain usable design data or descriptions of special testing equipment and techniques. Progress reports and government papers which are restricted or otherwise difficult to obtain have been omitted. Referring to Table 18.1 columns 2, 3 and 4 give the year of publication, and country of origin of the paper and indicate whether it is a literature review or contains a substantial literature review. The majority of papers do not contain a literature review of any substance. Columns 5 to 18 inclusive give details of laminate materials. Columns 19 to 24 give details of test speed and mode of stressing. It is not always clear whether four-point or three-point bending was used, whether flexural stressing was fully reversed, and in the case of three-point bending, whether the failures were flexural or interlaminar shear. Columns 25 to 35 indicate the type of data available and the failure criterion adopted. Columns 36 to 39 indicate the environmental conditions and columns 40 to 44 indicate the extent of other information in the paper which might be of value to designers. Virtually all papers give the

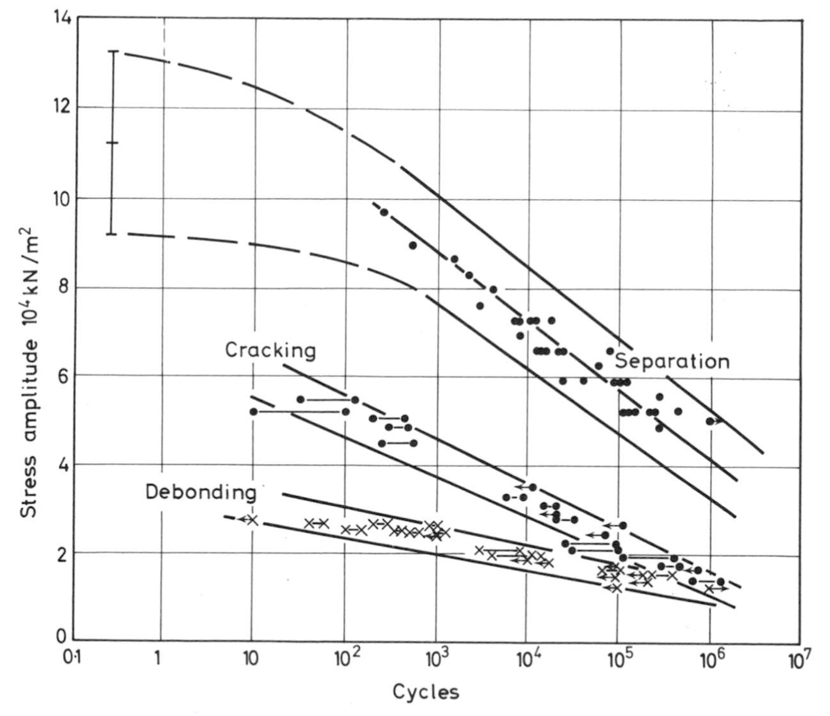

Figure 18.2. S–N Diagram showing the various stages of failure of chopped strand mat/High reactivity polyester resin. Zero mean stress: 20°C *and* 40–42 *per cent r.h.*

Table 18.1 AVAILABILITY OF TEST DATA

				Resin matrix					*Reinforcement material*				*Reinforcement type*				
1	2	3	4	5	6	7	8	9	10	11	12	13	14	15	16	17	18
Reference number	*Year of publication*	*Country of origin*	*Literature review*	*Polyester*	*Epoxide*	*Phenolic*	*Silicone*	*Thermo plastic*	*A-Glass*	*E-Glass*	*S-Glass*	*Asbestos*	*Mat*	*Fabric*	*Woven roving*	*Non-woven roving*	*Filament wound*
1	1968	UK		×						×				×			
2	1952	US		×						×				×			
3	1954	US		×						×				×			
4	1956	US			×					×				×			
5	1958	US			×					×				×			
6	1956	US		×	×	×	×			×			×	×			
7	1958	US			×	×				×		×	×	×		×	
8	1961	US				×				×				×			
9	1964	US				×				×				×			
10	1965	US	×		×	×						×	×	×			
11	1962	US	×		×					×						×	
12	1963	US	×	×	×	×	×			×			×	×			
13	1964	US		×						×						×	
14	1964	US		×	×	×	×				×			×			
15	1967	UK	×														
16	1965	UK		×									×				
17	1963	US			×					×	×			×		×	×
18	1963	US			×					×						×	
19	1965	US			×						×						×
20	1968	US			×						×						
21	1964	US	×		×					×						×	
22	1966	Japan		×						×				×		×	
23	1966	US			×					×	×			×			×
24	1957	US		×									×	×			
25	1966	US			×						×					×	
26	1958	UK	×														
27	1961	UK		×						×				×			
28	1965	UK															
29	1967	US	×		×						×			×			×
30	1956	US			×									×			
31	1968	US			×									×			
32	1966	US															
33	1958	UK				×						×	×				
34	1952	UK				×						×	×				
35	1966	US			×						×			×			
36	1957	US		×	×											×	
37	1967	UK															
38	1967	UK	×														
39	1968	UK		×						×			×				
40	1966	UK		×						×			×				
41	1968	UK		×	×					×			×			×	
42	1962	US			×					×							
43	1957	US		×	×	×				×			×	×			
44	1966	Switzerland			×									×			
45	1967	US	×														
46	1968	US						×									
47	1962	UK		×	×					×			×	×			

Test speed	Mode of stressing					Type of data						Failure criterion					Environment				Other information				
19	20	21	22	23	24	25	26	27	28	29	30	31	32	33	34	35	36	37	38	39	40	41	42	43	44
Cyclic speed: c/min	*Axial*	*3-pt Bending*	*4-pt Bending*	*Inter-laminar shear*	*Biaxial*	*Replicates*	*Fully reversed stress*	*Zero-tension*	*Zero-compression*	*Other mean stress*	*Master diagram*	*Cumulative damage*	*Total separation*	*Resin cracking*	*Debonding*	*Other*	*Normal ambient*	*Low temp.*	*High temp.*	*Wet*	*Notch effects*	*Stress rupture data*	*Loading at angle to reinforcement*	*Applications*	*Special testing equipment*
2000, 6000	×						×			×	×		×				×		×			×			
900	×						×			×	×		×				×			×	×		×		×
900	×										×		×									×			
900	×						×						×				×			×	×		×		
900	×						×						×								×				
900	×						×			×	×		×				×		×	×	×	×	×		
900	×						×						×				×		×	×			×		
5	×						×						×				×		×						×
5, 25, 900	×						×			×			×				×		×						
10, 900	×						×						×				×								
900	×						×			×	×		×				×		×	×	×	×	×		
900	×					×	×					×	×				×								
—	×							×									×	×							
—																									
0.3, 10, 60								×					×									×			
20	×				×				×				×									×		×	×
—	×								×				×	×	×							×			
—					×				×							×									
60, 2400	×					×		×								×									
—	×	×																						×	
1850			×				×									×	×		×				×		
—	×			×	×				×																
1800			×				×						×				×								
1				×		×										×	×			×		×			
—																									
1000, 400						×	×			×	×		×				×					×			
—																									×
6, 50, 1800	×	×	×			×	×	×	×			×	×			×	×			×		×			×
1250	Cantilever																×								
—																									
							—																		
—																									×
1000, 1500, 1740	×					×				×	×						×								
1500	×							×									×				×	×			
	×	×				×	×										×	×	×				×		
1800				×											×		×		×						
—																									×
74	×					×	×	×		×	×		×				×					×			
74	×					×	×	×		×	×		×	×	×		×					×			
74	×					×	×	×		×	×		×	×	×		×					×			
4, 33																								×	×
1800			×			×	×						×				×								
5000		×						×	×							×									
—																								×	
1800	Cantilever																					×			
100, 1000	×						×	×		×	×		×								×	×	×		

ultimate strength of a material and most give an indication of the glass content. In many of the better papers there is much useful detail which cannot be included in a summary of this kind.

The bulk of the American work emanates from two sources, the U.S. Forest Products Laboratory and the Illinois Institute of Technology Research Institute. The majority of the work from the former is under the authorship of Boller and his co-workers[2–13], and with the exception of one or two recent papers not listed in the references, most of the publications are readily available. Much of the IITRI work forms part of the United States Navy's deep submergence programme. There have been numerous progress reports although many of them are difficult to obtain. Cornish *et al.*,[17] Broutman[18] and Cole *et al.*[19] cover much of the deep submergence work and the work of Hofer *et al.*[29] coming from the same laboratory relates to aerospace materials. Further fatigue information arising from the deep submergence programme is given by Freund and Silvergleit[23] and Fried *et al.*[25].

British work has been more fragmentary. A number of reports appeared from the Royal Aircraft Establishment[26–28, 33, 34] between 1952 and 1961. In the last 2 years, a number of papers[37–41] have arisen from the University of Nottingham where a substantial programme is in hand. There are several smaller programmes in hand both in the U.K. and the U.S.

6. EFFECT OF MATERIALS

Some typical fatigue strength data are given in Table 18.2. The original references should be consulted before attempting to use these data for design.

For similar reinforcements phenolic laminates give slightly better fatigue performance than epoxy laminates which, in turn, give better fatigue performance than polyester laminates[12, 43]. Unidirectional laminates give better fatigue strength than cross-plied non-woven laminates which, in turn, give better fatigue strength than fabric laminates which give better properties than mat reinforced laminates. Laminates reinforced with S-glass reinforcements are generally superior in strength to E-glass reinforcements of the same type and quality[35].

Polyester laminates reinforced with glass fabric reinforcements (of widely differing weaves) give surprisingly uniform properties. Fourteen S–N curves are available[6, 27], covering a number of fabric constructions and employing different resins of British and American manufacture which, when superimposed on the same graph, form a band whose width does not exceed the normal sort of scatter obtained in fatigue testing[15]. Rather more variation is obtained with epoxy laminates. Schmid[44], in particular, has pointed out that the properties of epoxy resin laminates vary widely with both base resin and curing agent. Other workers testing epoxide laminates, either in flexure or under axial loading conditions controlled by tensile stress, have not appeared to find such large differences. Schmid's findings may be due to his method of test. His results are reported as three point flexural tests whereas from the specimen dimensions and the descriptions of failure given, they were either interlaminar shear tests or else the results were influenced by the presence of high interlaminar shear stresses.

Boller[6] reported a very low fatigue strength for a chopped strand mat/polyester laminate containing 42 per cent glass by weight. Owen and Smith[39] found better properties for mat laminates containing only 32 per cent glass by weight. It is well known that the conventional ultimate strength of this type of laminate is very sensitive to glass content. Owen and Smith[39] found that large differences in u.t.s. associated with glass content were not reflected in the fatigue strength at 10^6 cycles. Unpublished work by the author on a series of mat type laminates, press cured and with various glass contents, led him to believe that the properties of press cured mat laminates are inferior to wet lay up, room temperature cured laminates under fatigue conditions,

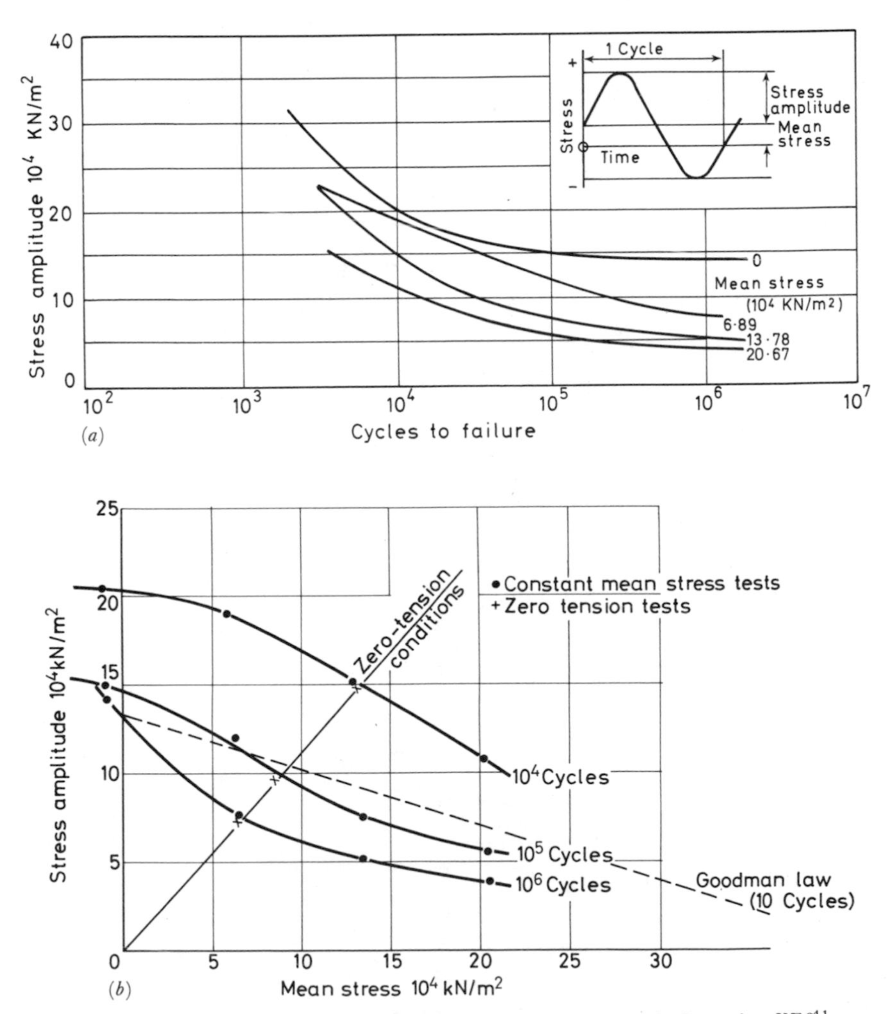

Figure 18.3 (a) S–N curves for Permaglass XE6 (b) Master diagram for Permaglass XE6[41]

Table 18.2 SAMPLE FATIGUE STRENGTHS (10^7 CYCLES AND ZERO MEAN STRESS) FOR TYPICAL GLASS FIBRE REINFORCED PLASTICS[15]

Reference	*Glass type*	*Finish* (a)	*Resin type* (a)	*Resin* % *by wt*	*Test conditions*	*Fatigue strength* 10^7 *cycles*		*Ultimate tensile or flexural strength*		*Fatigue ratio* (c)
						lbf/in^2	$kN/m^2 \times 10^4$	lbf/in^2	$kN/m^2 \times 10^4$	
MAT REINFORCED LAMINATES										
Boller[6]	E	—	Polyester (Paraplex P.43)	58	Direct Stress. Unnotched. 73°F. 50% R.H. 900 c/min	±3500	±2.41	13900	9.58	0.25
Boller[6]	E	—	Polyester (Paraplex p.43)	58	Direct Stress. Notched. 73°F. 50% R.H. 900 c/min	±4500	±3.10	14100	9.72	0.32
Owen & Smith[39]	E	—	Polyester	68	Direct Stress. 68°F. 42%. R.H. 74 c/min	±6000	±4.14	16200	11.15	0.37
Pusey[43]	E	Chrome	Polyester (PLL5147)	—	Uniform Bending. Unnotched. 1800 c/min	±8800	±6.07	36300	25.03	0.24
Pusey[43]	E	Chrome	Polyester (PLL5193)	—	Uniform Bending. Unnotched. 1800 c/min	±8300	±6.41	33000	22.75	0.28
Pusey[43]	E	Chrome	Polyester (PLL4264)	—	Uniform Bending. Unnotched. 1800 c/min	±7200	±4.96	39300	27.10	0.18
Pusey[43]	E	Silane	Polyester (PLL5193)	—	Uniform Bending. Unnotched. 1800 c/min	±8000	±5.52	—	—	—
Pusey[43]	E	Chrome	Epoxide	—	Uniform Bending. Unnotched. 1800 c/min	±10000	±6.89	—	—	—
8-SHAFT SATIN WEAVE FABRIC REINFORCED LAMINATES										
Boller[6]	E	Volan A	Polyester (Paraplex) (P43)	34.3	Direct Stress. Unnotched. 73°F. 50% R.H. 900 c/min	±10400	±7.17	46000	31.72	0.23

Boller[6]	E	Volan A	Epoxide (Epon 828)	37.7	Direct Stress. Unnotched. 73°F. 50% R.H. 900 c/min	±16100	±11.10	41200	28.41	0.39
Boller[6]	E	Volan A	Epoxide (Epon 828)	37.7	Direct Stress. Notched ($\frac{1}{8}$ in. hole) 50% R.H. 900 c/min	±14200	±9.79	32400	22.34	0.34
Boller[6]	E	Volan A	Phenolic (BV 17085)	28.0	Direct Stress. Unnotched. 73°F. 50% R.H. 900 c/min	±12500	±8.62	45100	31.10	0.28
Boller[6]	E	Volan A	Phenolic (BV 17085)	28.0	Direct Stress. Notched ($\frac{1}{8}$ in. hole) 50% R.H. 900 c/min	±10000	±6.89	31900	22.00	0.21
Boller[6]	E	Heat Cleaned	Silicone (DC2106)	31.0	Direct Stress. Unnotched. 73°F. 50% R.H. 900 c/min	±7900	±5.45	35300	24.34	0.22
Boller[6]	E	Heat Cleaned	Silicone (DC2106)	31.0	Direct Stress. Notched ($\frac{1}{8}$ in. hole) 50% R.H. 900 c/min	±6900	±4.76	30400	20.96	0.19
Heywood[27]	F	Garan	Polyester	—	Direct Stress. Unnotched. 400 c/min	±12000	±8.27	50700	34.96	0.24
Heywood[27]	E	Volan	Polyester	—	Direct Stress. Unnotched. 400 c/min	±12000	±8.27	53300	36.75	0.23
Thompson[47]	E	Silane	Epoxide	42	Direct Stress. Unnotched. 100 c/min	±12000	±8.27	44300	30.54	0.27
Mettes & Lockwood[35]	S	HTS	Epoxide	—	Flexural Stress. Unnotched. 1800 c/min	±18500	±12.76	92000	63.43	0.20
UNWOVEN REINFORCEMENTS										
With ±5° Bias:										
Boller[13]	E	Prepreg	Epoxide (1002)	35	Direct Stress. Unnotched. 73°F. 50% R.H. 900 c/min	±34000	±23.44	118300	81.57	0.29
Boller[13]	E	Prepreg	Epoxide (1009)	24	Direct Stress. Unnotched. 73°F. 50% R.H. 900 c/min	±36000	±34.82	127500	87.91	0.28
With alternative layers cross-plied 0° and 90°:										
Boller[13]	E	Prepreg	Epoxide (1002)	35	Direct Stress. Unnotched. 73°F. 50% R.H. 900 c/min	±22000	±15.17	67300	46.40	0.33
Boller[13]	E	Prepreg	Epoxide (1009)	22	Direct Stress. Unnotched. 73°F. 50% R.H. 900 c/min	±24000	±16.55	96500	66.54	0.25

(*a*) For further details consult the original references.
(*b*) SI Units: 1 lbf/in^2 = 6894.76 N/m^2; 1 kN/m^2 = 10^3 N/m^2
(*c*) Fatigue ratio = Fatigue strength ÷ Ultimate strength. The unnotched tensile strength is used for calculating the fatigue ratio for notched specimens.

and that the effect of the elevated temperature press cure dominates the effect of glass content. Further investigation of this point would be very worth while.

Compressive stresses of large magnitudes are probably rare in GRP structures, a notable exception being deep submergence applications involving filament wound thick-walled cylinders. Intensive effort has led to substantial improvements in compressive properties under uniaxial and biaxial loading conditions. Basically, stiffer fibres give material less prone to buckling and hence S-glass fibres are preferred to E-glass fibres. For a given fibre type, compressive strength appears to be dominated by the interfacial properties of the glass resin system. Hence the compressive properties are closely related to the interlaminar shear properties which are easier to measure. Both compressive and interlaminar shear properties are dependent on the finish, the resin and the void content. Sophisticated epoxide resins and finishes and good quality control in manufacture and laminating have produced very good compressive properties. However, scatter is still high and the fatigue strength is only about half the ultimate strength at 10^4 cycles[23].

7. EFFECT OF MEAN STRESS

It has turned out that the investigation of the relationship between alternating stress and mean stress is even more important for GRP than it is for the common structural metals. The ferrous metals are relatively insensitive to the effect of mean stress. Consequently, on the plot of stress amplitude against mean stress, a line drawn from the endurance limit at zero mean stress to the ultimate strength at zero stress amplitude (the Goodman line) gives a conservative estimate of the relationship for design purposes. The Goodman law is also a reasonable approximation for aluminium alloys at ambient temperature. Boller[6] found that at long lives, i.e. 10^6 or 10^7 cycles, the results for GRP fall well below the Goodman line. Boller proposed a modification of the Goodman law for use with GRP.

$$\frac{S_A}{S_E} = 1 - \frac{S_M}{S_U} \quad \text{(GOODMAN)}$$

$$\frac{S_A}{S_E} = 1 - \frac{S_M}{S_C} \quad \text{(BOLLER)}$$

where S_A is the stress amplitude at mean stress S_M, S_E is the fatigue strength at stated life and S_U is the ultimate strength of the material. S_C is the stress rupture strength at a time equivalent to the stated number of fatigue cycles. Unfortunately, it has been found in a number of cases that the data fall even below this line in the tensile mean stress field. A better approximation in the tensile mean stress field has been suggested by Smith and Owen[41] and that is the following rule originally proposed by J. O. Smith for cast irons.

$$\frac{S_A}{S_E} = \frac{1 - (S_M/S_C)}{1 + (S_M/S_C)}$$

A serious limitation in the usefulness of master diagrams is that the true extent of scatter is hidden. If replicate tests are conducted at any stress level and plotted on the conventional S–N diagram, then they are likely to cover a range of between one or two orders of magnitude. In this respect GRP are no worse than and, in fact, may be better than the common structural metals. If replicate tests are not used then the curve obtained is an estimate of the median life. A family of such curves are used to produce the master diagram. Indeed, it is often necessary to carry out a slight smoothing operation in drawing curves through the points used to construct the master diagram. These factors are often glossed over in presenting design information.

8. OTHER DATA OF INTEREST TO THE DESIGNER

The designer has many factors to take into consideration in producing a reliable design and the uniaxial stress S–N curve based on total separation of specimens, obtained at ambient temperature and humidity, often may be of little greater use than the conventional ultimate tensile strength. Some indication is required of the margins which have to be allowed to avoid incipient damage and the effects of scatter, and to take account of biaxial stresses, and the effects of variable amplitude or variable frequency loading, etc. Relatively few test programmes give any indication of scatter although in the last year or two, there has been a tendency to introduce six to ten replicates at a number of stress levels. Column 34 of Table 18.1 indicates the papers where replicates have been used.

In many applications, incipient damage must be avoided and there is a growing awareness of the need to take into account progressive failure in presenting test results. The IITRI test programme is making use of ultrasonics to chart the progress of damage, and Hofer and Olsen[29] have shown residual interlaminar shear strengths in relation to ultrasonic attenuation produced by fatigue damage. They also commented on the need to correlate results with observations of microstructural damage. Smith and Owen[41] have shown that it is possible to use conventional tensile modulus measurements to indicate the onset of damage. From the limited amount of information available, it seems that there is only a slight loss of ultimate strength during the early stages of damage although the loss of other properties, such as modulus or electrical resistance may be more marked. Smith and Owen[41] found that in a tensile test a wide variety of laminates commenced to debond at a strain of approximately 0.3 per cent and that fully reversed strains of 0.14 per cent produced debonding after 10^6 cycles. Their paper includes a master diagram for chopped strand mat/polyester resin laminates at the onset of debonding. The same paper gives a comparable diagram based on total separation of the specimens.

Although components and structures are rarely, if ever, subjected to simple uniaxial stresses, the effects of notches, complex stresses, and fibre direction have received relatively little attention. Boller[6] considered the effect of small round holes drilled through the neck of fatigue specimens to produce a stress concentration and concluded that, compared with metals, GRP are relatively unaffected by notches at the longer lives. This is not really surprising when total separation of the specimens is taken as the criterion of failure since progressive damage will effectively remove the stress concentration effect. It does not follow that large apertures in structures are not fully effective stress raisers especially if incipient damage is important. Apart from the work on compressive properties of thick-walled cylinders relatively little work has been carried out under biaxial stress conditions. Freund and Silvergleit[23] compared data on the basis of percentage ultimate strength showing that compressive biaxial stress (2:1) fatigue appeared to be only slightly more damaging than uniaxial compressive stress fatigue. There does not seem to have been any systematic investigation of fatigue under a range of biaxial stress ratios.

With orthotropic reinforcements, either fabric or non-woven roving, it is not possible to ensure that filament orientation is favourable at every detail. Some attempts have been made to conduct uniaxial fatigue tests at an angle to the filaments (usually 45 degrees). Generally the fatigue properties under these conditions are very poor and are unfavourable even when compared with mat laminates of relatively low glass content. Recently, Endo and his co-workers[22] have pointed out that there is a substantial size effect and the properties depend partly on the continuity of the fibres across the neck of the test specimens.

GRP find favour in preference to metals when they have a favourable combination of properties or a price advantage. This often means that they are required to operate in a special environment. A certain amount of fatigue testing has been carried out at elevated and low temperatures as indicated in Table 18.1. Brink[14] reports that the fatigue properties of a variety of laminates are as good, or better than, those at ambient conditions right down to cryogenic temperatures. At elevated temperatures the properties fall off and depend very much on the resin selected. Many workers have tested specimens under wet conditions (tap water or distilled water, or in saturation humidity). The general conclusion is that at longer lives the effects of moisture under fully reversed stress are not as serious as they are at short term tests. No results are known for more damaging chemical environments.

In a real application, it is unlikely that a component would be subjected to constant amplitude stressing at constant frequency. Very little work has been done on GRP under cumulative damage or random loading conditions. Boller[13] and Hofer[29] report the beginnings of cumulative damage studies and report tests begun at one stress amplitude and completed at another. At the present stage the results are inconclusive. Much further work is required before cumulative damage laws can be formulated for GRP. It is well known that there is a frequency effect under laboratory testing conditions. This appears to be associated with heating effects at the more damaging stress levels. It is not proven that there is a frequency effect at practical stress levels in the absence of heating. It is, however, certain that repeated loading is far more damaging than static loading under normal ambient conditions.

REFERENCES

N.B. 'S.P.I. CONF' denotes the Annual Technical and Management Conference of the Reinforced Plastics Division of the Society of the Plastics Industry Inc., New York.

1. Armstrong, D. J. and Beveridge, A. A. *N.E.L. Report* No. 355, May 1968
2. Boller, K. H. *U.S. Dep. Agric. Forest Prod. Lab. Rep.* No. 1823, 1952
3. Boller, K. H. *U.S. Dep. Agric. Forest Prod. Lab. Rep.* No. 1823A, 1954
4. Werren, F. *U.S. Dep. Agric. Forest Prod. Lab. Rep.* No. 1823B, 1956
5. Kimball, K. E. *U.S. Dep. Agric. Forest Prod. Lab. Rep.* No. 1823C, 1958
6. Boller, K. H. *Modern Plastics* **34**, June 1957, p. 163. (Also *WADC Tech. Rep.* 55–389, 1956). (Also *S.P.I.* 12*th Conf.*, 1957, Section 5-B)
7. Stevens, G. H. and Boller, K. H. *WADC Tech. Rep.* 59–27, 1958
8. Stevens, G. H. *U.S. Dep. Agric. Forest Prod. Lab. Rep.* No. 1884, 1961
9. Stevens, G. H. *U.S. Dep. Agric. Forest Prod. Lab. Rep.* F.P.L.-027, 1964
10. Stevens, G. H. *U.S. Dep. Agric. Forest Prod. Lab. Rep.* F.P.L.-37, 1965
11. Boller, K. H. *Tech. Docum. Rep.* No. ASD-TDR 62–464, 1962
12. Boller, K. H. *Mod. Plast.* **41**, June 1964, p. 145. (Also *Tech. Docum. Rep.* ASD-TDR 63–768, 1963).
13. Boller, K. H. *Mod. Plast.* **42**, April 1965, p. 162. (Also *ML-TDR* 64–168)
14. Brink, N. O. *S.P.E. Journal*, October 1964, p. 1123
15. British Plastics Federation, 'Fatigue and Creep-Reinforced Plastics', *Publication No.* 64/1, 1967
16. Carswell, W. S. and Borwick, G. R. *Trans. Plastics Inst.* **33**, Oct. 1965, p. 169. (Also *NEL Rep.* No. 210, December 1965)
17. Cornish, R. A., Nelson, H. R. and Dally, J. W. *S.P.I.* 19*th Conf.* 1964, Section 9E
18. Broutman, L. J. *S.P.I.* 19*th Conf.* 1964, Section 9C
19. Cole, C. K., Cornish, R. H. and Elliott, J. P. *S.P.I.* 21*st Conf.* 1966, Section 17C
20. Dally, J. W. and Broutman, L. J. *J. Compos. Mater.* **1**, No. 4, 1967, p. 424
21. Davis, J. W., McCarthy, J. A. and Schurb, J. N. *Mats. Des. Engng.* **60**, December 1964
22. Endo, K., Yano, A. and Okuday. 9*th Japan Congr. Testing Materials—Non-Metallic Materials* 1966, p. 104–8
23. Freund, J. F. and Silvergleit M. *S.P.I.* 21*st Conf.* 1966, Section 17B
24. Fried, N. *S.P.I.* 12*th Conf.* 1957, Section 5A. (Also *Reinf. Plast.* **2**, 1958, No. 7, p. 19)
25. Fried, N., Kaminetsky, J. and Silvergleit, M. *S.P.I.* 21*st Conf.* 1966, Section 14A
26. Heywood, R. B. *R.A.E. Tech. Note Chem.* 1337, 1958
27. Heywood, R. B. and Sage, J. A. *R.A.E. Tech. Note Chem.* 1376, 1961
28. Heywood, R. B. *Symposium on Developments in Materials Testing Machine Design*, Sept. 1965, Manchester I.Mech.E. (Also *R.A.E. Tech. Note Chem.* 1380, 1961)
29. Hofer, K. E. and Olsen, E. M. *IITRI Report* (DDC Accession No. AD 652415)
30. Hooper, R. C. *S.P.I.* 11*th Conf.* 1956, Section 8B. (Also *Plast. Tech.* **3**, 1957, p. 644–9)
31. James, T. K., Appl, F. J. and Bert, C. W. *Expl. Mech.* **8**, No. 7, 1968, p. 327
32. Kinna, M. A. and Prosen, S. P. *S.P.E. Journal* Nov. 1966, pp. 38–43. (Also *NDL TR*65–191, 1966)
33. McCue, D. J. *R.A.E. Tech. Note Chem.* 1334, 1958
34. McMullen, P. L. *R.A.E. Tech. Note Chem.* 1166, 1952
35. Mettes, D. G. and Lockwood, P. A. *S.P.I.* 21*st Conf.* 1966, Section 4G
36. Nara, H. R. *S.P.I.* 12*th Conf.* 1957, Section 5D
37. Owen, M. J. *Trans. Plast. Inst.* Feb. 1967, p. 353
38. Owen, M. J. and Dukes, R. *J. Strain Analysis* **2**, No. 4, 1967, p. 272
39. Owen, M. J. and Smith, T. R. *Plastics and Polymers* **1**, 1968, p. 33
40. Owen, M. J., Dukes, R. and Smith, T. R. *S.P.I.* 23*rd Conf.* 1968, Paper 14–B
41. Smith, T. R. and Owen, M. J. *British Plastics Federation* 6*th Int. Conf.* London, Nov. 1968, Paper 27
42. Pflederer, F. R. *S.P.I.* 17*th Conf.* 1962, Section 1-D
43. Pusey, B. B. *Plast. Tech.* **3**, 1957, p. 721–6
44. Schmid, R. *Brit. Plast. Fed.* 5*th Int. Conf.* 1966, Paper 10
45. Soltysiak, D. J. and Toth, J. M. *S.P.I.* 22*nd Conf.* 1967, Section 14E
46. Theberge, J. F. *S.P.I.* 23*rd Conf.* 1968 (Separate)
47. Thompson, A. W. *Trans. Plast. Inst.* **30**, 1962, pp. 39–47

19

Design Theory of GRP Boats

G. KINGSLAKE

1. INTRODUCTION

In this chapter the reinforced plastics boat hull is considered from the point of view of its structural strength. Although boats have been built for thousands of years we have very little detailed knowledge of the forces acting on them at sea, and the design criteria we use are, unfortunately, still rather empirical. However, applying even empirical loading conditions to a hull helps to produce a balanced design, and should lead to a more efficient structure than one designed entirely 'by eye'.

GRP presents a number of special problems in boat-building, and in this chapter an attempt is made to show how some of these may be solved and to point out known pitfalls.

At the present time most reinforced plastics boats are small, i.e. less than about 25 m long (80 ft) and this chapter deals with boats in this range. Far larger boats are in the design study stage, but the structural analysis of these craft is too complex to include here.

2. THE FORCES ACTING ON A BOAT HULL

For convenience we can divide these into three groups:

(1) The overall forces which tend to bend the complete hull.

(2) The pressures on the bottom, sides and decks.

(3) Local forces from shrouds or stays, propeller brackets, engine feet etc.

The first group is most important for large craft, where the bending moments induced by the action of waves provide the main design cases. It is however, seldom necessary to consider these cases in detail for boats less than about 25 m long, as with such craft, when the other design criteria have been met, the complete hull is usually adequate to withstand these overall loading cases. The calculation of the overall shear forces and bending moments is rather complicated, and can be found in books on Naval Architecture. However, in this chapter we shall consider the longitudinal strength qualitatively not quantitatively.

The second group usually provides the critical design cases for the hull in normal GRP boats. The bottom of a displacement boat must be able to withstand the water pressures caused by waves reaching at least to the gunwale level. In addition, if the boat is to venture out in rough seas it is likely to experience slamming pressures, especially on the bottom near the bows and perhaps just under the stern. The actual slamming pressures are difficult to specify as they are dependent on hull shape and size as well as the sea state. The pressures are very local, and may vary between about 70 kN/m^2 for small boats up to 350 kN/m^2 or more for large sea-going craft. On a sailing boat the worst slamming area may be on the topside forward, as a result of the boat heeling in a 'choppy' sea.

Fast, hard-chine boats are supported at speed by planing on the surface of the water, and slamming is very frequent and can be severe. The bottom is made with a vee section to relieve slamming pressures as much as possible, but still the slam dominates the design of the hull structure. The peak impact pressures depend on the speed and the size of the boat, together with the dead-rise angle of the bottom and, of course, the sea state. As a guide to the peak pressure to assume, it may be said that a small race boat (7–10 m long) would experience about 140 kN/m^2, whilst a typical fast patrol boat (30 m long) would experience pressures in excess of 350 kN/m^2.

These pressures do not have to be supported over large areas of the bottom; the peak pressure probably only acts on less than a tenth of a square metre, and only exists for a very small fraction of a second. As the area in contact increases the pressure drops rapidly. These high pressure zones can occur anywhere over the so-called slamming area of the bottom which effectively starts at about a quarter of the waterline length from the bows, and for a large or a relatively slow small boat, finishes at about amidships. Fast small boats sometimes jump clear of the water, and the main slamming area extends back to the transom.

The design pressures for the decks and topsides of any type of boat are necessarily somewhat arbitrary. It is advisable to make sure the topsides can withstand about half the bottom pressure, to allow for sea-slap forces and, in the case of a fast boat, to allow for the possibility of the boat temporarily riding on its chine in a rough sea. Decks should be designed to carry 7–10 kN/m^2 (1–$1\frac{1}{2}$ lbf/in^2), with the foredeck taking the highest pressure. For larger sea-going boats it would be advisable to strengthen the foredeck to withstand14–20 kN/m^2.

Watertight bulkheads should be designed to support a head of water up to the main deck level, acting in either direction. If the bulkhead forms part of an integral tank, the head of fuel or water should be taken at least to the top of the vent pipe when estimating the design pressure.

The third group of forces, that is the forces from local mountings and attachments. is difficult to deal with quantitatively as we do not usually know the magnitude of the loads concerned. In the case of the attachments for standing rigging in a sailing boat, however, past experience is used to decide rope sizes, and the strength of the rope can be used to design its attachment. To avoid the risk of damage to the hull, the strength of the attachment should be at least 50 per cent higher than that of the weakest link in the rigging.

Mountings for heavy items should withstand the weight of the item times the acceleration factor. This factor should be two or three for a normal displacement boat, rising to a maximum of about ten in the bows of a fast planing boat.

The design of propeller shaft brackets is a complicated matter for a fast boat where weight-saving is important, and is beyond the scope of this book. Here again the hull locally should be stronger than the bracket itself. In the case of 'P' brackets, at least, it should be possible to bend the bracket under a sideways load at the shaft before breaking the reinforced plastic.

3. MAIN TYPES OF HULL STRUCTURE

The simplest hull structure is a pure monocoque, that is a shell without additional stiffening. In practice this form is only used for very small craft, but the term monocoque is used in a loose sense when the amount of stiffening is small, and the inherent strength of the shell contributes significantly to the strength of the boat.

In a monocoque structure the shell is most effective if it can support most of the loading applied to it by means of 'membrane' stresses in the plane of the shell, rather than by shear and bending stresses across the thickness. This is theoretically only possible if the curvature of the shell bears a specific relationship to the distribution of pressure on it, but provided the hull has a reasonable degree of curvature, membrane stresses can account for a large proportion of the applied pressure. In general however, some form of stiffening is needed to support the skin, and this usually takes the form of longitudinal and/or transverse members fixed to the inside of the shell. Longitudinals also contribute to the overall bending strength of the hull.

Another way of stiffening a skin is to build up a 'sandwich' composed of two strong skins with a low density core material between. This very greatly increases the strength of the shell across its own thickness for a given shell weight, and enables it to carry shear and bending in itself, and reduces the need for separate stiffeners.

4. THE HULL CONSIDERED AS A BEAM

The external pressures acting on the hull must be balanced by the weights and inertia of all the items that make up the complete boat. The complete hull can be treated as a beam in equilibrium under these forces. The deck and the bottom shell (together with the associated longitudinal stiffeners) form the top and bottom booms of this beam, and the two sides form the web. The vertical forces must be fed into the sides of the boat and carried along them as shear forces, while the booms resist the bending moments resulting from these shear forces. Thus when considering any particular vertical force on a boat hull from pressure or inertia, we must ensure that it can be satisfactorily transferred into the topsides. The bulkheads play a large part in this, and as they are usually very stiff and strong for forces in their own plane, they are well suited to support secondary structures such as girders, carrying the loads directly into the sides of the boat.

The water pressure on the bottom of the boat is applied partly via the keel and girders to the bulkheads, and partly via the frames to the topsides. If the boat has a vee bottom, the vee can itself act as a girder. It is easy to appreciate this if the two panels are at right angles to one another, as each then becomes in effect a very deep girder supporting the edge of the other and carries the loads to the bulkheads by forces in the plane of the skin panels. This effect still operates when the vee is not a right angle, though the shear stresses and hence the deflections increase, as the angle becomes more obtuse. Hard chines can similarly be treated as girders supporting the bottom and the sides.

When the bottom or the bilges are rounded, the same effect exists to a lesser extent, though the actual location of the effective girder is indeterminate.

The various items that make up the total weight of the boat are supported in a variety of ways, but in all cases there must be a path for the load into the main structure of the boat, and thence into the topsides, to balance the forces from the water pressure.

5. MATERIALS

The term 'reinforced plastic' when applied to boatbuilding usually means glass fibre reinforced polyester resin. There are of course other reinforcing fibres that could be used and other resins. Unfortunately at the present time the higher strength fibres such as carbon or boron are far too expensive for boat hulls, whilst the thermoplastic fibres such as nylon or Terylene are too flexible. About the only alternative resin that could be used at room temperature is cold setting epoxy, which is more expensive than polyester, more difficult to use and, in general, has a lower wet strength retention. On the other hand, its adhesive properties are generally better, and its curing is less likely to be inhibited by moisture and other substances to which the polyesters are sensitive.

Glass fibre is available in several forms, but for boat building only two are significant: chopped-strand mat and woven rovings (together with surfacing cloths or tissues which do not contribute to the strength of the hull). The choice between these two forms depends on a number of factors, perhaps the most important of which is ease of manufacture. Chopped strand mat can be easily draped over complex shapes when wetted with resin, or it can even be produced *in situ* on the mould by spray techniques. Woven rovings, on the other hand, retain their full strength when wetted with resin and can be laid continuously over large areas using a dispenser which pre-wets the resin if the size of the boat justifies this. Woven material cannot be made to lie smoothly on a surface with double curvature, nor can it be bent to form a curved angle (except to a limited extent by distorting the weave), and it becomes necessary to 'tailor' the material by cutting. This leads to butt joints or even gaps at times, which produce lines of weakness in the finished laminate.

Obviously these cuts should be staggered between layers, but even so there is a loss of strength.

More skill is needed in rolling down chopped strand mat than woven rovings to ensure that an even thickness is maintained, especially over sharp corners. Where unskilled labour is involved woven rovings are probably more reliable than mat.

Laminates made from woven rovings have a higher glass content than those made from mat, leading to a higher tensile strength. In compression the difference is not so great. As glass is more expensive than resin, the material cost per unit laminate thickness is greater for woven rovings, which offsets the strength advantage in this respect. When a panel or strip of material is loaded in flexure, the strength is proportional to the square of the thickness and the stiffness to the cube of the thickness. Thus in spite of the lower strength and Young's modulus for mat laminates, they have the advantage over woven rovings in that the thickness is greater for the same amount of glass. The result is that for a given total weight, mat and woven rovings have about the same flexural strength, whereas the flexural stiffness of the mat laminate is superior. This means that chopped strand mat panels are less likely to buckle under shear or compressive stress, again for a given panel weight.

The adhesion between layers is nominally better with chopped strand mat than with woven rovings because the irregularities of surface allow the fibres of one layer to penetrate the previous layer to some extent. However with normal good workmanship, woven rovings can be made to produce a perfectly satisfactory laminate in this respect. Some boatbuilders use alternate layers of mat and rovings to give good interlaminar strength.

Perhaps we could summarise by saying that chopped strand mat is most suitable for small boats where curvatures are high and where the flexural stiffness of the skin is more important than its tensile strength, whilst woven rovings are more suitable for larger boats where stresses are higher and where mechanical dispensing of cloth is worth while.

A special type of woven rovings which should be mentioned is the so-called unidirectional cloth or tape. In this material the heavy rovings all lie in the warp direction, and the weft is formed of light yarns widely spaced. This is very useful where high stresses in one direction only have to be carried, for example in the 'crown' of a top-hat section stiffener. In such cases the saving over plain weave material is considerable, as the unwanted transverse fibres are eliminated.

If sandwich construction is to be used the core material must possess the following properties:

(1) Its density must be considerably lower than that of the basic laminate.
(2) It must be capable of carrying shear stress between the two skins.
(3) It must adhere to the laminate sufficiently strongly to carry this shear stress into the skins.
(4) It must withstand the maximum temperature range required without loss of properties.
(5) It must withstand the effects of water vapour diffusing through the skin and of liquid water in case of minor damage to the skin.

The commonest core material is rigid plastics foam. For the best strength properties it is usual to use polyvinyl chloride foam, which is tough and durable, though rather more expensive than most. This foam can be obtained in a range of density, and with a fine or a coarse cell structure. The best shear strength is obtained with the higher density material (as would be expected) and the fine cell material is somewhat stronger than the coarse for the same density. Unfortunately, PVC foam softens significantly above about 50°C, and it should be used with caution where high temperatures might be experienced (e.g. on exposed decks, especially if they are dark in colour, or for use in the tropics). Polyurethane foam is cheaper than PVC and has the added advantage that it can be foamed *in situ*, but its strength properties are far lower, and it tends to be brittle and may crumble under vibrating loads. Polystyrene foam, though cheap, is not suitable for reinforced plastic sandwich construction, as it is rapidly attacked by the uncured polyester resin. Phenolic foam is too weak and 'crumbly' to use structurally, but it is more fire resistant than the others mentioned.

Another effective core material is balsa wood, preferably laid so that the skins are applied to the end grain of the wood. This core is harder than normal foams, and has excellent shear carrying properties. Its strength is not affected by high atmospheric temperatures. Its main disadvantage is the danger of waterlogging if the outer skin of a sandwich becomes accidentally punctured, though it is not usually affected by water vapour migrating through the laminate.

Where high shear strength is required, especially on large boats, it is suggested that some form of fabricated reinforced plastics core would be structurally more suitable, either laminated *in situ* or pre-formed and built into the sandwich. Typical of this type of construction is a corrugated core with inner and outer skins attached to it.

Honeycomb cores have not been used to any great extent for boat hulls, though their strength properties are excellent. The durability of the core material under continuous exposure of the sandwich to sea-water is somewhat doubtful, and flooding of the honeycomb seems to be a possibility. There may be an application for this material in small race boats.

6. BASIC STRUCTURAL DESIGN

The choice between single skin construction (monocoque or with stiffeners) or sandwich construction depends on a number of things. From the manufacturing point of view the orthodox method of laying up inside a female mould has the great advantage that the outside surface is controlled, leading to good appearance and good hydrodynamic shape, and it enables us to mould in sea inlets and outlets, propeller bosses, platforms for shaft brackets etc. It also allows all the internal structure to be fitted while the hull is accurately held in the mould. With this process the simplest method is to produce a thick, single-skin monocoque hull with the minimum of internal stiffening (perhaps only bulkheads). Such a hull is heavy and material costs are likely to be high, but labour costs are correspondingly low. Chopped strand mat would be the best reinforcement for such a hull to give the maximum shell thickness for a given weight.

To make a lighter hull, the skin would have to be thinner and stiffeners would have to be fitted inside. The lightest structure is usually attained by having the thinnest practicable skin with closely spaced stiffeners, but this tends to be costly. A compromise solution is usually the best. If the skin is curved, woven rovings could well be used for this type of hull since much of the support is given by membrane stresses, but if the bottom is largely flat, mat would still probably have a slight advantage.

If the shell is made of sandwich construction, in certain circumstances this results in a very light structure. However, if the thickness of the outer skin is to be made sufficient to withstand the usual bumping and scoring that a boat suffers, and the core thickness sufficient to carry the shear forces across the skin, it often results in a heavier structure than the stiffened shell. The simplicity and inherent buoyancy of a sandwich construction, and the fact that minor damage to the outer skin does not cause a leak are in favour of this type of construction.

On the manufacturing side sandwich construction presents special problems; using the orthodox moulding process the core has to be laid inside the outer skin. With foam this is difficult to achieve without trapping air underneath, which results in lack of shear strength with the possibility of a progressive breakdown of the sandwich under slamming forces for example. Incidentally, end-grain balsa is better in this respect as air can easily escape along the grain of the timber. Another method of moulding sandwich hulls which avoids this difficulty is to drape heated PVC foam over a simple male mould and cover first the outside, then the inside with GRP; this leads to a sound laminate, but the advantages of a moulded outside surface are lost.

Although with smaller boats it is not usually difficult to provide sufficient strength to carry the bending moment in the hull as a whole, it is worth keeping in mind during the course of a design that longitudinal strength is necessary. In this respect the primary consideration is usually continuity of strength, that is, the avoidance of stress concentrations. Wherever possible longitudinals and girders should be continuous, at least over the middle half of the boat's length. This is not always practicable, but a lot can be achieved by a common-sense approach during the early stages of a design. For example, the engine foundations and the sides of integral tanks can be incorporated into the main longitudinal stiffening of the bottom, provided the joints between them and the girders are at least as strong as the girders themselves. It is not sufficient just to 'mat' them separately to opposite sides of a bulkhead; they must be joined directly by laminating them together in the plane of the vertical webs and flanges.

Large deck openings are inevitable, especially in smaller boats, and the corners are almost certain to be points of high local stress. Trouble on this account can be minimised by providing as large a radius as possible in the corners, and if it is at all possible, by continuing the carling members as deck beams well beyond the opening.

The shear strength of the sides of a normal boat is likely to be ample to carry the loads applied to them, but it must be remembered that the complementary shear forces are carried through the joint between the topsides and the deck. For this reason an adequate amount of shear-carrying material must be provided to form a bridge across this joint.

When deciding on the layout of the stiffening to support a loaded surface, it should be remembered that the bending moment in a uniformly loaded beam is proportional to the square of the span between supports, and for this reason it is generally better to arrange stiffeners across rather than along a rectangular area. For example, if we wish to use only one set of stiffeners on the bottom of a small boat, if the distance between the vee bottom and the chine is less than the bulkhead spacing, transverse stiffening is more efficient than longitudinal.

The main longitudinal girders in the bottom of the boat are often very heavily-loaded members, and every effort should be made to keep their span as short as possible by avoiding a wide bulkhead spacing. In a large engine compartment it is sometimes worth fitting a deep floor (that is a shallow bulkhead) between the main engine room bulkheads to give extra support to the girders in this critical area. To help reduce the bending moments and more particularly, the deflections in the main girders, it is desirable that they should be flexurally continuous through bulkheads and deep floors.

Long girder spans under decks cannot be supported by the equivalent of the deep floor, and such members can be held up by means of pillars passing down to the bottom of the boat. Pillars may be of any suitable material, but metal tubes are very convenient for this purpose. It is obviously necessary to check that the pillar locations fit in with the accommodation. If they can be designed to lie in minor bulkheads this is the best arrangement. Pillars cannot be assumed to provide any significant support to the bottom of the boat.

Bulkheads for reinforced plastics boats are generally made from plywood or GRP boards, attached to the shell all around their edges. As their primary structural function is to support the main girders, they must be capable of carrying shear forces from these to the sides of the boat. This presents no problem unless there are doorways or other openings in the bulkheads. In a small boat a doorway through a bulkhead may reach from the deck to the bottom. In such cases the girders should be arranged to run outboard of the opening and attach to the effective parts of the bulkhead, but at the same time robust transverse beams should run across the opening above and below to maintain some shear strength across the centre of the boat to allow for asymmetrical loads on the bottom. Openings in bulkheads should always have well rounded corners.

When a bulkhead is to act as a watertight partition in case of flooding, it may require further stiffening on this account. Such stiffening can be vertical or horizontal (depending on which gives the shorter span), or the bulkhead may be made strong enough in itself by the use of sandwich construction.

7. DETAIL STRUCTURAL DESIGN

The components of the hull are usually joined together by layers of GRP applied wet across the joint. In many cases these are laid between surfaces at right angles to one another, to form jointing angles. This applies where bulkheads meet the skin of the boat, at the corners of tanks, at the ends of stiffeners, and so on. The plastic can be reinforced with mat or woven rovings, though mat is easier to apply where awkward shapes are involved. These jointing angles have two weaknesses which must be guarded against:

(1) Due to the time-gap between laying up the main skin and fitting the internal stiffening, the skin is often cured before the joints are made. This leads to a low interlaminar strength unless special precautions are taken (surface roughening and degreasing).

(2) If any tension is applied, the angle is likely to peel off, starting from the 'heel'. The adhesive offers little resistance to this peeling when the joint is good; a poor joint offers almost no resistance at all.

Even when pure shear force is applied there is usually a considerable amount of stress concentration, so average 'glue' stresses should be kept as low as possible. Suggested overlaps are 40 mm ($1\frac{1}{2}$ in) per layer, either of 0.6 kg/m^2 (2 oz/ft^2) mat or 0.8 kg/m^2 (24 oz/yd^2) woven rovings. It is very desirable that the thickness of the jointing angles should be graded from a maximum at the heel to a minimum at the toe, to reduce stress concentrations in the joint and to save material. (*See Figure 19.1*).

Stiffeners are usually made of 'top hat' section, by laying strips of GRP over a suitable former (e.g. polyurethane foam) *in situ*, allowing the strips to spread over the skin and form their own attachment angles. It eases the lay-up process if the core is narrower at the 'crown' than at the base. Additional

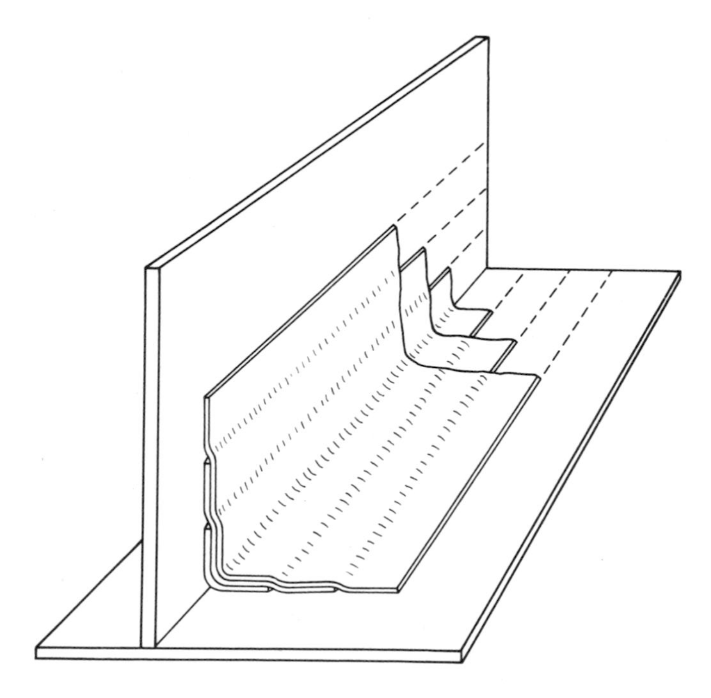

Figure 19.1. GRP jointing angle

bending strength can be provided by reinforcing the crown of the hat with unidirectional tapes (*Figure 19.2*). Sometimes it is convenient to laminate the stiffener formers from GRP on a separate mould, and then to lay up over these, in place of the foam core suggested above. This has the advantage that the former contributes to the strength of the stiffener. This idea can be extended to the pre-moulding of complete stiffeners, which are then fixed in place by GRP angles. (This is only worth while for deep stiffeners such as engine foundations.)

The top hat type of stiffener is efficient structurally because its box form almost entirely eliminates torsional instability. Sometimes single-web stiffeners

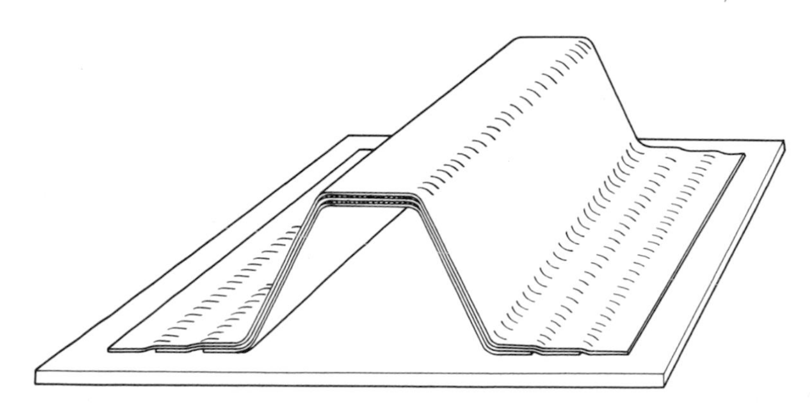

Figure 19.2. Typical 'Top Hat' stiffeners

are desirable, for example where it is required to bolt components to the web. In these cases it may be necessary to provide 'tripping brackets' to support the inner boom laterally at intervals. If the web is very deep, vertical stiffeners may be required on it to prevent shear instability.

Stiffeners may be of wood attached by GRP angles. These are often preferred by boatbuilders, especially for engine foundations, as subsequent bolted or screwed fastenings are made more easily, and the foundation can be trimmed if necessary. These are quite satisfactory in practice, though they tend to be heavier than GRP stiffeners for the same strength. It is best to bed the bottom of wooden stiffeners into some form of plastics filler to prevent water collecting underneath and rotting the wood. Plywood is often used for light stiffeners. It should be remembered however, that only about half the thickness of the plywood is effective in bending (the rest having its grain across the member).

All forms of stiffener require their ends to be anchored so that the shear forces can be carried from the webs of the stiffeners into the shell or bulkheads. This is usually done by means of GRP angles each side of the stiffener (so that they are only loaded in shear). If there is any risk of bending forces being carried across the joint the boom should also be supported, for example by means of a small 'knee' (*Figure 19.3*). Where possible stiffeners should end on

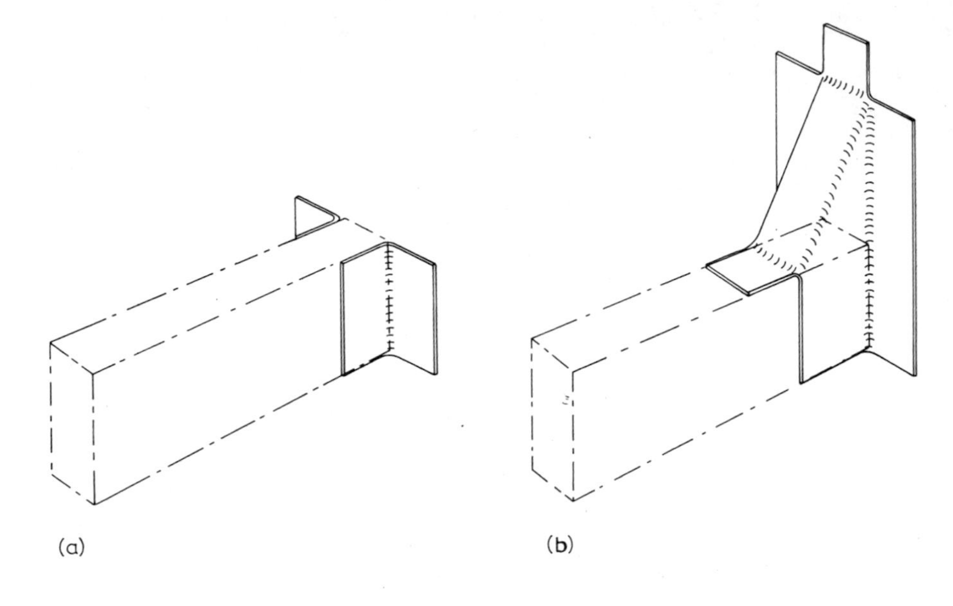

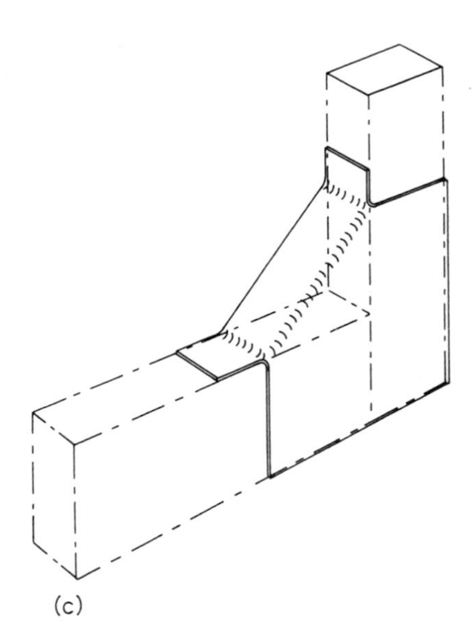

Figure 19.3 Typical stiffener and attachments
(a) side angles only
(b) 'knee' joint
(c) 'knee' joint between stiffeners

other stiffeners at right angles to themselves; for example, deck beams should pick up on vertical frames in the topsides.

It is particularly important to provide adequate shear connections between the webs of main longitudinal girders and the bulkheads that support them. A simple calculation would show that the loads to be transferred are surprisingly large.

When sandwich shell is used, the core material has to carry the shear forces from the water pressure. To ensure that it can do this it is necessary for all slabs of foam or balsa to be edge glued together. If this is not done it is possible for the shear forces to exert a peeling action on the skins (*Figure 19.4*).

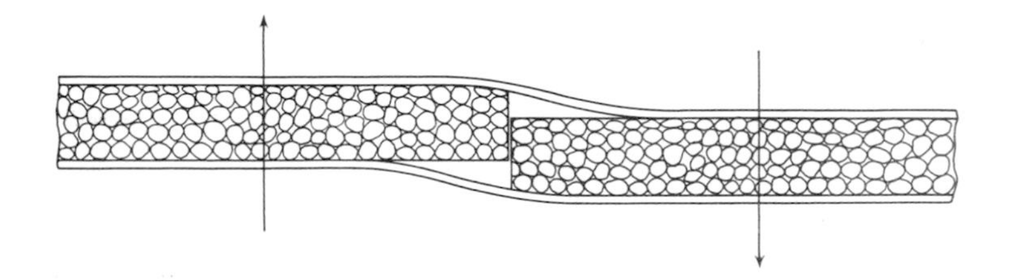

Figure 19.4. Peeling effect of shear discontinuity in a sandwich core

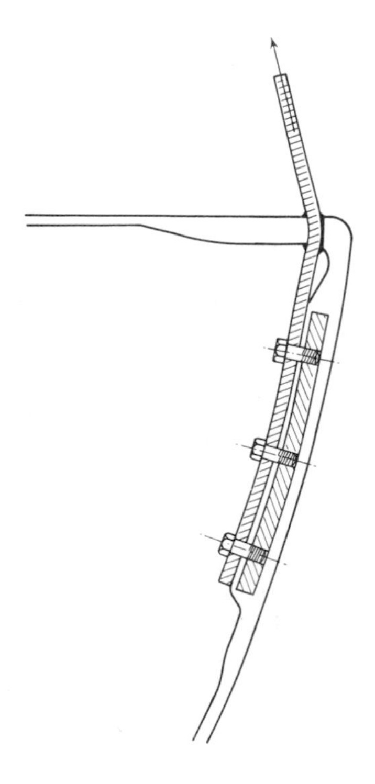

Figure 19.5. Typical stay or shroud attachment

When a stay is attached to the hull of a boat, the local structure should be so arranged that the tension is carried in the plane of the laminate, and does not have a component at right angles to the surface. Where a shroud plate passes through the deck to pick up on the topside, the inwards component can be reacted by the deck itself (stiffened if necessary) while only the vertical component is carried into the skin. This avoids the tendency to peel off the attachment fitting from the skin. A forestay can be attached to a metal stem fitting which carries the load directly into the sides of the boat, or a small bulkhead can be provided under the deck aligned with the stay. The same principle can be applied to other strong points, always bearing in mind the dangers of GRP peeling under tensile stress.

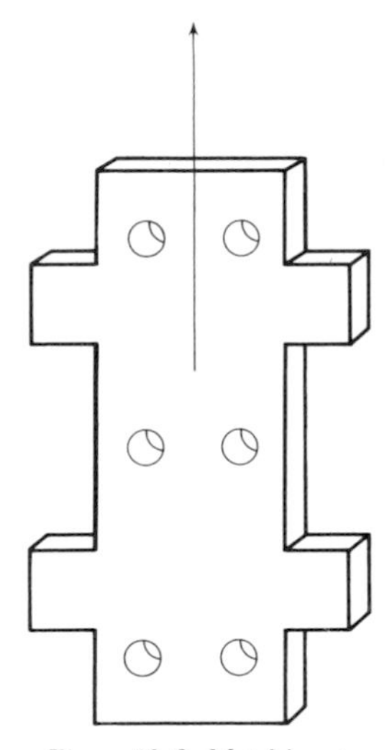

Figure 19.6. Metal insert

It is frequently desirable to incorporate metal inserts into the structure to carry concentrated local forces. In such cases it is not safe to assume any adhesion between the metal and the GRP, and the joint should be so designed that the metal butts against the laminate internally and carries the load in this way. The bearing area can be increased if necessary by providing sideways projecting tongues on the insert (*Figure 19.6*). When the GRP has been built up against the tongues to carry the shear force, the insert can be covered to trap it in position. The insert should not, of course, be loaded in such a way as to pull it out of the shell. Small metal insert plates to carry studs should preferably be star-shaped so that the edges of the plate bear on as large an area as possible. This also leads to better adhesion between layers than when a circular plate is used, with or without holes.

8. STRENGTH CALCULATIONS

In this section it must be assumed that the reader understands the stress

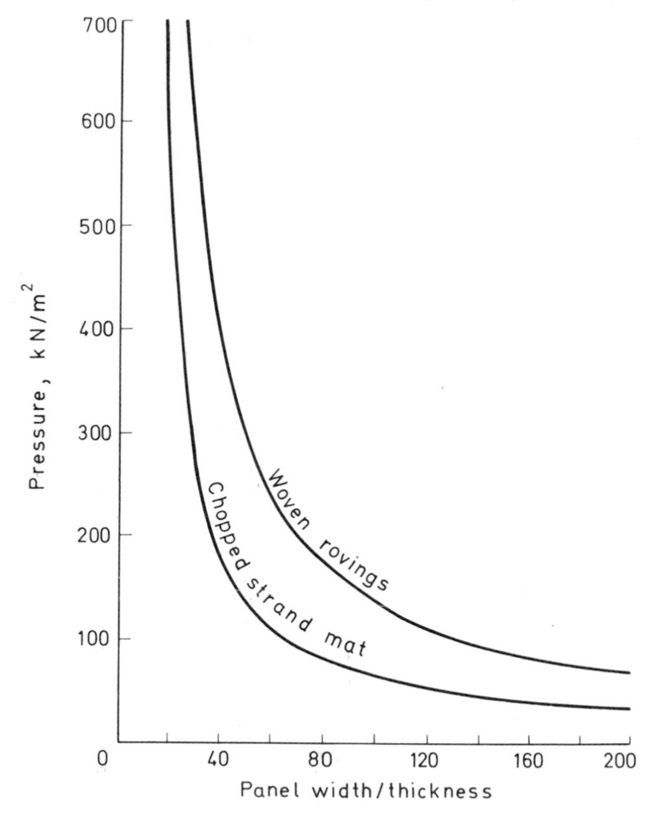

Figure 19.7. Ultimate strength of flat panels under normal pressure (edges ensastré)

analysis of normal structures, and a few points will be put forward which are perhaps more specific to GRP boats.

When considering the strength of a panel of skin, the degree of initial curvature is significant. If the panel is flàt, the 'classical' formulae can be used to calculate the maximum stresses. A number of such formulae are listed by Roark[1] for various types of loading and various edge conditions. For a flat panel whose edges are capable of resisting tension in the plane of the skin, large deflections can introduce membrane stresses which give a significant improvement in strength. *Figure 19.7* shows curves of maximum permissible pressure on a panel plotted against the breadth to thickness ratio for the two main types of reinforced plastics, taking account of membrane effect.

When the original panel is curved, membrane effects become more powerful, and simple theory can no longer be applied. In these cases, except for such simplifying assumptions as treating the bottom as a part of a cylinder under uniform pressure, it is not generally practicable to calculate stresses in the shell. When curvature is small and the pressure is on the outside, there is a risk of the panel 'oil-canning'; in such cases it is safest to ignore the effects of membrane compression and treat the panel as flat.

Each stiffener on a fairly flat area is assumed to support a strip of skin bounded by imaginary lines drawn half way between the member concerned and the two adjacent stiffeners (*Figure 19.8*). From the pressure loading on this area the maximum shear force and bending moment in the stiffener can be calculated, taking the appropriate degree of end fixity. The frames which support light longitudinals, or the girders which support frames are, strictly speaking, not loaded evenly but at discrete points where the lighter stiffeners intersect. It can, however, be shown that for three or more intersecting members the assumption of evenly distributed loading is sufficiently accurate for our purposes.

When the pressures are due to slam it is not necessary to assume the peak pressure operates over large areas. Consequently in such cases, transverse framing need only be designed to carry between a half and three quarters of the peak pressure, and longitudinal girders between a third and a half, depending on the unsupported span (the shorter the span the greater the mean pressure). Continuous beams loaded by slam pressures cannot be assumed fully fixed at the ends. In the absence of a more accurate assessment of fixity, it is probably advisable to use the formula:

$$BM = \frac{Wl^2}{10} \qquad \text{Instead of} \frac{Wl^2}{12}$$

where W = loading and l = bay length.

When calculating the section modulus for a stiffener, account must be taken of the different Young's modulus for the various component parts of the section. For example, where unidirectional tape is used to strengthen the crown of a top-hat section of chopped strand mat, its effective area relative to mat is about four times its actual area. Where timber is incorporated in a stiffener,

Table 19.1 TYPICAL ULTIMATE STRENGTH FIGURES FOR REINFORCED PLASTICS

	Chopped strand mat	*Woven rovings*	*Unidirectional rovings*
Tension (kN/m^2 × 10^4)	10	24	48
Compression (kN/m^2 × 10^4)	10	17	35
Bending (kN/m^2 × 10^4)	12	24	48
Shear (in plane of laminate) (kN/m^2 × 10^4)	7	10	–
Interlaminar shear (kN/m^2 × 10^4)	1	1	1
Young's modulus (kN/m^2 × 10^6)	6	14	24
Shear modulus (kN/m^2 × 10^6)	3	4	–

again its Young's modulus must be taken into account. For plywood, a value about half that of solid wood of the same species should be used.

The effective width of skin to take as working with a stiffener is always

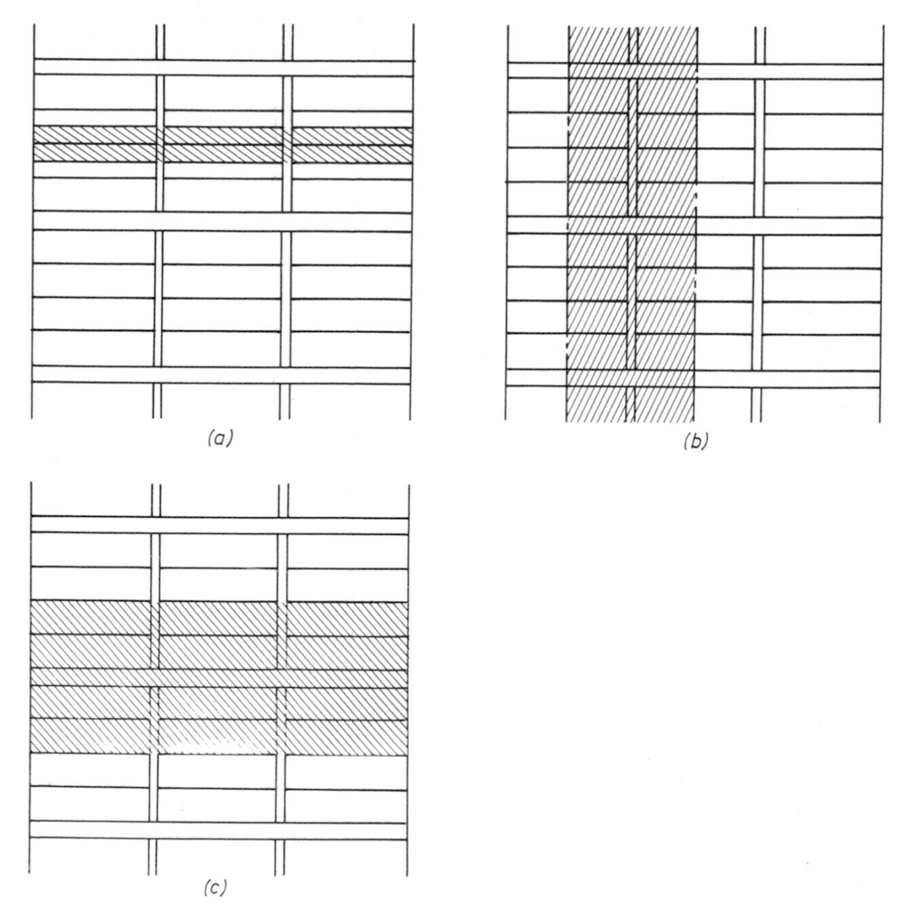

Figure 19.8. Area of skin supported by stiffeners
(a) area supported by light longitudinal
(b) area supported by frame or deck beam
(c) area supported by girder

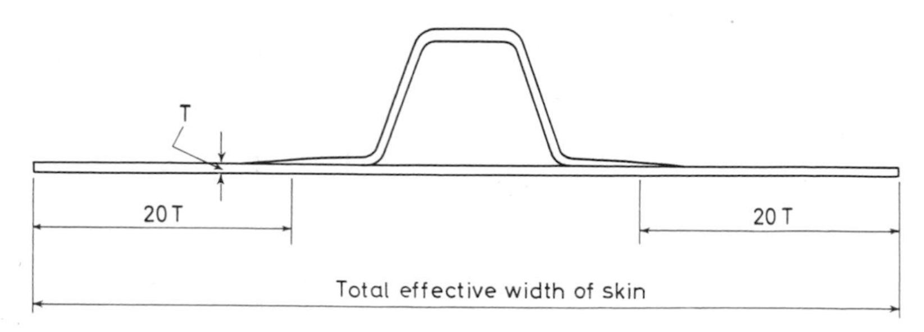

Figure 19.9. Effective width of skin working with a top hat stiffener

difficult to assess. A reasonable figure is perhaps twenty times the skin thickness each side of the stiffener, plus that width which lies directly under the stiffener itself (*Figure 19.9*).

The strength values for GRP laminates show a considerable variation between different authorities. For accurate work the standard of workmanship should be taken into account, as well as the type of material used and the glass fibre treatment. Best of all, calculations should be based on a comprehensive series of tests on material produced under the actual conditions envisaged. As a rough criterion, however, the figures quoted in Table 19.1 can be taken as nominal ultimate values.

For safe working stresses a safety factor is required, which varies according to conditions. Where loading is applied for relatively long periods (for example, static fluid pressures), this factor should be about 4. On the other hand, slamming pressures which act for very short periods of time could work closer to the ultimate stress, with a safety factor of say $1\frac{1}{2}$ to 2.

Reinforced plastics as a material suffers from both fatigue and creep. The effects of these properties are not very well known as yet, though there has been some work done on this subject (Chapter 18). The safety factors quoted above are intended to make allowances for both effects as they apply to a normal boat. Nevertheless where conditions are likely to be more severe than usual, due allowances should be made. The 'Marine Design Manual for Fiberglass Reinforced Plastics'[2] has some data on this subject.

It is hoped that this chapter will be of some use in the structural design of GRP boats, though everything stated from applied loads to permissible stresses is only approximate. This is unavoidable, as this design theory is intended to apply to any type of boat. It would be possible to design a given boat more accurately, if a programme of testing could be carried out, but this is beyond the scope of this chapter.

REFERENCES

1. Roark: *Formulas for Stress and Strain*. McGraw Hill Book Company. 1966.
2. Gibbs and Cox: *Marine Design Manual for Fiberglass Reinforced Plastics*. McGraw Hill Book Co. 1960.

20

Design Logic for GRP Buildings

R. P. BOUVERIE

1. INTRODUCTION

The momentum of the growth of the whole reinforced plastics industry has over the last 10 years been conspicuously slower than might have been indicated by its early explosive development. Since its fairly rapid encroachment into the boatbuilding industry, it seems to have failed to have made any real impact upon other industries and save for minor isolated successes the initial promise of this, in many ways, remarkable material remains unfulfilled.

The reasons for this apparent tendency of the industry to stagnate after its early success are multitudinous, but four reasons are pre-eminent and they are mentioned here because an understanding of them is of positive assistance to the designer.

Firstly, the extremely rapid growth of the use of the material in the boatbuilding industry in the United States was spurred by a real need for the material. United States waters contain the Torredo worm which bores holes in wooden hulls but which will not attack GRP. Also the appearance of this material coincided with an acute shortage of skilled labour in the American boatbuilding industry and its advent allowed an ailing industry to take advantage of an extremely rapid growing demand from an affluent society, by using unskilled labour to fabricate repetitive units. It is notable that its introduction into the U.K. boatbuilding industry was far less rapid. Here there was no real need for it. U.K. waters do not commonly harbour the Torredo, and initially there was a sufficiency of skilled boatbuilders to cater to the not very great demands of a society growing affluent at a far slower rate than its American counterpart.

Secondly, the fact that a GRP fabricator can start up a business with very little capital indeed, a feature once cited as a positive advantage of the material, has resulted in the bulk of the fabricating arm of the industry being grossly under-financed. This has inevitably led to a conspicuous lack of work done on research and development, and the condition is acute enough for there to be apparent a real inability to pay for the many outside skills necessary for a healthy growth industry, market research, design, skilled salesmen, public relations and so forth. One of the direct and pernicious results of this deplorable lack of capital funds has been that it has led to the excessive reliance on archaic and inefficient methods of fabrication and this has held fabrication costs at a consistently high level.

Thirdly, there appears to be a very real paucity of reliable correlated technical information regarding the true nature of the material and its physical properties. Experience has shown that the published physical properties of samples prepared and tested in the laboratories of the raw materials suppliers are not always accurate indications of the physical properties of the same laminate fabricated as part of a production run in a factory. Extravagant claims from the fabricating part of the industry contrast strangely with the rather cautious vagueness of the material suppliers, and the material's behaviour and durability over a long length of time still seems to be largely a matter of conjecture. The only standard engineering manual on the subject was compiled primarily for naval architects and even this is now over 10 years old.

Lastly, much of the fault lies with design. The tendency of early design and manufacture in a new material is to use the new material to duplicate, with almost loving exactness, a product or component which has previously been designed for and produced successfully using another material. This is understandable as it is all part of the process of learning about the new material, and it has been true of every new material throughout history. It can, however, be argued that in no instance has this initial approach to design ever taken full advantage of the physical properties and economics of the new material. This straightforward material substitution rather than proper design for the material has persisted for an inordinately long time as far as GRP is concerned and it is the cause of some concern that grass roots design in the material is still a comparative rarity, with the result that considerable production capacity is still tied up turning out sham.

It is a prerequisite of successful design in any material that the material, its

Table 20.1

(NOMINAL THICKNESS AND AVERAGE PRICES.
GLASS F.G.E. 2000. RESIN/GLASS RATIO 2.5–3/1)

Laminate specification	*Thickness mm*	*Price per sq. metre laminate*	*Materials content*	*Labour, overheads, profit*
Gel coat 1 ply 1 oz mat	1.6	£2.70	£0.76	£1.94
Gel coat 2 ply 2 oz mat	2.4	£4.30	£1.48	£2.82
Gel coat 3 ply 2 oz mat	3.2	£6.45	£2.15	£4.30
Gel coat 4 ply 2 oz mat	4.8	£8.60	£2.82	£5.78
Gel coat 5 ply 2 oz mat	6.4	£10.75	£3.50	£7.25
Gel coat 6 ply 2 oz mat	8	£12.92	£4.17	£8.75

(Note that the above prices do not include pattern and mould write off).

properties, its methods of production and shaping and its economics are all understood by the designer. He must be familiar enough with the material to be able to feel its possibilities and limitations, and for what cost. Only with this familiarity with the material can he hope to accomplish successfully a simultaneous design for the function and the material. Too often this essential background is lacking. Successful design in GRP for the building industry presupposes that the designer understands fully the material, has more than a passing knowledge of the materials or buildings he is competing with, and is reasonably familiar with the building industry as a whole.

Much has been written about the properties and fabrication techniques of GRP and it is not intended to repeat any of it here. Much recent theoretical work has also been published proving the material's structural viability when used in certain forms and shapes. Little consideration seems to have been given, however, to the down-to-earth practical considerations which must be dealt with in detail before successful design of buildings or building components in this material can be accomplished.

2. BASIC DESIGN CRITERIA

Given that the strength and properties of the material are such that formed into the right shapes it can be made to be structurally viable, the design of any building product in GRP, or indeed in any material, involves detailed consideration of five factors of which the first two are of paramount importance.

2.1. FUNCTION

Every successful design of building in reinforced plastics must be primarily functional and regardless of aesthetic embellishments or trimmings it must adequately perform all the functions its end user desires.

2.2. ECONOMICS

Compared with more common building materials GRP as a material is extremely expensive. Considered solely as a cladding or covering material where its use is confined to this function and where aesthetic effect is not the primary consideration, it is often difficult to justify its use economically.

GRP imitation of clap board facing for the elevations of a house is, for instance, some three to four times as expensive in first cost as the original painted wood. Even if it is argued that it is 'maintenance free' the money saved on the initial capital outlay by using the original wood is more than sufficient for its repainting and maintenance over quite a long life.

The nature of the material is such that with proper design it can, by itself, provide aesthetic appearance and finish, weathering surface, and the necessary structural strengths for certain walls and roofs and it can materially assist in thermal insulation. To be a truly economically viable proposition in building, the material must be designed for in such a manner that as many as possible of these functions are performed simultaneously by the material.

Assuming fabrication by the hand lay-up method, the selling price from a fabricator of 1 m^2 GRP of different thicknesses is shown in Table 20.1.

It will be seen from the figures in Table 20.1 that assuming a modest profit margin, the labour/overhead content of the total price is high. This is accounted for by the basic inefficiency of the use of labour inherent in the hand lay-up method. An indication of just how inefficiently labour is used in the hand lay-up technique, is seen when it is realised that most manufacturers cost labour on the basis of one man laying-up and rolling out 1.5/1.7 m^2 of single ply laminate in one hour. A good bricklayer can be expected in one hour to build 1.7 m^2 of a single-course brick wall.

More mechanized forms of laminate production, by reducing the labour cost, can produce laminates of significantly lower price, but these techniques at their present stages of development have certain major disadvantages and limitations. One of the most promising of these techniques, catalyst injection spraying, can reduce labour costs by about one third and this technique is of growing importance to the designer since it shows promise of permitting him to consider the use of laminates thicker than those that are economic when laid-up by hand, a fact of great structural importance. The major drawback at the moment of spray lay-up, is that when the required laminate is thin, for example below 1/8 in (3 mm) thick, there is some difficulty in maintaining quality control.

Table 20.2 shows guide prices of various traditional wall and roof constructions and certain common types of low cost buildings.

It should be noted that a direct comparison between Table 20.1 and Table 20.2 is not possible since they are not on an equivalent basis, but taken together they do clearly show that for whatever use in a building GRP is intended, the designer must concentrate on using as thin a laminate as is functionally possible, and of necessity he must improvise shapes for the material which permit an economical laminate to be used.

Just how thick an economical laminate is likely to be is dependent to a certain extent on the particular project. An intelligent appraisal of Tables 20.1 and 20.2 will show that in the field of low and medium cost building the designer, to be economic, should try to keep the weight of his laminates below 6 oz/ft^2 (1.8 kg/m^2), and should recognise that thicknesses greater than this should be confined to small areas, jointing flanges, stiffening ribs, and points of high stress.

Although the material can be moulded into virtually any shape, with limits on size governed only by transport and handling considerations, an important consideration concerning the choice and shape, and the economics, of any design in GRP is the basic pattern and mould costs. Any component made in GRP requires a pattern and at least one mould, although this statement should be qualified by saying that in certain instances the pattern and the mould can be one and the same thing. Therefore, as well as limits on shape imposed by other considerations, the designer must also bear in mind that the fabricated prices of each component will contain the cost of pattern and mould write-off. If this pattern and mould cost is not to have an economically

Table 20.2

TYPICAL COSTS OF CERTAIN KINDS OF TRADITIONAL BUILDING

Item	*Finished cost* *Cost per meter super*
WALLS	
Brick—single course fair faced both sides	£2.96
Brick—cavity brick wall fair faced exterior internal render and set; 3 coats emulsion	£5.52
Breeze—loadbearing; plastered with 3 coats emulsion both sides	£4.04
Concrete block—fair finish; 3 coats emulsion both sides	£4.30
Site cast 216 mm concrete panel—exposed aggregate exterior finish; fine finish interior ready to accept paint	£4.08
ROOFS	
Flat roof—timber beam supported covered 25 mm sawn boarding; built up; 3 layer felt with chipping; includes insulation but excludes ceiling	£4.98
Pitched roof—timber trusses; tiles on battens felt backed; inclusive insulation, ceiling, eaves and soffits, rainwater goods	£4.71
Pitched roof—steel trusses (span 12 m or less); steel purlins; asbestos cement roofing includes rain goods, excludes insulation	£3.23
Typical 2-storey detached house—89 m^2; cavity brick gable walls; timber elevations with vertical tiling. Tiled roof on timber trusses. Timber 1st floor, on steel R.S.J. Finished shell above slab inclusive of partitioning and internal doors; excluding plumbing, electrics and internal decoration	£19.37
Concrete or steel portal frame construction with concrete, steel or timber purlins clad with corrugated asbestos cement. Not insulated. Finished shell above slab; spans of 15 m or less 4.6 m or less height to eaves from	£8.07
Concrete or steel portal frame construction with concrete, steel or timber purlins clad with insulated corrugated asbestos cement. Cavity brick infill to eaves—18 m span. Total finished cost inclusive of ground slab and foundations, drainage, etc.	£18.33

crippling affect on the cost of the whole project, either the number of exactly similar components must be sufficient to absorb the total cost of patterns and moulds without affecting the overall economics of the building severely, or the component must be designed to a shape which minimizes the fabricator's capital outlay on tooling.

Thus, a long run of similar components can absorb high initial pattern and mould costs permitting the designer's consideration of complex shapes, whereas a smaller number of similar components necessitates the consideration of only those shapes for which the pattern and mould costs are minimal.

All of the foregoing makes it clear that the design of buildings or components in GRP, if they are to be competitive with the current, comparatively low cost of traditional types of construction, requires radicalism and originality.

Simplification of the building and components is the essence of economic design in this material. Effort put into the standardisation of the joints of a building, or indeed the elimination of them, the moulding-in of rainwater drainage systems, the fast frames of windows, door architraves and other time consuming, skilled site labour features of conventional building must be considered in the light of comparative economics and included if they reduce the cost of the building considered as a whole. At the present moment it is both technically and economically possible to think in terms of designing certain common buildings, houses and some kinds of industrial buildings, so that their structures can be made out of two or three different major components in GRP, with the joints standardized to two, the joint between components and the joint to the ground. Such structures can be, even today, erected as cheaply as their traditional counterparts, and with traditional building costs constantly rising at a fast rate, such structures will become increasingly acceptable.

2.3. BUILDING REGULATIONS

The current sets of building regulations, model by-laws and codes of practice were all of them written before it was envisaged that GRP, or indeed any plastics, would be used structurally or in any marked degree in building.

Where regulations or by-laws are unclear concerning the use of GRP in a building, those in authority whose job it is to interpret the regulations are apt, due to unfamiliarity with the material, to apply the regulations to the letter. To the designer this is often extremely discouraging. To have, for instance, to make a material which can deflect considerably in perfect safety without spoiling its finish, conform to code of practice B.S. 499 (deflection in structural steel), can be heartbreaking and sometimes cripplingly expensive. The designer is well advised at the outset of a project to seek out and talk to the local County or Borough Building Regulations Officer, Fire Officer and County Engineer with the view of acquainting them with the material and its properties and obtaining their prior advice on the project.

Although it can by no means be guaranteed, prior consultations of this nature are sometimes advantageous in that the regulations and by-laws may then subsequently be applied to the material and project with knowledge, intelligence and common sense, and it will at least ensure that the project receives a sympathetic hearing, by no means always the case when the interpreters are asked to make decisions on projects which use materials they are unfamiliar with.

2.4. MARKET RESISTANCE TO CHANGE AND THE CONSIDERATION OF TOTAL DEVELOPMENT COST

Because the economic design solution of a building, or buildings, in GRP will, as often as not, give an appearance and shape very different from any traditional alternative, certain psychological considerations must be taken into account during design. This is because in this material shapes dictated primarily by function and economics, and adequately satisfying these two factors, may nevertheless produce a product which will not find acceptance for psychological reasons. What is technically and economically possible is not necessarily desirable, and this is especially so if it lays a burden of radical change on people who are not ready or willing to change. In simple terms the originality and radicalism necessary to produce viable designs of buildings in this material must always be tempered with a strong measure of common sense if the design is to prove acceptable in a practical sense to the end user. In a very real sense life must be made more easy for the client, not more difficult.

A much more practical aspect of this consideration of change with regards to design concerns the approach made by the designer himself and impinges upon the economics of the design with regards to total development cost. Consider mass housing for instance. Although the outside appearance of a house may influence a house buyer, it is of minor importance compared to his habitual concern with the inside. Structurally and economically the ideal shape for a cheap GRP house may call for a basic oval shape with compound curved walls and primary furniture moulded in as part of the structure, the dwelling being largely open plan. It is improbable that such a structure is acceptable currently in outside appearance, but even if it were, at this moment in time, the inside of such a house would have to be largely cuboid or near cuboid with vertical flat internal walls, the space being partitioned off into rooms conventionally. All of the industries subordinate to the building industry or householders, from baths and kitchen stoves to fenestration and heating units and from furniture and carpets to picture hanging and wall paper, are geared for, and cater to, flat vertical or horizontal internal surfaces. The acceptance of mass housing with radically untraditional internal shapes presupposes change on a massive scale. The total development cost of bringing such a building to a point where it could adequately cater to, and be accepted by, a mass market will run into millions of pounds, not just for the development costs of the building itself, but the development costs of all the industries subordinate to house owning. To expect the short term acceptance change on this sort of scale is unrealistic.

In present day practical terms, giving due regard to all factors, there is strong argument for saying that as far as mass housing in GRP is concerned design should start from the inside arrangement of the house. This would be largely conventional and, because of the economic factor, in conventional materials. To a very real extent the inside of the house will dictate the outside appearance and shape of the house and although local shaping to obtain the correct balance between function and economics will be necessary, the overall shape of an economic house whose main structure is GRP will be surprisingly traditional in appearance.

To a great extent these same arguments apply to all new buildings in the low-priced range, and although in due course radically different user habits and building shapes may well prove acceptable, in the short term this will almost certainly not be so. Change, to gain ready acceptance, must be introduced gradually.

2.5. AESTHETICS

Consideration of aesthetics as a factor in the design of medium and low cost buildings in GRP has deliberately been left until last. Unless the budget for any project includes money which can be spent solely on aesthetic effect, a rare occurrence except in the case of expensive prestige architectural projects, the prime factors of design, function and economies, will largely dictate the shape and appearance of the building or component. For this reason the designer's room for manoeuvre is strictly limited by these two disciplines. This does not mean that GRP structures whose parameters are dictated solely by function and economics are necessarily aesthetically unpleasing. The reverse is more often true, but it will almost always be the case that where function and economics have been equated well, the final appearance will generally be one of great simplicity.

It is when colour and light and shadow effect is used to dramatize or give aesthetic appeal to a functional/economic design that the designer in this material has a most free hand. The nature of the material is such that with negligible cost increase a laminate can be made to transmit any amount of light from complete opacity to about 80 per cent transmission, and by the addition of pigment to the resin prior to fabrication it can be made in virtually any colour without the need for painting.

It follows that it requires no extra trade or cost to vary at will the colour of components, over the full British Standard range if desired. In addition, any portion of the structure or component may be translucent, and this also may be colour tinted. Used intelligently these two very real advantages of the material not only produce buildings of excitement and appeal, but they can sometimes be used to save significant cost on the building as a whole, since they can be used to provide all, or a large measure of, the internal decorations of a building. *Figure 20.1* shows the internal light and shadow effect from a building whose structure is a thin laminate of extremely pale blue, translucent GRP.

3. SMALL LOW COST CHURCH

Figure 20.2 is a sketch drawing illustrating a small low cost church designed primarily for economy and it is interesting to consider this design in the light of the five major factors affecting design previously outlined.

Since, under the building regulations a church, being a place of assembly, requires a half-hour fire rating on the load-bearing exterior walls and since the prospect of obtaining ministerial waiver of the regulation for a building of this nature is negligible, the practical use of structural GRP is of necessity confined to the roof only where the properties of the material with regard to fire will meet the regulations completely.

By making the roof the main feature of the building, the exterior walls can be left plain and low and brick or concrete block construction would be envisaged for these, each being economical and uncomplicated.

Because it is a 'one off' building it is desirable to choose a functional shape of roof which requires only a small pattern and mould cost. The shape chosen

Figure 20.1. Internal light and shadow effects of a building whose main structure is a thin translucent GRP laminate

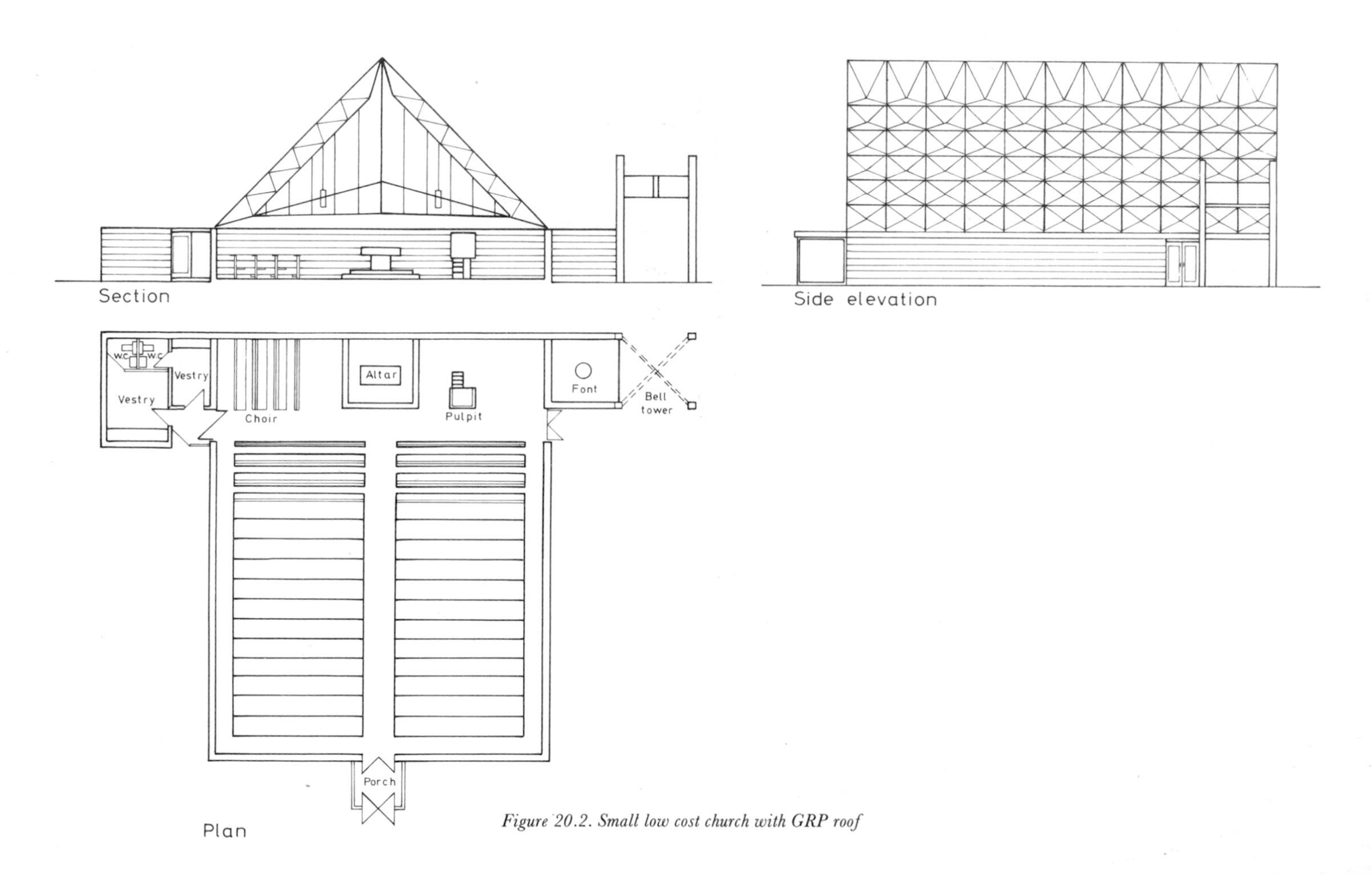

Figure 20.2. Small low cost church with GRP roof

is a compounded folded plate of plain vee-section made up in the form of diaphragm beams, each 11 m long, 1.8 m wide and with a 0.9 m depth of web. In this case the pattern can also be the mould, and the pattern can be made out of ten sheets of standard hardboard minimally stiffened with timber. Total cost of pattern would not exceed £70.

The main folded plate of this roof is compounded on both sides, inside and out and the resultant shape, that of interlocking, irregular but similar, tetrahedra, is extremely strong and stiff both laterally and longitudinally. The overall specification for the laminate for this roof structure would call for a gel coat backed by 2 plies of 2 oz/ft^2 (0.6 kg/m^2) glass mat and resins meeting the fire standard B.S. 476 Part III S.A.A. would be used. The sections of this roof each 1.8 m wide and 11 m long are convenient to transport, and would be shipped to the site where they would be joined together on the ground into larger sections prior to being erected by crane. It is likely that this roof could be erected in only two crane lifts, each covering half of the building full width, each half of the roof having its fenestration and stainless tie wires fixed prior to erection. The joint between the two halves is easily made good after erection by riveting and bonding the closing pieces from the outside.

The structural shape of the roof is such that it is effectively double skinned, giving, by itself, a reasonably acceptable measure of thermal insulation. The shape on the inside, exactly similar to the shape on the outside, is also good from the point of view of acoustic damping.

The windows along the top apex of the roof and along each side of the base are all similar repetitive units, and lighting, if arranged along the top edge of the supporting walls internally will not only adequately light the interior by reflection off the opposite side of the roof internally, but will also produce an interesting external effect, tending to light up the whole of the external base and the apex of the roof.

The internal wall surfaces of the building would be plain rough cast and all of the internal surface decoration beyond this would be done by the natural shadow effects cast by the roof and windows assisted by the discreet use of colours and translucency in the roof itself.

In this instance, therefore, it is seen that one material only, GRP, provides this building above the walls with:

an outer weathering surface,
an outer finish and appearance,
a main structural roof,
thermal insulation,
adequate acoustic insulation,
an inside finish and appearance.

It also assists materially in reducing costs by saving erection time and provides a large measure of the internal decoration. Roof maintenance costs would also be extremely low. The roof, erected but excluding fenestration would cost approximately £12.50/m^2 covered.

The functional and economic advantages given by using GRP in this case, will materially reduce the overall cost of the building below a comparable traditional alternative, and it is entirely probable that the finished, decorated cost of this building, excluding pews, would not exceed £30.00 per meter super.

INDEX